Twelve Problems in Physics

First Edition

By Ian H. Redmount
Saint Louis University

Bassim Hamadeh, CEO and Publisher
Michael Simpson, Vice President of Acquisitions and Sales
Jamie Giganti, Senior Managing Editor
Miguel Macias, Graphic Designer
Angela Schultz, Senior Field Acquisitions Editor
Michelle Piehl, Project Editor
Alexa Lucido, Licensing Coordinator

First published in the United States of America in 2016 by Cognella, Inc.

Trademark Notice: Product or corporate names may be trademarks or registered trademarks, and are used only for identification and explanation without intent to infringe.

Cover images: Shuttle launch source: http://st.depositphotos.com/1011061/1360/i/110/depositphotos_13600465-Launching-from-Houston-Texas.jpg. Copyright in the Public Domain.
Jumbo jet copyright © 2012 by Depositphotos/Ivan Cholakov.
Billiard game copyright © 2010 by Depositphotos/Dmitrijs Dmitrijevs.
Earth and moon source: https://archive.org/details/GPN-2000-001437. Copyright in the Public Domain.
Spring copyright © 2011 by Depositphotos/Giuseppe Ramos.
Wave copyright © 2010 by Depositphotos/Clément Levet.
Atom vector copyright © 2011 by Depositphotos/Ion Popa.
Active neuron copyright © 2010 by Depositphotos/Sebastian Kaulitzki.
Magnet copyright © 2013 by Depositphotos/ClassyCatStudio.
Old electric device copyright © 2012 by Depositphotos/Aleksandr Volkov.
Nuclear submarine copyright © 2012 by Depositphotos/Andreus.
Binoculars copyright © by Depositphotos.

Interior images: Figure II.1 copyright © 2013 by Depositphoto/Arashaqar.
Figure III.1 copyright © 2010 by Depositphotos/Krzysztof Grzymajlo.
Figure IV.1 copyright © 2010 by Depositphotos/Tristan3D.
Figure VIII.1 copyright © 2012 by Depositphotos/Sebastian Kaulitzki.
Figure XI.1 copyright © 2012 by Depositphotos/Andreus.

Printed in the United States of America

ISBN: 978-1-62661-137-5 (pbk) / 978-1-63487-075-7 (br)

www.cognella.com   800-200-3908

Contents

Preface		vii

I Flight of the Rocket   **1**
A   Kinematical quantities—the language of motion   2
B   Kinematics of a rocket in vertical flight—no drag   13
C   Kinematics of angled flight—no drag   20
D   Theory of Relativity (Galileo)   27
E   Flight dynamics—thrust and drag   30
F   Relativity and Newton's Laws   49
G   Engine performance: momentum conservation   50
H   The Einsteinian rocket   54

II Nonstop Service to Anywhere   **67**
A   Dynamics of straight, level flight   69
B   Work and energy   72
C   Flight range   76
D   Other flight-performance characteristics   81
E   Potential energy and energy conservation   83

III "Trouble in River City"   **93**
A   Collisions: interactions of short duration   94
B   Collisions in two and three dimensions   100
C   Center of mass and the CM frame   103
D   Rotation about a fixed direction   111
E   Rotation and billiard-ball dynamics   126
F   Rotation in three dimensions   131
G   A few words about static equilibrium   139

IV Shoot the Moon   **145**
A   Newton's Law of Universal Gravitation   146
B   Kepler's Laws of Planetary Motion   159
C   Orbital dynamics   164
D   Einsteinian gravitation   168

V   The Harmonic Oscillator   175
   A   Universality of the harmonic oscillator 176
   B   The Initial-Value Problem 180
   C   Solving the *linear* SHO IVP 182
   D   The damped harmonic oscillator 186
   E   Driven oscillator and resonance 189
   F   Coupled oscillators and normal modes 193
   G   The quantum harmonic oscillator 198

VI   Catch a Wave (A Knotty Problem)   201
   A   Waves defined 203
   B   Wave kinematics 204
   C   Wave dynamics 212
   D   Scattering at a knot 217
   E   Sound waves 220

VII   The Classical Electron   227
   A   Electrostatic "amber" force 228
   B   The Coulomb Law 230
   C   Electric fields 235
   D   The Gauss Law 242
   E   Electric potential energy and potential 255
   F   Classical radius of the electron 262
   G   Capacitance and field energy 263

VIII   The Nerve as Electrical Network   279
   A   The neuron—axon, dendrites, and synapses 280
   B   Electric current—charge in motion 282
   C   The Ohm "law"—a property of materials 285
   D   Power (energy, work) and electric currents 290
   E   DC Circuits: closed paths for current 291
   F   Axon as resistance ladder: electrotonus case 301
   G   RC Circuits: time-dependent currents 305
   H   Telegraphers' Equation: time-varying signals 308
   I   Action potentials on the axon 309
   J   Semiconductor devices 310
   K   Electronic logic 320

IX   Return of the Classical Electron   329
   A   Magnetism: another force known in antiquity 330
   B   Magnetic poles 331
   C   Operational definition of **B**: Lorentz force law 332
   D   Magnetic forces on moving charges 334
   E   Sources of magnetic fields: moving charges 344
   F   Magnetic moment of the electron 357
   G   Magnetic effects in bulk matter 363

X The Solenoid **375**

   A   The Faraday Law of Induction 377

   B   Inductance 385

   C   Force exerted by a solenoid 392

XI Submarine Radio Communication **397**

   A   The Maxwell Equations 399

   B   Electromagnetic-wave solutions 400

   C   Electromagnetic wave dynamics 411

   D   Submarine communication 416

   E   Quantum wave equations 420

XII A Pair of Binoculars **425**

   A   Geometrical optics 427

   B   Optical images 432

   C   Operation of binoculars 439

   D   Physical optics: interference 443

   E   Lens coatings 449

   F   Physical optics: diffraction 449

   G   Resolution of binoculars 455

   H   Matter waves and quantum mechanics 456

Preface

This is a text for a calculus-based introductory physics course. It is based on an approach to the two-semester Engineering Physics course at Saint Louis University that I have explored and developed over the last fifteen years. Rather than attempt comprehensive coverage of all the topics appropriate to such a course, I select a small number of more-or-less realistic problems to be explored in as much depth as the preparation of the students will allow. Along the way, the students become familiar with the physics concepts and principles necessary to address the problems, the framework of ideas I deem essential to an understanding of (classical) physics as it stands today.

This approach is based on four observations. First, the introductory university physics course is very unlikely to be the students' first course in physics. (According to the American Institute of Physics, some 98% of American high school students have access to a physics course, a figure that accords with my own experience.) So it is no longer necessary to teach the course as if it were. Second, for many students the introductory university physics course is very likely to be their *last* physics course. Most of my students are in engineering or premedical fields; physics majors are thin on the ground. If students are to appreciate the intellectual heights that the science of physics reaches and to which it aspires, they must see some of them here. Third, students invariably complain on course evaluations that the instructor does not illustrate the material with enough examples. Very well: Here is a course that is all examples. Fourth, it is my experience that the defining characteristic of the modern introductory physics course is *lack of time*. The topics illuminated here can be covered in a two-semester, four-credit-hour course. For a three-credit-hour course, some squeezing is necessary.

The choice of the Twelve Problems is shaped in part by the fact that many students in the calculus-based course at Saint Louis University are affiliated with Parks College of Engineering, Aviation, and Technology. (As was the author, until a reorganization in 2006.) This school, founded in 1927 and donated to the university in 1946, holds the first federally issued certification for a flight-training school in the United States. Some 10% of America's World War II pilots were trained here. The aviation and aerospace traditions of the school run deep. Similarly, the major topics not covered here—for lack of time, not significance—are determined by the curricula of these students. They see thermodyamics in their chemistry courses, fluid mechanics in specialized aerodynamics courses, and alternating-current circuits in circuits courses offered by specialists. "Modern physics" has its own two-semester course here.

Calculus is used freely in this text. It is assumed that students are conversant with the ideas and techniques of their introductory calculus course—which they may be taking concurrently—or know where to look them up. Notions beyond the introductory calculus sequence are explained as needed.

The problems offered at the end of each chapter are not, for the most part, intended as exercises. Those are plentiful. Rather, they are intended as vehicles for the students to explore further the concepts presented and applied in the chapters.

The references listed in each chapter are not intended to document every assertion in the text. Nor are they a comprehensive review of relevant texts, which would require an entire book in itself. They are intended as guides to further and deeper exploration. The choice of sources is shaped by those texts with which I have personal experience.

The biographical footnotes are drawn mostly from *Webster's Biographical Dictionary* [Sta69]. Appendix J of Halliday, Resnick, and Krane [HRK02] and Wikipedia (wikipedia.com) supplied some of the more recent entries.

This text was created with LaTeX 2_ε, using the amsmath, amssymb, bm, and graphicx packages. Five of the figures with which each chapter begins— Figs. II.1, III.1, IV.1, VIII.1, and XI.1—are the work of the artists cited in the captions, provided by Depositphotos.com. The other seven are my own photographs. The line drawings are my work, using the Oracle OpenOffice.org 3 Draw utility.

The list of people to be thanked for making it possible for me to undertake this work, including my family, my own teachers, my mentors, and my colleagues at Saint Louis University, is very long. But I should like to thank in particular the hundreds of SLU students who suffered the torments of hell, so that this book might come to be. And the team at Cognella Academic Press, for their interest and industry: Bassim Hamadeh, Michael Simpson, Angela Kozlowski, Jamie Giganti, Jessica Knott, Jess Bush, Jennifer Bowen, Ivey Preston, Chelsey Rogers, Brian Fahey, Miguel Macias, Angela Schultz, Michelle Piehl, Alexa Lucido, Allie Kiekhofer, Danielle Menard, Arek Arechiga, Sean Adams, Jennifer Levine, copyeditors Sharon and LeeAnn, and all the others working behind the scenes but not named to me. And my wife, Hisako Matsuo, for her patience and understanding.

The Cognella team has put considerable effort into reviewing and correcting the manuscript. The errors that remain are mine alone.

Let the quest begin.

Ian H. Redmount
Associate Professor of Physics
Saint Louis University

Chapter I

Flight of the Rocket

- *Evaluate the performance of a model rocket in vertical flight without drag.*
- *Evaluate the performance of the rocket in angled flight without drag.*
- *Analyze the effects of aerodyamic drag on the rocket's flight.*
- *Analyze rocket engine performance.*
- *Analyze the performance of an Einsteinian rocket.*

PHYSICS is a Greek word. Its use goes back at least as far as the *Physics* [McK41] of Aristotle.[1] It means, literally, *the study of nature,* and that ambitious project is the subject of this text. This author will not attempt to describe here every facet of the vast array of knowledge and understanding that is physics in its current state. Rather, he will attempt to illuminate the field through a sequence of straightforward examples, explored in as much depth as a calculus-based introductory course will allow.

Motion is a feature of nature of great interest to the ancient Greek thinkers and their successors, as it is a prominent characteristic of life. The branch of physics that deals with motion is called *mechanics.* The science of mechanics is divided into three principal branches: *kinematics,* the description of motion; *dynamics,* causes and effects of motion; and *statics,* the study of *non*motion. If this last seems trivial, it is not. Statics is the oldest and longest-developed of the three branches. You cannot build a pyramid or a ziggurat or a Gothic cathedral without some considerable understanding of the principles of statics.

The model rocket pictured in Fig. I.1 is an easily visualized example through which to explore the mechanics of a single body. Though constructed of cardboard and balsa wood, it operates in the same way as the large rockets that have carried humans to the Moon and robot spacecraft to every planet in the solar system and to the threshold of interstellar space. The mechanics of such grand journeys is the subject of Ch. IV; this chapter will treat flights of the model rocket near the surface of the Earth.

A   Kinematical quantities—the language of motion

To ask and answer questions about the flight of the rocket requires the language of kinematics. Although the science of mechanics has roots going back to ancient Egypt, Babylonia, and Greece (and India and China as well), the kinematics we use was first properly formulated by Galileo[2]. It is a language of quantities; the descriptions of motion offered by the science of mechanics are quantitative. Some thinkers have gone to considerable lengths to understand why this should be so. Perhaps the simplest explanation is that it is the quantitative aspects of nature that we find conform to simple rules. The beauty of a sunset can be discussed endlessly, but its spectral energy distribution can be measured and understood.

Quantities

As quantities are our stock in trade, it is useful to examine in some detail how we describe and represent them. Consider an example:

$$y = (4.55 \pm 0.02) \times 10^2 \text{ m} . \tag{I.1}$$

---

[1] Aristotle, 384–322 BCE
[2] Galileo Galilei, 1564–1642

Figure I.1: This model rocket, powered by solid-fuel engines, illustrates the same laws of physics that govern its larger counterparts. (Photograph by author.)

This is a distance, expressed in scientific notation, with which it is assumed the student is familiar. The components of this representation have names: y is the *symbol* for this quantity; the digits 4.55 are the *mantissa* or *significant figures*; the 0.02 is the *uncertainty*; the factor 10^2 is the *order of magnitude* of the quantity, and the m is its *unit*. Each of these is important: If the mantissa is wrong, you have a (possibly minor) numerical error. If the order of magnitude is wrong, you have a major numerical error; you are off by factors of ten. If the units are wrong, you have garbage; if the units are not right, the rest is meaningless.

The symbol represents the entire quantity; hence, it is superfluous to write y meters. Some symbols have traditional meanings: x, y, and z are often distances; m is usually mass; θ and ϕ usually angles, and so on. Other symbols are used as needed. It would be ideal to have a distinct symbol for each quantity, but there are not enough symbols in the combined Latin, Greek, and Hebrew alphabets for that. So circumstances arise in which the meaning of a symbol must be understood from context: p might represent momentum or pressure, for example; V might represent velocity or volume.

The mantissa or significant figures, combined with the order of magnitude, represent the best estimate of the value of the quantity. The number of digits presented is not a matter of convenience. It indicates how well we know what we claim to know about the quantity. The general rule is that we display all the digits we know and the first digit that is uncertain by "a few." Thus 4.55 represents a number somewhere between 4.52 and 4.58. Trailing zeros are signifcant figures, but leading zeros are not—they are simply place holders. For example, 0.0020 has two significant figures. Dropping trailing zeros for convenience is an error: 0.5 is a number somewhere between 0.2 and 0.8, but 0.500 is a number somewhere between 0.497 and 0.503. (And neither is the same as 1/2, which is *exact,* of infinite precision.) The numbers 0.497 and 0.500 are the same, but 0.4970 and 0.5000 are not. To display too few significant figures is to discard hard-won information, while to display too many is a lie—a claim to more than one actually knows.

The number of significant figures offers a guide to how well a value is known, but actual measurements are properly displayed with an explicit statement of their uncertainty. Usually (but not always), one shows one standard deviation of a distribution of measurements: For the example above, 68.3% of measurements of this quantity would lie between 453 m and 457 m, 95.5% between 451 m and 459 m, et cetera, assuming the errors are distributed according to a Gaussian[3] distribution, the familiar bell curve.[4] In actual experimental work, nearly as much effort is expended on the uncertainty of a measurement as on the value itself, since the uncertainty determines whether two measurements, or a measurement and a calculation, agree or disagree. In this text, however, we shall not often display explicit uncertainties, but shall rely on significant figures to indicate precision.

---

[3] Karl Friedrich Gauss, 1777–1855

[4] This is the language of *statistics,* the science of data analysis. Students who have not taken an introduction to this subject should remedy that.

The use of powers of ten to specify the order of magnitude of a quantity allows us to handle both very large and very small numbers without having to contend with inconveniently long decimal expressions. Some powers can also be indicated by prefixes on the units, e.g, mm (millimeters, 10^{-3} m) or MW (megawatts, 10^6 W). These prefixes extend from yocto- (10^{-24}) to yotta- (10^{+24}) by factors[5] of 10^3, plus traditional prefixes for $10^{\pm 1}$ and $10^{\pm 2}$. It is assumed here that the student is familiar with these or will look them up.

In this text we shall use units almost exclusively from the *Système International des Poides et Mesures* (International System of Weights and Measures) or *SI* subset of the metric system, maintained by the International Union of Pure and Applied Physics (IUPAP) and the International Union of Pure and Applied Chemistry (IUPAC). This system is based on units for seven basic quantities: length (meter), time (second), mass (kilogram), amount (mole), temperature (kelvin), electric current (ampere), and luminous intensity (candela), and combinations of these. The choice of seven quantities is rooted in human experience of the world. These quantities cannot be defined except circularly, in terms of synonyms of themselves; they can only be experienced. The units for these quantities are defined *operationally,* via instructions for measuring them. The names for units derived from those for the basic quantities will be introduced here as we encounter them.

Displacement r—the prototype vector

The kinematical quantities we need are those required to address the questions we might ask about a rocket's motion: How high or how far does it fly? How longdoes it take to get there? How fast does it go? and so on. The first of these requires a description of *position* or *displacement*. More precisely, displacement is change in position, but since position is usually specified as displacement from a chosen origin, we need not emphasize the distinction.

The mathematical notion of a *vector* is the appropriate tool to describe displacement. Or conversely, spatial displacement is the inspiration for the vector concept. In introductory physics it is customary to define a vector as a quantity that: has magnitude $|\mathbf{r}|$ or r; has direction, perhaps specified by angular coordinate θ (or θ and ϕ, in three dimensions); and adds or combines in the manner suggested by spatial displacements. Illustrated in Fig. I.2, this is described as "tail-to-tip" or "parallelogram" addition. This last is important: There are quantities that possess magnitude and direction but do not combine in this way and are not vectors. Electric currents, for example, have magnitude and direction but add according to the "junction rule" illustrated in Fig. I.3. Consequently, electric current—to be described in detail in Ch. VIII—is *not* a vector quantity.

We should note that the standard notation used here glosses over important distinctions. The magnitude of a vector, $|\mathbf{r}|$, looks like the absolute value of a

---

[5]This grouping of numbers by powers of 10^3 is a Westernism. In China and Japan, traditionally, numbers are grouped by powers of 10^4, e.g., *man* (10^4), *oku* (10^8), and *cho* (10^{12}) in Japanese.

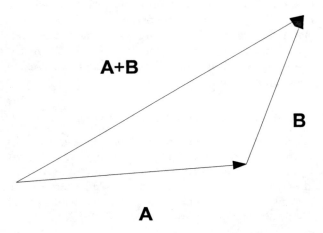

Figure I.2: "Tail-to-tip" or "parallelogram" method of vector addition. The combination of spatial displacements is the prototype for this mathematical operation.

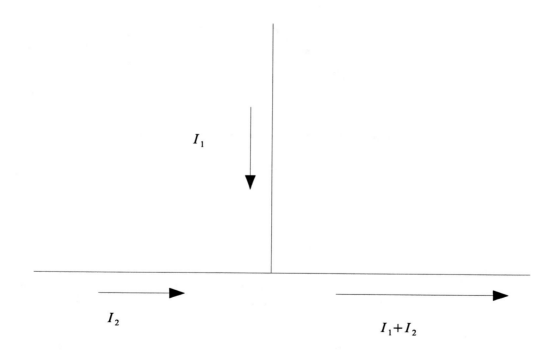

Figure I.3: Junction rule for electric currents. This is *not* vector addition; electric currents are scalar quantities.

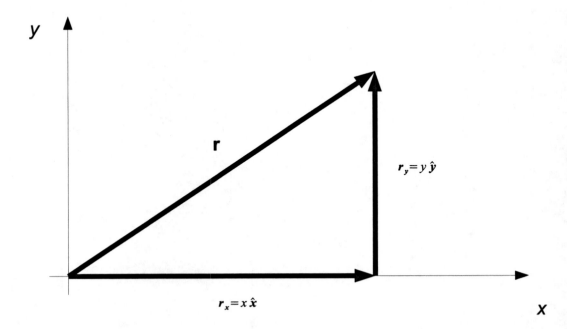

Figure I.4: Resolution of vector **r** into components.

number, $|x|$, but these are distinct concepts. (Mathematicians, wishing to be more specific, might use $\|\mathbf{r}\|$ for vector magnitude.) Likewise, the vector sum $\mathbf{A}+\mathbf{B}$ uses the same symbol as real-number addition $a+b$, but these are different operations on different classes of objects.

If we can combine vectors, we can also take them apart. The displacement vector **r** in Fig. I.4 can be represented as the sum of *vector components* $\mathbf{r}_x$ and $\mathbf{r}_y$ in the x and y directions, which we are at liberty to choose. These in turn can be written as *scalar-component* multiples of vectors in the chosen x and y directions of unit, dimensionless magnitude, the *unit vectors* $\hat{\boldsymbol{x}}$ and $\hat{\boldsymbol{y}}$:

$$\mathbf{r} = \mathbf{r}_x + \mathbf{r}_y$$
$$= x\,\hat{\boldsymbol{x}} + y\,\hat{\boldsymbol{y}} \ . \tag{I.2}$$

Most textbooks use the notation **i**, **j**, and **k** for unit vectors in the $+x$, $+y$, and $+z$ directions, respectively. Here, we shall use instead $\hat{\boldsymbol{x}}$, $\hat{\boldsymbol{y}}$, and $\hat{\boldsymbol{z}}$, as this notation is more explicit about the directions of the unit vectors, is readily generalized to unit vectors in other directions, and facilitates the use of **i**, **j**, and **k** for other purposes. This disassembly of a vector is properly called *resolution of* *a vector into components.* It is not required that the directions of the components be perpendicular (orthogonal), though this greatly simplifies calculations. In crystallography, for example, it might be useful to resolve vectors into components along nonperpendicular crystal axes. In this text, however, we shall not find it necessary to depart from the use of perpendicular axes.

The scalar components of a vector are readily calculated from its magnitude and direction, and vice versa, i.e., the two sets of quantities represent the same information. In two dimensions, a displacement vector **r** with magnitude r, at angle θ counterclockwise from the $+x$ axis, has scalar components

$$x = r \cos\theta \quad \text{and} \quad y = r \sin\theta . \tag{I.3a}$$

Conversely, a position vector with components x and y has magnitude and direction angle

$$r = \left(x^2 + y^2\right)^{1/2} \quad \text{and} \quad \theta = \tan^{-1}(y/x) \tag{I.3b}$$

respectively. We note that while the components x and y can be positive, negative, or zero, the vector magnitude is always non-negative, and positive for any nonzero vector. Also, the expression for θ does not contain the principal inverse tangent function; the appropriate branch of the inverse tangent (appropriate quadrant for θ) is determined from the signs of x and y. Similar but slightly more elaborate relations apply in three (or more!) dimensions, where more angles are needed to specify directions.

As numerical examples, consider a displacement $r = 200.$ m at $\theta = 45°.0$ to the $+x$ axis. The corresponding x and y components are

$$x = 141. \text{ m} \quad \text{and} \quad y = 141. \text{ m} . \tag{I.4a}$$

Conversely, components $x = 50.0$ m and $y = 86.6$ m correspond to a displacement of magnitude and direction

$$r = 100. \text{ m} \quad \text{and} \quad \theta = 60°.0 \tag{I.4b}$$

respectively. The arithmetic is left as practice for the student. These results are given consistently to three significant figures; the decimal point in the r values indicates the significance of the zeros.

The utility of the resolution procedure is illustrated in Fig. I.5. This shows that tail-to-tip vector addition proceeds "componentwise":

$$\begin{aligned}(\mathbf{A} + \mathbf{B})_x &= \mathbf{A}_x + \mathbf{B}_x \\ &= (A_x + B_x)\,\hat{\boldsymbol{x}} ,\end{aligned} \tag{I.5}$$

and similarly for the y (and z) components. In terms of magnitudes and directions, tail-to-tip addition involves the trigonometric "solution of triangles"; it becomes very cumbersome if more than two vectors are involved, and it is difficult even to envision in three dimensions. Componentwise addition is much more straightforward.

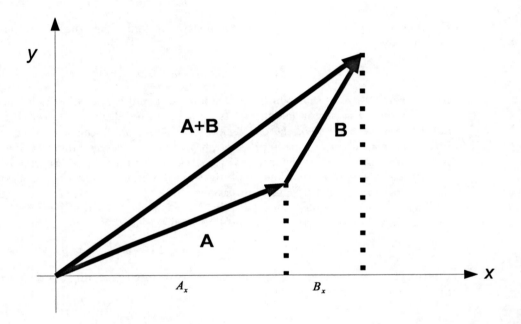

Figure I.5: Component of a vector sum is the sum of components.

In contrast to vectors, quantities that are described by a single number are called *scalars.* The name derives from the fact that multiplying a vector by a scalar multiplies its magnitude or scale while leaving its direction unchanged (if the scalar is positive) or reversing it (if negative). The magnitude and direction angle of a vector, or its scalar components, are all individually scalar quantities. Scalar and vector are not the only two possiblities; we shall encounter others at the end of Ch. III.

It should be emphasized that this approach to vectors in physics is rather pedestrian. Mathematicians start with a more fundamental, wide-ranging definition of vectors, treating magnitude and direction as lagniappes. Their approach has great power. But we shall not have immediate need of it, so we shall start from this physical definition.

Students are cautioned to avoid careless or haphazard use of vector notation. In this text, vectors are identified in boldface: **r**; by hand, it is customary to write an arrow over the symbol: $\vec{r}$. But this must be done consistently. Vectors cannot be set equal to scalars or added to scalars. Vectors can be added together or multiplied by scalars, and we shall see that there are several ways to multiply them together, but it is not possible to divide by a vector. Statements in which vector designations are misused are false.

Velocity **v**—*rate* of displacement

To address questions about how fast the rocket goes, we employ *velocity,* the time rate of change of position. In common parlance the terms velocity and speed are often interchanged, but in the code language of physics they are distinguished: Velocity is a vector quantity, while speed is a scalar.

When the speedometer of your car reads 60 miles per hour, say, how does it know where you will be in an hour? It doesn't, of course. To appreciate what it does know, i.e., what that reading means, we introduce the distinction between *average* and *instantaneous* velocities. Average velocity is the displacement $\Delta\mathbf{r}$ over a finite time interval Δt divided by that interval:

$$\langle \mathbf{v} \rangle \equiv \frac{\Delta\mathbf{r}}{\Delta t} \, , \tag{I.6}$$

the angled brackets denoting average over the time interval Δt. (The three-bar equals sign signifies equality by definition or identical equality, as constrasted with equality by coincidence.) As displacement is a vector and the time interval is, in Galilean kinematics, a scalar, this is a vector quantity. At first glance, the notion of velocity *at an instant* is nonsense: In zero time an object covers zero distance, and zero divided by zero is undefined. We give it meaning by taking the ratio over time intervals so short that the result would not change if we make the interval shorter. This is the notion, central to calculus, of a *limit,* and the ratio becomes the *derivative* of position:

$$\mathbf{v} \equiv \lim_{\Delta t \to 0} \frac{\Delta \mathbf{r}}{\Delta t}$$
$$\equiv \frac{d\mathbf{r}}{dt} \, . \tag{I.7}$$

Although the abstraction of a limit may get short shrift in modern calculus courses, it is a profound concept essential to physics, as it has been conceived since Newton[6] and Leibniz[7] introduced limits and derivatives in the seventeenth century.

We should note that the limit and derivative in Eq. (I.7) are of vector quantities, rather than of scalars as presented in introductory calculus. But since vectors can be subtracted as scalars can, and the notion of "closeness" can still be applied (two vectors are "close" if the magnitude of their difference is small), it is legitimate to extend limits and derivatives to vectors.

All measured velocities are average velocities, since all measurements take some finite time interval. Instantaneous velocities can only be approached. For example, the speedometer of your car actually knows nothing about where your car is at any time. It measures the speed at which the drive shaft (or transaxle for front-wheel drive) is turning and is calibrated to convert this into a linear speed for the car. This measurement is made over some milliseconds (for an old-fashioned speedometer that senses the twisting of a cable) or microseconds (for a modern electronic speedometer). Since the velocity of the car does not change appreciably during such intervals, this is a reasonable approximation to the instantaneous value. The speed reported by your GPS, by the way, is a different measurement entirely.

Speed, to be precise, is the magnitude of the velocity vector. Instantaneous speed is the magnitude of instantaneous velocity: $s \equiv |\mathbf{v}|$. The term average speed might be interpreted in two different ways: as the average of the magnitude of the (instantaneous) velocity or as the magnitude of the average velocity. The first interpretation is chosen because it is more useful: $\langle s \rangle \equiv \langle |\mathbf{v}| \rangle$. This represents the total distance—regardless of direction—covered divided by the time interval. For any motion that returns to its starting point, the average velocity over that interval is zero, since the net displacement is zero. To define the average speed of such motion to be zero would not represent the nature of the motion. All speeds, being magntitudes, are non-negative.

The distinction between velocity and speed is important. It is a much more serious offense to be driving at the wrong velocity than at the wrong speed: The latter is speeding; the former is driving east in the westbound lanes!

Acceleration—rate of change of velocity

Acceleration in physics refers to the rate of change of the velocity vector, not merely to increasing speed as in its ordinary usage. This quantity is key to

---

[6]Isaac Newton, 1642–1727
[7]Gottfried Wilhelm von Leibniz, 1646–1716

our understanding of motion. Galileo was apparently the first to grasp its significance as a kinematic variable.

As we do with any rate of change, we distinguish between average and instantaneous acceleration. Average acceleration is the vector change in velocity divided by the time interval:

$$\langle \mathbf{a} \rangle \equiv \frac{\Delta \mathbf{v}}{\Delta t} \, . \tag{I.8}$$

Instantaneous acceleration is the limit of this, as the time interval is made so small that the ratio doesn't change if we make it smaller:

$$\mathbf{a} \equiv \lim_{\Delta t \to 0} \frac{\Delta \mathbf{v}}{\Delta t}$$
$$\equiv \frac{d\mathbf{v}}{dt} \, . \tag{I.9}$$

It is this quantity that appears in the relationships which describe motion in Newton's formulation of mechanics—even though it is a limit, an abstraction.

Acceleration is a vector. It is helpful to resolve it into components parallel and perpendicular to the velocity vector:

$$\mathbf{a} = a_{\parallel} \, \hat{\boldsymbol{v}} + \mathbf{a}_{\perp} \, , \tag{I.10}$$

where $\hat{\boldsymbol{v}}$ is a vector of unit magnitude in the direction of the velocity vector $\mathbf{v}$. The first term describes the rate of change of the magnitude of $\mathbf{v}$, i.e., speeding up (for $a_{\parallel} > 0$) or slowing down (for $a_{\parallel} < 0$). The second term describes the rate of change of the direction of $\mathbf{v}$, i.e., the rate of turning.

Jerk, et cetera

This is so much fun that we could keep going. The rate of change of (instantaneous) acceleration has a name: It is called *jerk*. As with velocity and acceleration, we identify an average jerk and an instantaneous jerk, the time derivative of $\mathbf{a}$. Although the mechanics problems we shall examine in this text do not require use of the jerk, there are contexts in which it is important. Detailed study of the operation of parachutes, for example, requires it. Also, jerk is a figure of merit in the design of roller-coaster rides, a nontrivial engineering challenge.

Yet higher-order derivatives of kinematic variables can be constructed. But these are not usually encountered in mechanics problems and do not have commonly used names of their own.

B Kinematics of a rocket in vertical flight—no drag

We shall begin our examination of the motion of the model rocket with the simple case of purely vertical motion, under the assumption that aerodynamic effects are negligible. It is a source of discomfiture to some that the treatment of a phenomenon in physics is not *exactly true* in every detail; this causes confusion in the description of physics as "an exact science." But the map must be smaller than the territory it covers. If it were necessary to understand *everything* in order to understand *anything,* we would never get anywhere. In fact, a physics problem is a sequence of approximations in a race between realism and difficulty. The goal is to include enough features in the description to enable understanding, and to make predictions that can be compared to observations, before the analysis becomes so difficult that no further progress is possible. We shall begin with this simplest case, include more realistic assumptions later, and attempt to evaluate the validity of our descriptions as we go.

The rocket flight we shall consider is illustrated in Fig. I.6. It begins at Liftoff, to which we assign time $t_0 = 0$, when the fuel/oxidizer mixture in its engines is ignited electrically. (These are self-contained solid-fuel engines in cardboard casings, with ceramic nozzles, a plug of fuel-oxidizer mixture, a quantity of smoke powder that burns during the coasting portion of the flight, and a small charge to separate the rocket and its nose cone, deploying a small parachute to return the rocket to Earth. We shall consider these engines in more detail in Sec. G, following.) The rocket rises under power until its fuel is exhausted at Burnout, at time $t = t_1$. It then coasts upward to its maximum altitude at Apogee—point of greatest distance from the Earth—at time $t = t_2$. Neglecting aerodynamic forces, we shall disregard the parachute deployment and return of the rocket and simply assume that it falls freely back to ground level at Impact, at time $t = t_3$. What questions might we wish to ask and answer about this flight? We shall try to determine the time t, the altitude y, and the velocity v_y of the rocket at each of the four key points just described.

This is *not* a constant-acceleration problem, such as is usually treated in high-school physics courses. The formulas used in such a course cannot blithely be applied here.[8] The rocket accelerates upward while its engines are running; thereafter it coasts, with only the downward acceleration provided by the Earth's gravity. If we approximate the acceleration under power as constant, however, then this is a *piecewise*-constant-acceleration problem. (We shall see what this approximation entails in Sec. E, following.) We can analyze the motion by dividing it into phases, in each of which the acceleration *is* constant.

Phase I: $0 \leq t \leq t_1$, and $a_y = a_0$

Phase I is the powered portion of the flight between Liftoff and Burnout. At Liftoff the kinematic variables have the values $t_0 = 0$, $y_0 = 0$, and $v_y^{(0)} = 0$. During the interval $0 \leq t \leq t_1$, the acceleration—the net value from all contributions—is taken to have the constant, positive value $a_y = a_0$. The variables y, v_y, and a_y are vector components. In this vertical flight, these are the only nonzero components of the vector quantities $\mathbf{r}$, $\mathbf{v}$, and $\mathbf{a}$. The

---

[8]Students are *strongly* cautioned against the mindless search for and applicaton of formulas, here and anywhere.

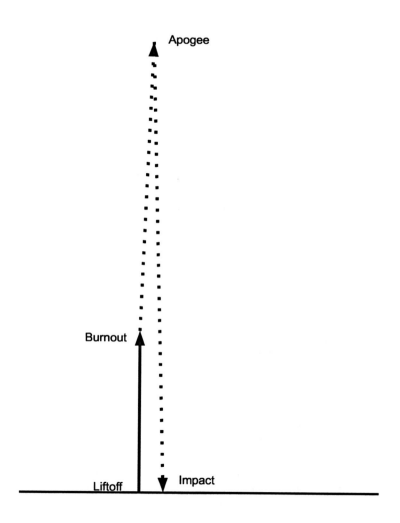

Figure I.6: Vertical flight of the rocket. The rocket accelerates under power between points Liftoff and Burnout; coasts upward, with fuel exhausted, between Burnout and Apogee; falls from Apogee to the ground at the Impact point.

acceleration a_0 and the burn time t_1 are determined by design parameters of the rocket's engines, i.e., they are inputs to our calculation.

Acceleration is the time derivative of velocity. Hence, the Fundamental Theorem of Calculus[9] implies that velocity is the time integral of acceleration:

$$
\begin{aligned}
v_y(t) &= v_y^{(0)} + \int_0^t a_y(t')\,dt' \\
&= v_y^{(0)} + \int_0^t a_0\,dt' \\
&= v_y^{(0)} + a_0 t \,,
\end{aligned} \tag{I.11}
$$

this last because a_0 is a constant, for all times t in Phase I. The reader should note that this author prefers the use of definite integrals, because constants of integration are icky. (This is a stylistic preference.) The reader should also note the distinction between the dummy variable of integration t' and the limit/argument t. (This is *not* a mere stylistic preference, although even some calculus textbooks are careless about it.)

Velocity is the time derivative of position. Hence, the same reasoning yields

$$
\begin{aligned}
y(t) &= y_0 + \int_0^t v_y(t')\,dt' \\
&= y_0 + \int_0^t (v_y^{(0)} + a_0 t')\,dt' \\
&= y_0 + v_y^{(0)} t + \tfrac{1}{2} a_0 t^2 \,,
\end{aligned} \tag{I.12}
$$

for all times t in Phase I. It is also useful to note that the time variable can be eliminated between Eqs. (I.11) and (I.12) to obtain the relation

$$
y - y_0 = \frac{v_y^2 - v_y^{(0)2}}{2a_0} \tag{I.13}
$$

among altitude, velocity, and acceleration.

Is it legitimate to integrate individual components of a vector relation, i.e., to integrate componentwise? Yes, because integration involves sums and limits. Vectors add componentwise, as we have seen. It is easy to show that two vectors are "close" if and only if their components are each "close," so limits, and derivatives and integrals, can be carried out componentwise—as long as the directions for the components are fixed, as here.

As a numerical example, to which we shall return throughout this chapter, let us take the values[10] $a_0 = +50.0$ m/s^2 and $t_1 = 3.00$ s. At burnout, the

---

[9]There is no reason not to plunge straight into the use of calculus at this point. It was for this, after all, that calculus was invented.

[10]For the cognoscenti, these values are approximately appropriate for a 300 g rocket of the type pictured, powered by three "C" engines.

velocity of the rocket is

$$
\begin{aligned}
v_y^{(1)} &= v_y^{(0)} + a_0 t_1 \\
&= 0 + (50.0 \text{ m/s}^2)(3.00 \text{ s}) \\
&= 150. \text{ m/s} .
\end{aligned}
\tag{I.14}
$$

Its altitude at burnout is

$$
\begin{aligned}
y_1 &= y_0 + v_y^{(0)} t_1 + \tfrac{1}{2} a_0 t_1^2 \\
&= 0 + 0 + \tfrac{1}{2}(50.0 \text{ m/s}^2)(3.00 \text{ s})^2 \\
&= 225. \text{ m} .
\end{aligned}
\tag{I.15}
$$

Students may wish to note the presentation style illustrated here. Equations are stacked: a symbolic expression or expressions, numerical values inserted at the end, followed by the final answer. Expressions connected by the symbol $=$ must be equal; the symbol does not mean "and the next step is." Once numerical values are inserted, correct significant figures and units are maintained throughout. Here, three-significant-figure values are multiplied; the precision of the product is that of the least precise factor, so the result is good to three significant figures. (The factor $1/2$ is exact, good to an arbitrary number of significant figures. If the trailing zeros in the value 3.00 s had been dropped, the results would then have been good only to one significant figure. The lost precision is not regained.) Units are combined as algebraic factors. The final result is presented with the correct number of significant figures and the correct units. These features are important—there is no reward for being careless with them.

We should also note that the only *physics* inputs in this analysis are the assertion $a_y = a_0$ and the values of a_0 and t_1. All else follows immediately from the *definitions* of velocity and acceleration.

Phase II: $t_1 \le t \le t_2$ and $a_y = -g$ (free fall)

At Burnout, engine thrust ceases. The rocket coasts upward, affected only—if aerodynamic effects are neglected—by the downward acceleration imposed by the Earth's gravity. Motion with only that acceleration is called "free fall." This is a more restrictive use of the term than, for example, that of skydivers, who describe the portion of their motion before the opening of a parachute as free fall. They, however, are subject to the accelerations both of gravity and of aerodynamic drag. (A skydiver for whom aerodynamic drag is entirely neglgible is in very big trouble.) Here we shall adhere to the gravity-only interpretation of free fall.

We shall examine the Earth's gravity in much more detail in Ch. IV. Here, we shall simply assign the acceleration a new constant value: $a_y = -g$. To avoid confusion, we take g to be a positive value, the *magnitude* of the gravitational acceleration. (Different texts employ different conventions for this symbol.) The rocket slows to a stop at the highest point in its flight, termed Apogee. The velocity at Apogee is $v_y^{(2)} = 0$ (exactly, and in any units). Relation (I.11), applied to the interval between Burnout and Apogee, implies

$$0 = v_y^{(1)} - g(t_2 - t_1) \ . \tag{I.16a}$$

This gives

$$t_2 = t_1 + \frac{v_y^{(1)}}{g} \tag{I.16b}$$

for the total time from Liftoff to Apogee. Relation (I.13), likewise applied to the interval between Burnout and Apogee, yields

$$y_2 = y_1 + \frac{v_y^{(1)2}}{2g} \tag{I.17}$$

for the altitude at Apogee, the maximum height the rocket reaches.

Using the previous numerical results and the approximate value $g = 9.81 \text{ m/s}^2$, we can calculate

$$\begin{aligned} t_2 &= t_1 + \frac{v_y^{(1)}}{g} \\ &= 3.00 \text{ s} + \frac{150. \text{ m/s}}{9.81 \text{ m/s}^2} \\ &= 18.3 \text{ s} \end{aligned} \tag{I.18}$$

for the time to Apogee. The matter of significant figures requires care. The precision of a sum like this is determined not by the number of significant figures of each term, but by the number of decimal places in the least precise term (in the same units). The second term is good to three significant figures, hence, only to tenths of a second. The sum is thus good only to tenths of a second, three significant figures here, despite the fact that the first term is known to hundredths of a second. The altitude at Apogee is given by

$$\begin{aligned} y_2 &= y_1 + \frac{v_y^{(1)2}}{2g} \\ &= 225. \text{ m} + \frac{(150. \text{ m/s})^2}{2(9.81 \text{ m/s}^2)} \\ &= 1.37 \times 10^3 \text{ m} \ . \end{aligned} \tag{I.19}$$

Similar considerations dictate that this value is uncertain to a few tens of meters. The scientific notation shows this clearly. To write this as 1370. m misrepresents the precision of the result.

Phase III: $t_2 \leq t \leq t_3$ and $a_y = -g$ (free fall)

With parachute deployment and descent disregarded, the rocket falls from Apogee, with the downward acceleration of gravity, striking the ground at the point labeled Impact. We note that a_y is negative both in Phase II and in Phase III (in fact, we could have analyzed them as a single phase), but that v_y is positive in Phase II and negative in Phase III. That is, the algebraic signs of these quantities can vary independently.

The Impact point is identified by the altitude value $y_3 = 0$. Relation (I.12), applied to the interval between Apogee and Impact, allows us to calculate the time t_3 at Impact:

$$y_3 = y_2 + v_y^{(2)}(t_3 - t_2) - \tfrac{1}{2}g(t_3 - t_2)^2$$

i.e., $\quad 0 = y_2 - \tfrac{1}{2}g(t_3 - t_2)^2$. $\hspace{3cm}$ (I.20a)

This implies

$$t_3 = t_2 + (2y_2/g)^{1/2} . \hspace{3cm} \text{(I.20b)}$$

Relation (I.13), similarly applied, yields the velocity at Impact:

$$y_3 - y_2 = \frac{v_y^{(3)2} - v_y^{(2)2}}{-2g} \hspace{3cm} \text{(I.21a)}$$

implies

$$v_y^{(3)} = -(2gy_2)^{1/2} , \hspace{3cm} \text{(I.21b)}$$

where the appropriate sign for $v_y^{(3)}$ is as shown.

Using our previous numerical results, we can determine the remaining values. The time of Impact is

$$t_3 = t_2 + (2y_2/g)^{1/2}$$

$$= 18.3 \text{ s} + \left(\frac{2(1.37 \times 10^3 \text{ m})}{9.81 \text{ m/s}^2}\right)^{1/2}$$

$$= 35.0 \text{ s} ; \hspace{3cm} \text{(I.22)}$$

here we must take the same care with significant figures as before. The velocity at Impact is

$$v_3 = -(2gy_2)^{1/2}$$

$$= -[2(9.81 \text{ m/s}^2)(1.37 \times 10^3 \text{ m})]^{1/2}$$

$$= -164. \text{ m/s} . \hspace{3cm} \text{(I.23)}$$

The speed of the rocket is not greatest at Burnout. The rocket regains Burnout speed as it reaches Burnout altitude on its descent, then continues to increase speed as it falls the rest of the way to Impact. In fact its speed in descent is the same as that in ascent at each altitude between Burnout and Apogee, i.e., at each altitude reached ascending in Phase II, then descending in Phase III. We may also note that Phases II and III are characterized by the same acceleration and could have been treated as a single phase.

Victory! We have completed the analysis of the rocket's flight—answered all the questions asked—under the assumptions we have made. Is the analysis valid? The proper scientific way to answer that question is to launch a model rocket with suitable characteristics, measure the kinematic features of its flight, and compare the measurements with our calculated results. That is not feasible here, though it would be an interesting challenge for a laboratory course. Instead, we shall refine our description of the flight in Sec. E following and compare the new results obtained with these. If the two sets of calculations agree (to within the precision of possible measurements), then the validity of the description here is affirmed.

C   Kinematics of angled flight—no drag

The use of vectors allows us to generalize our analysis to the case of rocket flight in any direction, still treating aerodynamic forces as negligible. This case would apply to a rocket launched to strike a distant target or to deliver a projectile over a wall. This goal—superior artillery—was one of Galileo's motivations for his studies in mechanics. This illustrates an important fact: Science does not exist in a vacuum. It is a part of human culture, undertaken to address human needs—benevolent or malevolent.

This more general flight path is illustrated in Fig. I.7. The rocket accelerates under power in Phase I, from Liftoff to Burnout. In the absence of aerodynamic forces on its tail fins, it does not fly in the direction it is pointing. It "sideslips" due to the combined effects of engine thrust and gravity. (We shall examine this in more detail in Sec. E following.) After Burnout it coasts under the influence of gravity, upward to Apogee and downward to Impact; we shall treat this entire portion of the motion as a single Phase II. Now our analysis must answer additional questions: In addition to the time, altitude, and vertical velocity of the rocket, its down-range distances and velocities, and the *shape* of its trajectory, are of interest.

Phase I: $0 \leq t \leq t_1$, and $\mathbf{a} = \mathbf{a}_0$ (net)

Anticipating results from Sec. E following, we take the net acceleration of the rocket under power to be a constant vector $\mathbf{a}_0$. The *vector* definitions of acceleration and velocity can be integrated, as we did with single components previously, to determine the rocket's velocity and position in Phase I:

$$
\begin{aligned}
\mathbf{v}(t) &= \mathbf{v}_0 + \int_0^t \mathbf{a}(t')\, dt' \\
&= \mathbf{v}_0 + \int_0^t \mathbf{a}_0\, dt' \\
&= \mathbf{v}_0 + \mathbf{a}_0 t \ ,
\end{aligned}
\tag{I.24}
$$

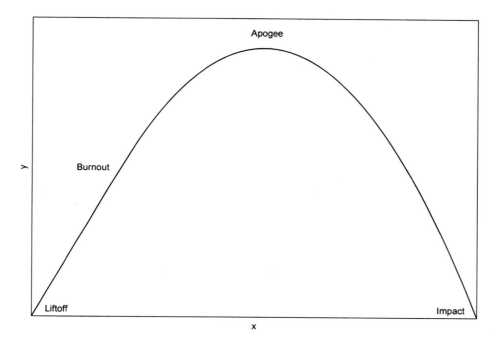

Figure I.7: Angled flight of the rocket. Launched from rest, the rocket accelerates under power until Burnout, then coasts under the influence of gravity through Apogee to Impact.

and

$$\mathbf{r}(t) = \mathbf{r}_0 + \int_0^t \mathbf{v}(t')\, dt'$$

$$= \mathbf{r}_0 + \int_0^t (\mathbf{v}_0 + \mathbf{a}_0 t')\, dt'$$

$$= \mathbf{r}_0 + \mathbf{v}_0 t + \tfrac{1}{2}\mathbf{a}_0 t^2 \,, \tag{I.25}$$

respectively.

The shape of the rocket's path follows from Eq. (I.25). If the rocket is launched from rest ($\mathbf{v}_0 = \mathbf{0}$), as here,[11] or if the initial velocity and acceleration vectors are parallel or antiparallel ($\mathbf{v}_0 \parallel \pm\mathbf{a}_0$), then this portion of the trajectory is *a straight line* through point $\mathbf{r}_0$. If not ($\mathbf{v}_0 \not\parallel \pm\mathbf{a}_0$), then distance traveled in the direction of $\mathbf{a}_0$ is a quadratic function of distance in the direction of the component of $\mathbf{v}_0$ perpendicular to $\mathbf{a}_0$. That is, the path is *a portion of a parabola*.

For a numerical example to use throughout this section, let us take

$$\mathbf{a}_0 = (52.5 \text{ m/s}^2)(\cos 45°.0\, \hat{\boldsymbol{x}} + \sin 45°.0\, \hat{\boldsymbol{y}}) \,, \tag{I.26}$$

$\mathbf{v}_0 = \mathbf{0}$, and $t_1 = 3.00$ s, with the direction of $\hat{\boldsymbol{x}}$ horizontally down range and that of $\hat{\boldsymbol{y}}$ vertically upward. One of the nice features of Cartesian[12] coordinates is that we are free to choose the axes for maximum convenience. We can also choose the origin at the Liftoff point, i.e., $\mathbf{r}_0 = \mathbf{0}$. The preceding results yield

$$\mathbf{v}_1 = \mathbf{v}_0 + \mathbf{a}_0 t_1$$

$$= \mathbf{0} + (52.5 \text{ m/s}^2)(\cos 45°.0\, \hat{\boldsymbol{x}} + \sin 45°.0\, \hat{\boldsymbol{y}})(3.00 \text{ s})$$

$$= (111. \text{ m/s})\, \hat{\boldsymbol{x}} + (111. \text{ m/s})\, \hat{\boldsymbol{y}}$$

$$= (158. \text{ m/s})(\cos 45°.0\, \hat{\boldsymbol{x}} + \sin 45°.0\, \hat{\boldsymbol{y}}) \,, \tag{I.27}$$

and

$$\mathbf{r}_1 = \mathbf{r}_0 + \mathbf{v}_0 t_1 + \tfrac{1}{2}\mathbf{a}_0 t_1^2$$

$$= \mathbf{0} + \mathbf{0} + \tfrac{1}{2}(52.5 \text{ m/s}^2)(\cos 45°.0\, \hat{\boldsymbol{x}} + \sin 45°.0\, \hat{\boldsymbol{y}})(3.00 \text{ s})^2$$

$$= (167. \text{ m})\, \hat{\boldsymbol{x}} + (167. \text{ m})\, \hat{\boldsymbol{y}}$$

$$= (236. \text{ m})(\cos 45°.0\, \hat{\boldsymbol{x}} + \sin 45°.0\, \hat{\boldsymbol{y}}) \,, \tag{I.28}$$

where both results are shown in x-y component form and magnitude-direction form.

Phase II: $t \geq t_1$, and $\mathbf{a} = \mathbf{g} = -g\,\hat{\boldsymbol{y}}$ (projectile)

In Phase II the rocket moves—again, with aerodynamic effects neglected—with only the acceleration due to the Earth's gravity: $\mathbf{a} = \mathbf{g} = -g\,\hat{\boldsymbol{y}}$. Care should be

---

[11]The reader should note the distinction between $\mathbf{0}$, the zero vector, and 0, the number zero. These are different objects.

[12]René Descartes, 1596–1650

taken with this notation. The vector **g** requires no negative sign. It is the vertical component that is negative (if the $+y$ direction is taken upward), written $-g$ with the symbol g a positive value. Motion with only this acceleration is often called "projectile motion." Again, this is physics code language. In ordinary usage, a projectile is an object that is projected, i.e., that moves without its own propulsion, but is subject both to gravity and drag. Here, we shall reserve the term "projectile motion" for gravity-only motion.

Starting from the Burnout point, the definitions of acceleration and velocity can be integrated to determine the remainder of the rocket's trajectory. Its velocity is given by:

$$
\begin{aligned}
\mathbf{v}(t) &= \mathbf{v}_1 + \int_{t_1}^{t} \mathbf{a}(t')\, dt' \\
&= v_1(\cos\theta_1\,\hat{\boldsymbol{x}} + \sin\theta_1\,\hat{\boldsymbol{y}}) + \int_{t_1}^{t} (-g\,\hat{\boldsymbol{y}})\, dt' \\
&= v_1\cos\theta_1\,\hat{\boldsymbol{x}} + [v_1\sin\theta_1 - g(t - t_1)]\,\hat{\boldsymbol{y}}\,,
\end{aligned} \tag{I.29}
$$

where we have introduced symbols v_1 for the magnitude and θ_1 for the angle, from the $+x$ axis, of the vector velocity $\mathbf{v}_1$ at Burnout. The position vector or trajectory is given by:

$$
\begin{aligned}
\mathbf{r}(t) &= \mathbf{r}_1 + \int_{t_1}^{t} \mathbf{v}(t')\, dt' \\
&= (x_1\,\hat{\boldsymbol{x}} + y_1\,\hat{\boldsymbol{y}}) + \int_{t_1}^{t} \{v_1\cos\theta_1\,\hat{\boldsymbol{x}} + [v_1\sin\theta_1 - g(t - t_1)]\,\hat{\boldsymbol{y}}\}\, dt' \\
&= [x_1 + (v_1\cos\theta_1)(t - t_1)]\,\hat{\boldsymbol{x}} + [y_1 + (v_1\sin\theta_1)(t - t_1) - \tfrac{1}{2}g(t - t_1)^2]\,\hat{\boldsymbol{y}}\,,
\end{aligned} \tag{I.30}
$$

with x_1 and y_1 the position coordinates at Burnout. The x and y coordinates of the rocket throughout Phase II can be read from this expression.

Result (I.30) indicates that the horizontal and vertical motions of the rocket are, *independently*, motion with constant velocity $v_1\cos\theta_1\,\hat{\boldsymbol{x}}$ and free-fall motion with "initial" velocity $v_1\sin\theta_1\,\hat{\boldsymbol{y}}$. This is a famous result attributed to Galileo. It implies, for example, that a bullet dropped from rest and a bullet fired horizontally from the same height will not only hit the ground at the same time, but will also be at the same height instant by instant the whole way down. The vertical motions of the two bullets, independent of their horizontal motions, are identical. *However,* it is very important to note that this is true if aerodynamic effects are negligible, as well as in other special cases. It is not true in general, as we shall see.

The shape of the trajectory is revealed by examining the y coordinate as a function of the x coordinate. Eliminating t from the two components of

Eq. (I.30) gives

$$
\begin{aligned}
y &= y_1 + \tan\theta_1\,(x - x_1) - \frac{g}{2v_1^2\cos^2\theta_1}\,(x - x_1)^2 \\
&= y_1 - \frac{g}{2v_1^2\cos^2\theta_1}\left((x - x_1)^2 - \frac{v_1^2\sin(2\theta_1)}{g}\,(x - x_1)\right. \\
&\qquad\qquad\qquad\qquad \left. + \frac{v_1^4\sin^2(2\theta_1)}{4g^2} - \frac{v_1^4\sin^2(2\theta_1)}{4g^2}\right) \\
&= y_1 + \frac{v_1^2\sin^2\theta_1}{2g} - \frac{g}{2v_1^2\cos^2\theta_1}\left[x - \left(x_1 + \frac{v_1^2\sin(2\theta_1)}{2g}\right)\right]^2 , \qquad (\text{I.31})
\end{aligned}
$$

where the trigonometric identities $\tan\theta = \sin\theta/\cos\theta$ and $\sin(2\theta) = 2\sin\theta\cos\theta$ have been used freely.[13] The addition and subtraction of the extra term $v_1^4\sin^2(2\theta_1)/(4g^2)$ in the second expression here, leaving the value of the expression unchanged, is an example of an old trick called "completing the square." It leaves the result in a form from which we can read off many features of the trajectory. This portion of the rocket's path is a downward-opening parabola, with its peak where the quantity in brackets is zero, at coordinates

$$
x_2 = x_1 + \frac{v_1^2\sin(2\theta_1)}{2g} , \qquad (\text{I.32a})
$$

and

$$
y_2 = y_1 + \frac{v_1^2\sin^2\theta_1}{2g} , \qquad (\text{I.32b})
$$

with subscript 2 for Apogee. The relationship between t and x in Eq. (I.30) then gives

$$
t_2 = t_1 + \frac{v_1\sin\theta_1}{g} , \qquad (\text{I.32c})
$$

the time at Apogee.

It remains only to calculate the time, down-range distance, and velocity at the point of Impact. The first two can be obtained from Eqs. (I.30) and (I.31), respectively, by imposing $y_3 = 0$ and solving. The time is given by

$$
\begin{aligned}
t_3 &= t_1 + \frac{v_1\sin\theta_1}{g} + \left(\frac{v_1^2\sin^2\theta_1}{g^2} + \frac{2y_1}{g}\right)^{1/2} \\
&= t_2 + \left(\frac{2y_2}{g}\right)^{1/2} , \qquad (\text{I.33})
\end{aligned}
$$

where the sign of the square root is chosen to obtain $t_3 > t_2$. The down-range distance is

$$
x_3 = x_2 + \left(\frac{2v_1^2\cos^2\theta_1}{g}\,y_2\right)^{1/2} . \qquad (\text{I.34})
$$

---

[13] It is assumed that students remember everything they learned in high school trigonometry or know where to find these things if they do not.

The velocity at Impact follows from Eqs. (I.29) and (I.33):

$$\mathbf{v}_3 = v_1 \cos\theta_1 \,\hat{\boldsymbol{x}} + [v_1 \sin\theta_1 - g(t_3 - t_1)] \,\hat{\boldsymbol{y}} \,, \tag{I.35a}$$

with corresponding speed

$$v_3 = \{v_1^2 \cos^2\theta_1 + [v_1 \sin\theta_1 - g(t_3 - t_1)]^2\}^{1/2} \,. \tag{I.35b}$$

As in vertical flight, the speed of the rocket is the same ascending and descending at each altitude above Burnout. The rocket regains Burnout speed when it returns to Burnout altitude y_1, then continues to gain speed, reaching the highest speed of its flight at Impact.

Using our results from Phase I, we can complete the analysis of our numerical example. The rocket reaches Apogee at time

$$\begin{aligned}
t_2 &= t_1 + \frac{v_1 \sin\theta_1}{g} \\
&= 3.00 \text{ s} + \frac{(158. \text{ m/s}) \sin 45°.0}{9.81 \text{ m/s}^2} \\
&= 14.4 \text{ s} \,,
\end{aligned} \tag{I.36a}$$

down-range distance

$$\begin{aligned}
x_2 &= x_1 + \frac{v_1^2 \sin(2\theta_1)}{2g} \\
&= 167. \text{ m} + \frac{(158. \text{ m/s})^2 \sin 90°.0}{2(9.81 \text{ m/s}^2)} \\
&= 1.43 \text{ km} \,,
\end{aligned} \tag{I.36b}$$

and altitude

$$\begin{aligned}
y_2 &= y_1 + \frac{v_1^2 \sin^2\theta_1}{2g} \\
&= 167. \text{ m} + \frac{(158. \text{ m/s})^2 \sin^2 45°.0}{2(9.81 \text{ m/s}^2)} \\
&= 799. \text{ m} \,.
\end{aligned} \tag{I.36c}$$

The unit km is used instead of scientific notation in Eq. (I.35b). The rocket strikes the ground at time

$$\begin{aligned}
t_3 &= t_2 + \left(\frac{2y_2}{g}\right)^{1/2} \\
&= 14.4 \text{ s} + \left(\frac{2(799. \text{ m})}{9.81 \text{ m/s}^2}\right)^{1/2} \\
&= 27.1 \text{ s} \,,
\end{aligned} \tag{I.37a}$$

and down-range distance

$$x_3 = x_2 + \left(\frac{2v_1^2 \cos^2 \theta_1}{g} y_2 \right)^{1/2}$$

$$= 1.43 \text{ km} + \left(\frac{2(158 \text{ m/s})^2 \cos^2 45°.0}{9.81 \text{ m/s}^2} (799. \text{ m}) \right)^{1/2}$$

$$= 2.85 \text{ km} ,$$
(I.37b)

and with speed

$$v_3 = \{v_1^2 \cos^2 \theta_1 + [v_1 \sin \theta_1 - g(t_3 - t_1)]^2\}^{1/2}$$

$$= \{(158. \text{ m/s})^2 \cos^2 45°.0$$

$$+ [(158. \text{ m/s}) \sin 45°.0 - (9.81 \text{ m/s}^2)(27.1 \text{ s} - 3.00 \text{ s})]^2\}^{1/2}$$

$$= 168. \text{ m/s} .$$
(I.37c)

A second victory: Kinematic definitions plus our assertions about the acceleration of the rocket have enabled us to analyze this more general example of the rocket's motion.

A Special Case

An important special case of angled flight arises if we assume that the acceleration $\mathbf{a}_0$ in Phase I is of very large magnitude and its duration t_1 is very small, such that velocity $\mathbf{v}_1$ is of finite magnitude but displacement $\mathbf{r}_1$ is infinitesimal, i.e., negligibly small.[14] As if, for example, the rocket were fired from a mortar. This is the case of a "ground-to-ground" projectile, such as is usually treated in introductory physics courses. In this approximation, Eq. (I.33) reduces to the ground-to-ground flight time

$$T_{gg} = \frac{2v_1 \sin \theta_1}{g} \; ;$$
(I.38a)

Eqs. (I.34), (I.32a), and (I.32b) reduce to the ground-to-ground range

$$R_{gg} = \frac{v_1^2 \sin(2\theta_1)}{g} \; ;$$
(I.38b)

and Eq. (I.32b) reduces to the peak height

$$H = \frac{v_1^2 \sin^2 \theta_1}{2g} \; .$$
(I.38c)

---

[14]Strictly speaking, these conditions are meaningless, because quantities with units cannot be said to be "large" or "small" except in comparison to other quantities with the same units. The actual approximations needed here are $a_0 \gg g$ and $t_1 \ll (gR_\oplus)^{1/2}/a_0$, where $R_\oplus$ is the radius of the Earth.

These familiar formulae imply famous results, e.g., for a fixed value of v_1, launch angle $\theta_1 = 45°$ yields the maximum ground-to-ground range. It is important to remember, however, that both formulae and results apply only under the assumptions and in the approximation from which they are derived.

D   Theory of Relativity (Galileo)

At this point it is appropriate to introduce the Theory of Relativity, which was incorporated into physics by ... Galileo. Albert Einstein[15] modified the theory in 1905 and 1915 to achieve new understanding of mechanics, but the idea was introduced by Galileo some three hundred years earlier. It is rooted in the relationships among kinematic descriptions of motions by different observers—more properly, different sets of observers, or "reference frames"—in motion with respect to one another.

Kinematic quantities relative to a moving observer

Strictly, a *reference frame* is an abstraction: an infinite collection of observers, all at rest with respect to one another, who fill space and measure time so that they can measure and describe the motion of any object. Suppose that my friends and I constitute one reference frame, establish our coordinate system, synchronize our clocks, and so on. Another observer—call her O, for "Observer"—and her friends do the same, and suppose that O is moving, in my frame of reference, on vector trajectory $\mathbf{r}_O(t)$. We both seek to describe the motion of some object—a bird in the sky, a star in the heavens, whatever. I shall assign position $\mathbf{r}(t)$, velocity $\mathbf{v}(t)$, and acceleration $\mathbf{a}(t)$ to this object. O assigns similar kinematic variables to the object, distinguished here by primes: $\mathbf{r}'$, $\mathbf{v}'$, and $\mathbf{a}'$. How are these two descriptions related?

Vectors make it easy to answer this question. The situation is as depicted in Fig. I.8. O assigns to the object *relative position* vector

$$\mathbf{r}'(t') = \mathbf{r}(t) - \mathbf{r}_O(t) \ . \tag{I.39a}$$

Since these expressions are equal at all times, their time rates of change are equal. Hence, O assigns the object *relative velocity*

$$\mathbf{v}'(t') = \mathbf{v}(t) - \mathbf{v}_O(t) \tag{I.39b}$$

and *relative acceleration*

$$\mathbf{a}'(t') = \mathbf{a}(t) - \mathbf{a}_O(t) \ , \tag{I.39c}$$

and since we all measure time the same way,[16] time

$$t' = t \ , \tag{I.39d}$$

which justifies the equating of time derivatives in Eqs. (I.39b) and (I.39c) here.

---

[15] Albert Einstein, 1879–1955

[16] Galileo and his successors, down to Einstein, had no reason to question this simple assertion. But it is not quite as innocuous as it looks.

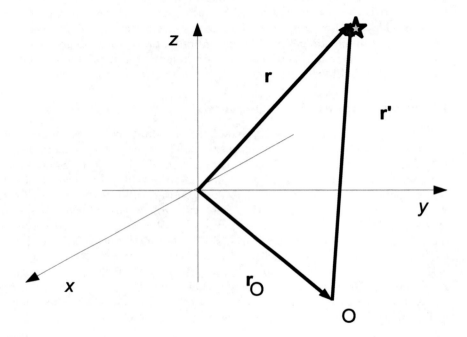

Figure I.8: Position vectors assigned to the same object by two different observers.

Galilean transformations

In the special case $\mathbf{a}_O(t) \equiv \mathbf{0}$, two sets of observers moving *uniformly*—at constant velocity—with respect to one another, Eqs. (I.39) define a *Galilean transformation*. Galileo's profound hypothesis was that *two reference frames related by a Galilean transformation are equally valid*. That is, any motion transformed via a Galilean transformation is another equally valid kinematic description, viz., the same motion as measured by a set of observers in uniform motion with respect to the first.

Galileo's purpose was to answer the question: *Can you play catch on a 747 jet in flight?* In this security-conscious age, it is not possible to answer this in true scientific fashion—by doing it. But the answer is yes: Space allowing, you can play catch on a jet in straight, level flight at constant speed, needing no more effort or skill than in a game played on the ground. Is this not remarkable, since both players and ball are actually moving in excess of 900 km/h? Certainly remarkable—and dangerous!—to an Aristotelian mechanician believing that objects return to a "natural state of rest" when released while plane and passengers continue to hurtle through the sky. But Galileo and experience assert that the game on board the plane proceeds exactly as does the game on the ground.

Of course, Galileo did not actually frame the problem in terms of a game of catch on a 747 jet. He wrote of a ship sailing steadily on calm waters. If a cannonball is dropped from the crow's nest, will it fall behind the mast while the ship sails out from underneath it, or will it fall right at the foot of the mast? The correct answer is the latter if aerodynamic effects can be neglected: The cannonball, once released, "remembers" its forward motion, moving with the ship as it descends straight down the mast. ("Remembering the motion" is the notion of inertia, which Galileo did not formulate explicitly; it took Newton[17] to do that. But Galileo was on the right track.) And the straight-down motion of the cannonball, as seen on the ship, is as valid a description as the parabolic trajectory described by observers on shore.

What was Galileo after? The fate of dropped cannonballs was not that urgent a scientific question at the time. But Galileo was enmeshed in the great controversy between the geocentric model of the solar system associated with Ptolemy[18] and the heliocentric model of Aristarchus,[19] reintroduced by Copernicus.[20] It was common in the youth of this author to scorn Galileo's contemporaries: They were fools; they didn't do scientific experiments. They weren't fools, but they didn't *trust* experiments, which dealt with corruptible matter and were prone to error. What did they trust? Logical arguments, as epitomized in the geometric theorems and proofs of Euclid.[21] Galileo was seeking a refutation of the argument that the Earth could not be orbiting the Sun, else its

---

[17] Isaac Newton, 1642–1727
[18] Claudius Ptolemaeus, second century CE
[19] Aristarchus of Samos, third century BCE
[20] Niklas Koppernigk, 1473–1543
[21] Euclid, fourth to third century BCE

motion would cause dropped objects to "fall to the back." No, argued Galileo, kinematics on the moving Earth—as on the ship or the jet—works just as well, and gives the same results, as that on some "stationary" platform.

Galilean relativity asserts that reference frames related by a Galilean transformation are all equally valid. This begs the question of how we identify that first valid reference frame. Consider, for example, two reference frames *not* connected by a Galilean transformation: one frame at rest relative to the ground, and one that "rides along" with some projectile. Suppose the motion of some other projectile is described in both reference frames, i.e., we seek the description of one projectile seen from another. In the ground frame, the target projectile has position and velocity

$$\mathbf{r}(t) = \mathbf{r}_1 + \mathbf{v}_1 t + \tfrac{1}{2}\mathbf{g}t^2 \tag{I.40a}$$

$$\mathbf{v}(t) = \mathbf{v}_1 + \mathbf{g}t \,, \tag{I.40b}$$

while the observer projectile has position and velocity

$$\mathbf{r}(t) = \mathbf{r}_{1,O} + \mathbf{v}_{1,O}t + \tfrac{1}{2}\mathbf{g}t^2 \tag{I.41a}$$

$$\mathbf{v}(t) = \mathbf{v}_{1,O} + \mathbf{g}t \tag{I.41b}$$

using results from Sec. C preceding. Transformation (I.39) then gives relative position and velocity

$$\mathbf{r}'(t) = (\mathbf{r}_1 - \mathbf{r}_{1,O}) + (\mathbf{v}_1 - \mathbf{v}_{1,O})\,t \tag{I.42a}$$

$$\mathbf{v}'(t) = \mathbf{v}_1 - \mathbf{v}_{1,O} \,. \tag{I.42b}$$

Seen from one projectile, any other projectile travels in a straight line with constant velocity! (It is required only that the projectiles be "nearby" compared to the size of the Earth, so that **g** is the same for both.) This is the true significance of the "Monkey and Hunter" demonstration shown in some introductory physics courses. It raises this issue: Our two reference frames are not equivalent in the Galilean sense. But which one is "valid"? Which one offers the simpler description of motion? These questions open a door, on the other side of which is Einstein's 1915 General Theory of Relativity.

It may occur to the student that all this discussion of reference frames is rather soporific. As will be seen, however, the matter of reference frames has a wider-reaching influence on the nature of physics than is yet apparent.

E   Flight dynamics—thrust and drag

To further explore the motion of the rocket, and to refine our analysis, we must move beyond kinematic description to *dynamics*—causes and effects. Our guide in this is the set of three Laws of Motion formulated by Isaac Newton in 1666.

Newton was a son of the rural gentry from Woolesthorpe in the north of England, an area from which both Stephen Hawking[22] and Margaret Thatcher[23] would later spring. Newton did not, apparently, fit the modern image of the "child prodigy," but he was bright enough to go up to Trinity College at the University of Cambridge.[24] There, he so distinguished himself that his tutor, Isaac Barrow,[25] resigned his endowed chair in 1669 so that Newton could have the position. That chair, the Lucasian Professorship of Mathematics, still exists— Stephen Hawking recently retired from it. The list of Lucasian Professors reads like a Who's Who of British science and mathematics. Barrow, by the way, did all right: He became head of Christ College, just down the street from Trinity.

In 1665 the plague broke out in Cambridge. In an age before antibiotics, Newton followed standard protocol—he got out of town. The university closed for the emergency, and Newton returned to the family farm to work on his research. That work would include Newton's Three Laws of Motion and his Law of Universal Gravitation, about which more will be presented. But it would not be until 1687, at the urging and at the expense of his friend Edmund Halley[26] (of Halley's Comet fame), that Newton would publish his results in his *Philosophiae Naturalis Principia Mathematica* (Mathematical Principles of Natural Philosophy) [New62], indisputably one of the most influential books ever published.

Newton's Laws of Motion

Newton's First Law is called the Law of Inertia: *A body at rest remains at rest; a body in motion remains in motion, in a straight line at constant speed, unless acted on by an outside force.* More concisely: *A free body moves with constant (or zero) velocity.*

The content of this deceptively simple assertion is multifaceted. First, it expresses the results of experiments carried out earlier by Galileo. By observing the motions of bodies allowed to slide down smooth inclined planes onto smooth surfaces, Galileo refuted the assertion of Aristotle that bodies come to rest when the force or impetus propelling them ceases. Galileo concluded, as an extrapolation of his results, that a body subject to no friction or other force would continue to move forever, in a straight line with unchanging speed.

Why did Galileo get this right and Aristotle get it wrong? Was Galileo smarter than Aristotle? It is difficult to concoct a valid comparison. Aristotle must have been describing objects of his experience, in which friction brings moving bodies swiftly to rest. Galileo was able to make the conceptual leap to a world without friction. Certainly, Galileo had better equipment with which to perform his experiments. But Galileo was also heir to fifteen centuries of a tradition that taught that the things of the Earth were perishable and imper-

---

[22]Stephen Hawking, 1942–
[23]Margaret Thatcher, 1925–2013
[24]You could visit Newton's rooms at Trinity, except they're still in use.
[25]Isaac Barrow, 1630–1677
[26]Edmund Halley, 1656–1742

fect, while Heaven was flawless and eternal. Galileo the astronomer may have been focused on the frictionless motion of celestial bodies.[27] The point, made apparent once again, is that science does not exist in isolation from the rest of the culture from which it grows.

There is more to the First Law. It is often taken to be a statement about reference frames. Many texts emphasize that the First Law may be applied only in what is called an *inertial reference frame*. Galileo's relativity principle asserts that any reference frame related to an inertial frame by a Galilean transformation is another inertial frame, but this begs the question of how to identify the first one. Prior to the twentieth century it was customary to define an inertial frame as any reference frame at rest or moving uniformly (with constant velocity) with respect to "the fixed stars," but this turns out to be problematic. Strictly, an inertial frame is a reference frame in which the Law of Inertia is obeyed, i.e., in which free bodies move with constant velocity. This is circular: The First Law is obeyed in those reference frames in which the First Law is obeyed. What can be the content of such a law? Simply this: *There are such reference frames.* The First Law asserts that it is possible to set up a frame of reference in which free bodies move with constant velocity.

The First Law is also an assertion about the measurement of time. When we say a body moves at constant speed, we mean it covers equal distances in equal times. Equal distances are easy to establish, e.g., by comparison with a single measuring device. But how do we know that the corresponding time intervals are equal? We measure them by means of some other phenomenon: dripping water, a swinging pendulum, an oscillating quartz crystal, a cesium atom. So the assertion is that a body moves equal distances in equal repetitions of some other motion. This again flirts with circularity. The content of the Law is: *It is possible to do this.* It is possible to define consistently a measure of time intervals, such that free bodies move equal distances in equal times.j

Newton's Second Law is the workhorse of the set. It is expressed in the equation:

$$\mathbf{F}_{\text{net}} = m\,\mathbf{a}\,, \tag{I.43}$$

where $\mathbf{F}_{\text{net}}$ denotes the net force on an object, m its mass, and $\mathbf{a}$ its acceleration.

Force describes the interactions of a body with its environment, traditionally pushes and pulls. The term "net force" here means *the vector sum of forces* acting on the body. This is not trivial; it expresses the experimental result that forces act as vectors, and that a body responds to the vector (tail-to-tip) sum of all applied forces. The Second Law establishes the units of force. The SI unit of force is:

$$1\,\frac{\text{kg}\,\text{m}}{\text{s}^2} \equiv 1\,\text{N} \qquad (\text{newton})\,, \tag{I.44}$$

named in honor of Isaac Newton. The size of this unit can be grasped from everyday experience, e.g., a 1.00 kg mass weighs 9.81 N; an adult human might

---

[27] There is friction in celestial motion, but its effects would not have been perceptible either to Galileo or to Aristotle.

weigh 700. N.[28]

The mass m in Eq. (I.43) denotes *inertial mass,* which measures the resistance of a body to changes in its velocity. This is one of four distinct concepts of "mass" we might expect to encounter in physics: inertial mass, as here; *passive gravitational mass,* which measures the extent to which an object *receives* gravity, i.e., its weight; *active gravitational mass,* which measures the extent to which an object *produces* gravity; and the chemists' notion of mass, a measure of "amount of stuff." Although these are distinct concepts, they are represented by the same symbol because they are found to be equal to the most amazing precision. Clearly, there is more to this story, which will be told later.

In the language of the biologist, the Second Law can be regarded as a "stimulus-response" relation: The net force on a body is the stimulus, its acceleration the response. Most modern texts describe the Second Law as the definition of force and the definition of inertial mass. Aside from the conundrum posed by trying to define two unknown quantities with a single relation, this interpretation fails to describe how Newton actually used the Second Law or how it is generally used in dynamics. That requires some specification of the forces acting on a body via a *force law* or laws. The Second Law then determines the acceleration. More precisely, it becomes a differential equation for the trajectory of the body:

$$m \frac{d^2\mathbf{r}}{dt^2} = \mathbf{F}(\mathbf{r}, \mathbf{v}, t) \; , \tag{I.45}$$

for example.

Newton's Third Law is called the Law of Action and Reaction. It is perhaps the most familiar of the Laws, often invoked in contexts in which it does not actually apply, e.g, psychology and politics. It states: *For every action there is a reaction, equal in magnitude, opposite in direction, and directed to contrary parts.* This last is often overlooked. The Third Law is a relation between forces that two bodies exert on each other. Two forces applied to a single body, even if equal in magnitude and opposite in direction, are never an action/reaction pair. A book sits on a table, at rest. The Earth exerts a downward force on the book, while the table exerts an equal upward force on it. But the reaction to the first force is an upward force exerted by the book *on the Earth,* while the reaction to the second is a downward force exerted by the book *on the table.* The Third Law reflects Newton's observations that motion (or changes in motion) cannot be gotten from nothing. An effect on one body must be accompanied by an opposite effect on another. We shall see further implications of the Third Law subsequently.

The reader is cautioned that these descriptions are not the only, or even the most common, interpretations of Newton's Laws. Alternatives can be found in many other texts. But these will serve our purposes.

---

[28] In this text, we shall not devote much effort to conversions between SI and "British" units, which the British no longer use. Only the United States and Chad still use them officially, and they are now defined in terms of SI units anyway.

Thrust

To apply Newton's Laws to our analysis of the motion of the rocket, we must examine the force laws for all of the forces involved. We start with the *thrust* force provided by the rocket's engines in Phase I of its motion. We shall later examine how the thrust force is produced and how it is determined by the design parameters of the engines. For now, let us assume that the rocket is powered by a cluster of three engines, producing a total force of magnitude $F_e = 18.0$ N. (This force is directed along the axis of the rocket.) If the mass of the rocket, including its engines, is $m = 0.300$ kg, then the Second Law implies that the thrust alone would impart to the rocket acceleration of magnitude

$$
\begin{aligned}
a_e &= \frac{F_e}{m} \\
    &= \frac{18.0 \text{ N}}{0.300 \text{ kg}} \\
    &= 60.0 \text{ m/s}^2 \ ,
\end{aligned}
\tag{I.46}
$$

where we have utilized the definition (I.44) of the newton of force. Students are urged diligently to practice the manipulation of units to complete calculations, a crucial skill.

The rocket engines expel fuel/oxidizer combustion products as they operate, so the mass of the rocket actually changes through Phase I. For the present calculations, however, we shall treat the change in mass as negligible.

Thrust plus weight—vector nature of force

In Phase I the rocket accelerates under the combined forces of thrust $\mathbf{F}_e$ and weight $\mathbf{W}$. The latter is described by the familiar force law

$$
\mathbf{W} = m\,\mathbf{g} \ ,
\tag{I.47}
$$

where m here is the passive gravitational mass of the rocket and $\mathbf{g}$ the vector acceleration of gravity, as previously. With aerodynamic forces still treated as negligible, Newton's Second Law implies

$$
\mathbf{F}_e + m\,\mathbf{g} = m\,\mathbf{a}_0 \ ,
\tag{I.48}
$$

where $\mathbf{a}_0$ is the net acceleration used in Sec. C. As described there, the rocket does not accelerate in the direction it is pointing. It "sideslips," its axis at an angle to $\mathbf{a}_0$ that, borrowing a term from aeronautics, we may call the "angle of attack" α. The geometry is illustrated in Fig. I.9. With a_0 the magnitude of the net acceleration and θ_0 the angle it makes with the $+x$ axis, Eq. (I.48) takes the form

$$
F_e \left[\cos(\theta_0 + \alpha)\,\hat{\boldsymbol{x}} + \sin(\theta_0 + \alpha)\,\hat{\boldsymbol{y}} \right] - mg\,\hat{\boldsymbol{y}} = a_0 \left(\cos\theta_0\,\hat{\boldsymbol{x}} + \sin\theta_0\,\hat{\boldsymbol{y}} \right) \ ,
\tag{I.49a}
$$

or

$$
ma_0 \cos\theta_0 = F_e \cos(\theta_0 + \alpha)
\tag{I.49b}
$$

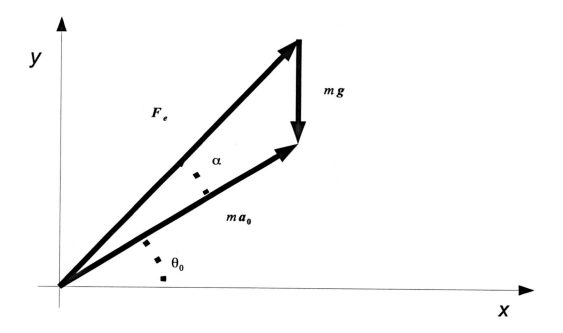

Figure I.9: Forces on the rocket under power, neglecting drag. Thrust is directed along the axis of the rocket. This makes angle α with the net acceleration because of the contribution of the rocket's weight.

and

$$ma_0 \sin\theta_0 = F_e \sin(\theta_0 + \alpha) - mg \ , \tag{I.49c}$$

because equal vectors must have equal components in each direction. The component expressions constitute two equations for the two unknowns a_0 and α, with thrust F_e and flight angle θ_0 specified quantities. That should be a guaranteed win, but what to do with these? It would be a mistake to insert numerical values at this point: The expressions would become more cumbersome and correct units and significant figures would have to be carried along. Instead, low cunning and trigonometry are needed. If we multiply Eq. (I.49c) by $\cos\theta_0$, multiply Eq. (I.49b) by $\sin\theta_0$, subtract equals from equals, and use the appropriate angle-difference identity, we obtain

$$0 = F_e \sin\alpha - mg\cos\theta_0 \tag{I.50a}$$

or

$$\alpha = \sin^{-1}\left(\frac{mg\cos\theta_0}{F_e}\right) \ . \tag{I.50b}$$

If we multiply Eq. (I.49b) by $\cos\theta_0$ and Eq. (I.49c) by $\sin\theta_0$ and add equals to equals, we obtain

$$ma_0 = F_e \cos\alpha - mg\sin\theta_0 \ , \tag{I.50c}$$

from which we can calculate a_0.

Continuing our numerical example from Sec. C, we find angle of attack

$$\begin{aligned}
\alpha &= \sin^{-1}\left(\frac{mg\cos\theta_0}{F_e}\right) \\
&= \sin^{-1}\left(\frac{(0.300\ \text{kg})(9.81\ \text{m/s}^2)\cos 45^\circ.0}{18.0\ \text{N}}\right) \\
&= 6^\circ.64 \ ,
\end{aligned} \tag{I.51a}$$

and net acceleration

$$\begin{aligned}
a_0 &= \frac{F_e\cos\alpha}{m} - g\sin\theta_0 \\
&= \frac{(18.0\ \text{N})\cos 6^\circ.64}{0.300\ \text{kg}} - (9.81\ \text{m/s}^2)\sin 45^\circ.0 \\
&= 52.7\ \text{m/s}^2 \ .
\end{aligned} \tag{I.51b}$$

To three significant figures, this is the same as the value used in Sec. C previously.

Drag

The most significant refinement of our analysis we can introduce at this point is to relax the assumption that aerodynamic forces are negligible. Then aerodynamic forces on the tail fins of the rocket would cause it to point in the direction it is moving, and aerodynamic drag might alter the characteristics of its flight.

Aerodynamic drag is a part of a class of forces termed *frictional forces*. These are exceedingly complicated interactions between one system of particles and another, e.g., between a solid body and a fluid, or between one solid body and another. To analyze these forces particle by particle and interaction by interaction would challenge the most sophisticated calculation scheme. Instead, we follow the traditional path and *approximate* these effects by greatly simplified force laws, the validity of which—over some range of circumstances—can be verified by experiment.

The most familiar such force laws are those describing *dry-surface friction,* in which molecular bonds are formed and broken between one solid body in contact with another. For example, if an external force is applied to one body in contact with another, bonds between the two bodies will engender a force, parallel to the contact surface, opposing the applied force and preventing the first body from moving. This is *static friction.* Only if the component of the applied force parallel to the contact surface exeeds a threshold will the bonds break and the body start to slide. The threshold is approximately proportional to the pressure (force per unit area) perpendicular to the contact surface times the contact area, i.e., the total perpendicular or *normal* force between the bodies. Hence, the magnitude f_s of the static frictional force is described by the inequality

$$f_s \leq \mu_s N \ , \tag{I.52}$$

with N the magnitude of the normal force and μ_s the *coefficient of static friction.* This is a dimensionless quantity characteristic of the surfaces involved, positive, typically of an order between 0.1 and 1, but not restricted to be less than unity.[29] Once the bonds of static friction are broken, the bodies slide against one another. The continuous forming and breaking of new bonds gives rise to a force, parallel to the contact surface and opposite to the motion, called *kinetic friction.* This too is proportional to the total normal force between the bodies:

$$f_k = \mu_k N \ . \tag{I.53}$$

This is an equation, not an inequality. The *coefficient of kinetic friction* μ_k is typically of the same order as μ_s, but somewhat smaller. There is a similar frictional force opposing the motion of one object rolling over another, as the contact "footprint" of the rolling object is continuously laid down and peeled up. This is called *rolling friction* or *tractive friction,* and it is approximated by a similar force law:

$$f_t = \mu_t N \ . \tag{I.54}$$

The *coefficient of tractive friction* μ_t is typically one or two orders of magnitude smaller than μ_k for the same surfaces. This is why it is usually much easier to transport a load on wheels or rollers than to drag or slide it. Most introductory physics texts describe the application of force laws (I.53)–(I.55) to a great variety of problems.

---

[29]It can be large. Two clean metal surfaces in vacuum can bond strongly to one another, a process called "contact welding."

Fluid friction or *drag* occurs when a solid body moves through a fluid, a liquid or a gas. It arises both from the displacement of the fluid from the path of the body, and from the flow of the fluid over the surface of the body. It is sensitive to the characteristics of the fluid and the nature of the flow. A general force law for drag relates the force $\mathbf{f}_D$ to some power of the speed of the body:

$$\mathbf{f}_D = -bv^n\,\hat{\boldsymbol{v}}\;, \tag{I.55a}$$

where b is a constant of suitable units, n a power, often but not necessarily an integer, v the speed of the body through the fluid, and $-\hat{\boldsymbol{v}}$ a unit vector opposite in direction to the velocity of the body. For "slow" flow, in which the fluid moves in well-defined layers, i.e., for *laminar* flow, the appropriate exponent is $n = 1$, and this takes the form sometimes called *Stokes*[30] *law of resistance*:

$$\mathbf{f}_D = -b\,\mathbf{v}\;. \tag{I.55b}$$

For faster flow, in which the fluid develops eddies and whorls, i.e., for *turbulent flow*,[31] the appropriate exponent is $n = 2$, and the force law is sometimes called *Newton's law of resistance*. This is usefully written in a form attributed to Ludwig Prandtl[32]:

$$\mathbf{f}_D = -\tfrac{1}{2}C_D\rho Sv^2\,\hat{\boldsymbol{v}}\;. \tag{I.55c}$$

Here ρ is the density of the fluid (*not* the body), S is the cross-sectional area of the body perpendicular to its velocity, and C_D is a dimensionless *drag coefficient*, which depends on the shape of the body but not its size.[33] This force law is appropriate for turbulent flow at speeds well below the speed of sound in the fluid. As the speed increases toward (transsonic) and past (supersonic) the speed of sound, drag force laws with other exponents n are needed; in the general case several such laws may be combined. We shall rely on law (I.55c) in our present analysis of the model rocket; the case of more powerful rockets must await a more advanced course.

Equation of Motion, including drag

The revised Equation of Motion[34] for the rocket is Newton's Second Law, with all relevant forces included:

$$
\begin{aligned}
m\,\frac{d\mathbf{v}}{dt} &= \mathbf{F}_e + \mathbf{f}_D + m\,\mathbf{g} \\
&= F_e\,\hat{\boldsymbol{v}} - \tfrac{1}{2}C_D\rho Sv^2\,\hat{\boldsymbol{v}} - mg\,\hat{\boldsymbol{y}}\;,
\end{aligned} \tag{I.56a}
$$

---

[30] George Gabriel Stokes, 1819–1903

[31] The transition from laminar to turbulent flow as the speed of flow increases is a complicated process, not fully understood even today.

[32] Ludwig Prandtl, 1875–1953

[33] Hence, for example, the drag coefficient of a large aircaft can be determined by measuring the forces on a scale model in a wind tunnel.

[34] Some texts apply the term *equation of motion* to an equaton for the trajectory $\mathbf{r}(t)$ of a body or system. This is a misnomer: The equation of motion is a differential equation describing the motion; the trajectory is a *solution* of the equation of motion.

where here the thrust force $\mathbf{F}_e$, assumed constant in magnitude as before, is in the direction the rocket is moving (identified by the unit vector $\hat{v}$), and we write the acceleration as $d\mathbf{v}/dt$ because the forces depend on the velocity of the rocket, but not on its location. The unit vector $\hat{v}$ is constructed by dividing the velocity vector $\mathbf{v}$ by its own magnitude. Hence, this equation can be separated into components to yield two differential equations for the velocity components v_x and v_y:

$$m \frac{dv_x}{dt} = \left(\frac{F_e}{(v_x^2 + v_y^2)^{1/2}} - \tfrac{1}{2} C_D \rho S \, (v_x^2 + v_y^2)^{1/2} \right) v_x \qquad \text{(I.56b)}$$

and

$$m \frac{dv_y}{dt} = \left(\frac{F_e}{(v_x^2 + v_y^2)^{1/2}} - \tfrac{1}{2} C_D \rho S \, (v_x^2 + v_y^2)^{1/2} \right) v_y - mg \; . \qquad \text{(I.56c)}$$

Our hearts soar with delight, because in physics *Differential Equations Are Our Friends.*

These particular differential equations, however, are not particularly friendly. The derivative of each velocity component depends on both components in a complicated and nonlinear way—in those square roots. The race between realism and difficulty is lost: For the general trajectory, analytic calculation can take us no further. We might proceed by solving the equations numerically, then integrating the velocity numerically to find the trajectory. There are a number of software packages, available to students, with facilities for solving systems of differential equations like this. (Some care would be needed with starting conditions, because the velocity components appear in denominators.) But to gain understanding this way, we would have to solve the equations for a wide variety of inputs, then search for patterns. Leaving such an exploration for another course, we can instead analyze the restricted case of vertical motion, for which v_x is identically zero and Eq. (I.56b) is trivial.

Solution for vertical flight

As in Sec. B preceding, we separate the flight of the rocket into phases. In Phase I, with $0 \le t \le t_1$, the rocket climbs under power, and v_y is positive. Equation of motion (I.56c) takes the form

$$\begin{aligned} \frac{dv_y}{dt} &= \frac{F_e}{m} - g - \frac{C_D \rho S}{2m} v_y^2 \\ &= a_0 - \frac{C_D \rho S}{2m} v_y^2 \; , \end{aligned} \qquad \text{(I.57)}$$

with $a_0 \equiv (F_e/m) - g$. At this point we do *NOT* attempt to obtain v_y by integrating the right-hand side of this equation with respect to t, and we do *NOT* compound the error by extracting the binomial expression from the integral as if it were constant. That it certainly is not, since dv_y/dt is nonzero. Such an integral could be evaluated only if we knew v_y as a function of t, in which case the problem would already be solved. Instead, we first clean up the equation: Defining

$$v_T^{(+)} \equiv \left(\frac{2ma_0}{C_D \rho S} \right)^{1/2} \tag{I.58}$$

—for the moment, just a collection of symbols, the meaning of which will be clear later—we can cast Eq. (I.57) in the form

$$\frac{dv_y}{dt} = a_0 \left[1 - \left(\frac{v_y}{v_T^{(+)}} \right)^2 \right] . \tag{I.59}$$

Fortunately, this is what is called a *separable* differential equation. We may not know v_y as a function of t, but we know it as a function of v_y. The form of this equation allows us to put all the v_y dependence on one side, and all the t dependence on the other, to obtain:

$$\frac{dv_y}{1 - \left(\dfrac{v_y}{v_T^{(+)}} \right)^2} = a_0 \, dt . \tag{I.60a}$$

This is a differential equation as it was originally conceived, i.e, a relation among differentials. Integrating both sides from zero to time t yields

$$a_0 t = \int_0^{v_y(t)} \frac{dv_y'}{1 - \left(\dfrac{v_y'}{v_T^{(+)}} \right)^2}$$

$$= v_T^{(+)} \int_0^{v_y(t)/v_T^{(+)}} \frac{du}{1 - u^2} , \tag{I.60b}$$

this last via the substitution $u \equiv v_y'/v_T^{(+)}$. This procedure is called *reduction to quadrature*: We have reduced the solution of a differential equation to the evaluation of an integral. ("Quadrature" is an old name for "integration.") As a rule, this is progress. It is usually easier, analytically and numerically, to evaluate an integral than to solve a differential equation.

We might evaluate integral (I.60b) with partial fractions and logarithms; the result would be correct, but unenlightening. Instead, we recognize that the integrand looks almost like that for the arctangent integral, except for the minus sign in the denominator. This indicates that the integral is the inverse *hyperbolic* tangent function:

$$a_0 t = v_T^{(+)} \tanh^{-1} \left(\frac{v_y(t)}{v_T^{(+)}} \right) . \tag{I.61}$$

This yields the velocity in Phase I:

$$v_y(t) = v_T^{(+)} \tanh\left(\frac{a_0 t}{v_T^{(+)}}\right) . \qquad (I.62)$$

The altitude of the rocket in Phase I requires only another quadrature:

$$
\begin{aligned}
y(t) &= y_0 + \int_0^t v_y(t')\, dt' \\
&= v_T^{(+)} \int_0^t \tanh\left(\frac{a_0 t'}{v_T^{(+)}}\right) dt' \\
&= \frac{v_T^{(+)2}}{a_0} \int_0^{a_0 t/v_T^{(+)}} \frac{\sinh u}{\cosh u}\, du \\
&= \frac{v_T^{(+)2}}{a_0} ln\left[\cosh\left(\frac{a_0 t}{v_T^{(+)}}\right)\right] , \qquad (I.63)
\end{aligned}
$$

here with substitution $u \equiv a_0 t/v_T^{(+)}$.

Students are encouraged to become familiar with the hyperbolic functions

$$\cosh u = \frac{e^u + e^{-u}}{2} , \qquad (I.64a)$$

$$\sinh u = \frac{e^u - e^{-u}}{2} , \qquad (I.64b)$$

and

$$\tanh u = \frac{\sinh u}{\cosh u} , \qquad (I.64c)$$

their identities, derivatives, and associated integrals. They are "kissin' cousins" of the circular or trigonometric functions and nearly as useful.[35]

The hyperbolic tangent function is equal to its argument for small ($\ll 1$) values of the argument and asymptotically approaches unity for large values. Result (I.62) shows that the velocity component of the rocket grows linearly with time immediately after Liftoff, as in the drag-free case of Sec. B, then grows more slowly as drag becomes significant. If Phase I were to go on long enough, with constant thrust, the rocket would asymptotically approach constant speed $v_T^{(+)}$, upward. This symbol, then, represents *terminal speed* for the upward, powered portion of the flight. It is the speed at which downward drag would equal the net upward force on the rocket (thrust minus weight), reducing the acceleration to zero. Many texts use the phrase *terminal velocity* for this notion, but velocity is a vector quantity. "Terminal speed" is the correct usage.

For comparison with our previous numerical results, we can take $a_0 = 50.0$ m/s^2, $t_1 = 3.00$ s, and $m = 0.300$ kg, and add parameters $C_D = 0.300$,

---

[35] Also, hyperbolic functions may rule the universe, but that is another story for elsewhere.

$\rho = 1.20 \text{ kg/m}^3$ for the density of air, and $S = 4.50 \times 10^{-3} \text{ m}^2$, reasonable values for the rocket pictured in Fig. I.1. The upward terminal speed takes the value

$$v_T^{(+)} = \frac{2ma_0}{C_D \rho S}$$

$$= \frac{2(0.300 \text{ kg})(50.0 \text{ m/s}^2)}{0.300(1.20 \text{ kg/m}^3)(4.50 \times 10^{-3} \text{ m}^2)}$$

$$= 136. \text{ m/s} . \tag{I.65a}$$

Result (I.62) then gives speed at Burnout

$$v_1 = v_T^{(+)} \tanh \left(\frac{a_0 t_1}{v_T^{(+)}} \right)$$

$$= (136. \text{ m/s}) \tanh \left(\frac{(50.0 \text{ m/s}^2)(3.00 \text{ s})}{136. \text{ m/s}} \right)$$

$$= 109. \text{ m/s} , \tag{I.65b}$$

and result (I.63) gives Burnout altitude

$$y_1 = \frac{v_T^{(+)2}}{a_0} \ln \left[\cosh \left(\frac{a_0 t_1}{v_T^{(+)}} \right) \right]$$

$$= \frac{(136. \text{ m/s})^2}{50.0 \text{ m/s}^2} \ln \left[\cosh \left(\frac{(50.0 \text{ m/s}^2 (3.00 \text{ s})}{136. \text{ m/s}} \right) \right]$$

$$= 190. \text{ m} . \tag{I.65c}$$

These are similar to results (I.14) and (I.15) in Sec. B preceding, but the differences are well outside the precision of the values.

The remaining portions of the flight can be analyzed in a similar way. In Phase II, with $t_1 \leq t \leq t_2$, thrust F_e drops to zero, and velocity component v_y remains positive. Equation of motion (I.56c) now takes the form

$$\frac{dv_y}{dt} = -g - \frac{C_D \rho S}{2m} v_y^2 . \tag{I.66}$$

Defining a *new* combination

$$v_T^{(-)} \equiv \left(\frac{2mg}{C_D \rho S} \right)^{1/2} , \tag{I.67}$$

we can cast this in the form

$$\frac{dv_y}{dt} = -g \left[1 + \left(\frac{v_y}{v_T^{(-)}} \right)^2 \right] . \tag{I.68}$$

Like Eq. (I.59), this is separable. The problem can be reduced to quadrature:

$$-g(t - t_1) = \int_{v_1}^{v_y(t)} \frac{dv_y'}{1 + \left(\dfrac{v_y'}{v_T^{(-)}}\right)^2} . \tag{I.69}$$

Here the character of the solution is different, because the gravitational and drag forces act in the same direction, decelerating the rocket to a stop. This integral is an inverse tangent:

$$-g(t - t_1) = v_T^{(-)} \left[\tan^{-1}\left(\frac{v_y(t)}{v_T^{(-)}}\right) - \tan^{-1}\left(\frac{v_1}{v_T^{(-)}}\right) \right] , \tag{I.70a}$$

implying

$$v_y(t) = v_T^{(-)} \tan\left[\tan^{-1}\left(\frac{v_1}{v_T^{(-)}}\right) - \frac{g}{v_T^{(-)}} (t - t_1) \right] \tag{I.70b}$$

for the velocity of the rocket in Phase II. (This expression can be cast in different forms using trigonometric identities. This is left as an exercise for the student.) This result is valid from time t_1 until the argument of the tangent function reaches zero. Hence, the time at Apogee, the end of Phase II, is given by

$$t_2 = t_1 + \frac{v_T^{(-)}}{g} \tan^{-1}\left(\frac{v_1}{v_T^{(-)}}\right) . \tag{I.71}$$

This reveals a curious feature: The inverse tangent function (which is a principal arctangent) will not exceed $\pi/2$ no matter how high the speed v_1 at burnout. The *duration* of Phase II has an upper bound:

$$t_2 - t_1 < \frac{\pi}{2} \frac{v_T^{(-)}}{g} . \tag{I.72}$$

This is more a mathematical curiosity than a physical result, however, because if v_1 is large enough, force law (I.55c) will not apply, and other approximations we have made may also fail. We also note that the constant $v_T^{(-)}$ does *not* play the role of a terminal speed in Phase II.

The altitude of the rocket in Phase II is obtained by integrating velocity (I.70b) with respect to time:

$$y(t) = y_1 + \int_{t_1}^{t} v_y(t') \, dt'$$

$$= y_1 + \frac{v_T^{(-)2}}{g} \ln \left(\frac{\cos\left[\tan^{-1}\left(\dfrac{v_1}{v_T^{(-)}}\right) - \dfrac{g}{v_T^{(-)}} (t - t_1) \right]}{\cos\left[\tan^{-1}\left(\dfrac{v_1}{v_T^{(-)}}\right) \right]} \right) . \tag{I.73}$$

Again, this result applies from time t_1 until the argument of the cosine in the numerator vanishes. This yields

$$y_2 = y_1 + \frac{v_T^{(-)2}}{g} \ln\left\{ \sec\left[\tan^{-1}\left(\frac{v_1}{v_T^{(-)}} \right) \right] \right\}$$

$$= y_1 + \frac{v_T^{(-)2}}{2g} \ln\left(1 + \frac{v_1^2}{v_T^{(-)2}} \right) \tag{I.74}$$

for the altitude at Apogee, the maximum height reached by the rocket. The last expression is obtained by applying trignometric and logarithmic identities.

Using the numerical values introduced previously, we calculate

$$v_T^{(-)} = \left(\frac{2mg}{C_D \rho S} \right)^{1/2}$$

$$= \left(\frac{2(0.300 \text{ kg})(9.81 \text{ m/s}^2)}{0.300(1.20 \text{ kg/m}^3)(4.50 \times 10^{-3} \text{ m}^2)} \right)^{1/2}$$

$$= 60.3 \text{ m/s} , \tag{I.75a}$$

time to Apogee

$$t_2 = t_1 + \frac{v_T^{(-)}}{g} \tan^{-1}\left(\frac{v_1}{v_T^{(-)}} \right)$$

$$= 3.00 \text{ s} + \frac{60.3 \text{ m/s}}{9.81 \text{ m/s}^2} \tan^{-1}\left(\frac{109. \text{ m/s}}{60.3 \text{ m/s}} \right)$$

$$= 9.55 \text{ s} , \tag{I.75b}$$

and altitude at Apogee

$$y_2 = y_1 + \frac{v_T^{(-)2}}{2g} \ln\left(1 + \frac{v_1^2}{v_T^{(-)2}} \right)$$

$$= 190. \text{ m} + \frac{(60.3 \text{ m/s})^2}{2(9.81 \text{ m/s}^2)} \ln\left(1 + \frac{(109. \text{ m/s})^2}{(60.3 \text{ m/s})^2} \right) \tag{I.75c}$$

$$= 459. \text{ m} .$$

Differences from results (I.18) and (I.19) of Sec. B are now apparent.

In Phase III, with $t_2 \leq t \leq t_3$, the rocket descends under the influence of gravitational and drag forces. These now oppose each other. We shall continue to disregard parachute deployment and descent. Instead, we assume that the rocket turns over at Apogee and descends straight down, so that its aerodynamic characteristics are the same as in ascent. In this phase v_y is negative, so care must be taken in evaluating the square roots in Eq. (I.56c); in particular, we must take $(v_y^2)^{1/2} = -v_y$. The equation of motion takes the form

$$\frac{dv_y}{dt} = -g + \frac{C_D \rho S}{2m} v_y^2$$

$$= -g \left[1 - \left(\frac{v_y}{v_T^{(-)}} \right)^2 \right] . \qquad (I.76)$$

As this is different from Eq. (I.66), it is *not* possible to treat Phases II and III as a single phase here. But separation still effects reduction to quadrature:

$$-g(t - t_2) = \int_0^{v_y(t)} \frac{dv_y'}{1 - \left(\dfrac{v_y'}{v_T^{(-)}} \right)^2}$$

$$= v_T^{(-)} \tanh^{-1} \left(\frac{v_y}{v_T^{(-)}} \right) , \qquad (I.77a)$$

which gives Phase III velocity

$$v_y(t) = -v_T^{(-)} \tanh \left(\frac{g}{v_T^{(-)}} (t - t_2) \right) . \qquad (I.77b)$$

In this phase the rocket *does* approach a terminal speed, here $v_T^{(-)}$.

The altitude of the descending rocket is given by the integral of $v_y(t)$:

$$y(t) = y_2 + \int_{t_2}^t v_y(t') \, dt'$$

$$= y_2 - v_T^{(-)} \int_{t_2}^t \tanh \left(\frac{g}{v_T^{(-)}} (t' - t_2) \right) dt'$$

$$= y_2 - \frac{v_T^{(-)2}}{g} \ln \left[\cosh \left(\frac{g(t - t_2)}{v_T^{(-)}} \right) \right] , \qquad (I.78)$$

in a manner similar to Eq. (I.63). The rocket reaches $y = 0$ at

$$t_3 = t_2 + \frac{v_T^{(-)}}{g} \cosh^{-1} \left[\exp \left(\frac{g y_2}{v_T^{(-)2}} \right) \right] , \qquad (I.79)$$

the time of Impact.

We can now complete the analysis of our numerical example. Using previous values, we obtain Impact time

$$
t_3 = t_2 + \frac{v_T^{(-)}}{g} \, \cosh^{-1}\left[\exp\left(\frac{gy_2}{v_T^{(-)2}}\right)\right]
$$

$$
= 9.55 \text{ s} + \frac{60.3 \text{ m/s}}{9.81 \text{ m/s}^2} \, \cosh^{-1}\left[\exp\left(\frac{(9.81 \text{ m/s}^2)(459. \text{ m})}{(60.3 \text{ m/s})^2}\right)\right]
$$

$$
= 21.3 \text{ s} \tag{I.80a}
$$

and velocity

$$
v_3 = -v_T^{(-)} \tanh\left(\frac{g}{v_T^{(-)}}(t_3 - t_2)\right)
$$

$$
= (60.3 \text{ m/s}) \tanh\left(\frac{9.81 \text{ m/s}^2}{60.3 \text{ m/s}}(21.3 \text{ s} - 9.55 \text{ s})\right)
$$

$$
= -57.7 \text{ m/s} . \tag{I.80b}
$$

This is nearly terminal speed in descent, but a substantially lower speed than that at Burnout [Eq. (I.65b)]. Again, differences from the drag-free results (I.22) and (I.23) are apparent.

The differences between the drag-free and turbulent-flow-drag (quadratic-in-v-drag) analyses of the rocket's motion in vertical flight are illustrated clearly in Figures I.10 and I.11. These are graphs of the altitude and velocity of the rocket, respectively, as functions of time. (Figure I.10 does *not* represent the trajectory of the rocket in angled flight.) The dashed curves show the drag-free results of Sec. B, "Flight in Vacuum," and the solid curves show the results of this section with drag included, "Flight in Air." The design parameters of the rocket are those used in the numerical calculations of those sections, the same for both cases. The curves were generated by numerical evaluation of the analytic results obtained in both sections, using a FORTRAN 77 program.

It is clear from these figures that the neglect of drag is a poor approximation indeed for this rocket. It might be acceptable for order-of-magnitude estimates, but it differs from the more realistic analysis by far more than the precision we might expect from actual measurements of the rocket. (The results of this section, incorporating drag, are in fact quite reasonable for a rocket of this type.) This is true, by the way, for a great many projectile-motion problems in most introductory physics texts: The neglect of drag makes the analyses tractable but is not very accurate. We have seen how a *succession* of approximations allows us to approach more closely an accurate description of nature.

It may have struck the reader that the more realistic analysis, encompassing Eqs. (I.56) through (I.80), involves rather a lot of work. That is a crucial point applicable to any important scientific issue: The questions that can be treated by recalling simple formulae have already been answered; indeed, they can be answered by machines. If we seek to understand more, we must work harder.

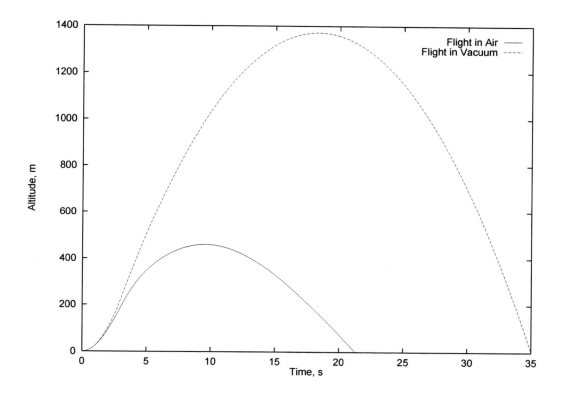

Figure I.10: Altitude of the model rocket as functions of time in the drag-free ("Flight in Vacuum") and quadratic-drag ("Flight in Air") approximations. The design parameters of the rocket are those used in the numerical calculations of Secs. B and E, the same for both cases.

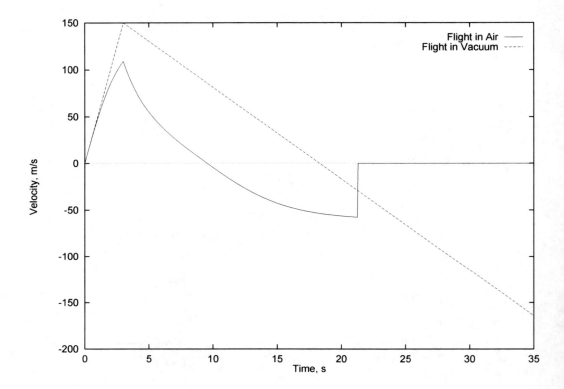

Figure I.11: Velocity of the model rocket as functions of time in the drag-free ("Flight in Vacuum") and quadratic-drag ("Flight in Air") approximations. The design parameters of the rocket are the same as for Fig. I.10.

F   Relativity and Newton's Laws

Galileo's Theory of Relativity is hard-wired into Newtonian mechanics, or in software parlance, comes bundled with it. Newton's Laws of Motion take the same form in any inertial reference frames related by Galilean transformations (Eqs. (I.39), with $a_O = 0$). Such transformations leave the constant velocities of all free bodies constant (First Law). They leave all accelerations unchanged (Eq. (I.39c)), so with the same force laws in play, $\mathbf{F} = m\mathbf{a}$ in one frame is equivalent to $\mathbf{F}' = m\mathbf{a}'$ in another (Second Law). And action/reaction pairs of forces in one frame remain action/reaction pairs in another (Third Law). This feature is called the *Galilean invariance,* or more properly, *Galilean covariance of Newton's Laws.*

Galilean covariance is more significant than it may appear at first glance. It actually constrains the laws of nature encompassed within Newtonian mechanics. For example, Newton's Laws can never tell us the velocity of anything. They can tell us the velocity of one thing *relative to another* (all the rocket velocities in Secs. B, C, and E are velocities relative to the ground), or they can tell us the *rate of change* of velocities—these are left unchanged by Galilean transformations. But the absolute, unqualified velocity of anything—no. Other far-reaching consequences of Gailean covariance will be seen in the chapters to follow.

Galilean covariance can be described in another way. Any solution of Newton's Laws, transformed by any Galilean transformation, becomes another solution—the same motion seen in a different inertial reference frame. That is, Newtonian mechanics is *relativistic.* It is common to label Newtonian mechanics, and the physical systems to which it applies, *nonrelativistic,* and to label Einstein's refinement of Newtonian mechanics, and the circumstances under which it is needed, *relativistic.* Although nearly universal, this usage is incorrect. Both theories are relativistic, the difference between them being the allowed group of transformations. (Aristotle's mechanics, with its absolute states of rest, might properly be termed nonrelativistic.) The appropriate diction would distinguish between *Newtonian* and *Einsteinian* mechanics or systems; this author will attempt to adhere consistently to this usage. Thus, for example, the quantum mechanics one learns in introductory chemistry, based on the Schrödinger[36] equation, is Newtonian quantum mechanics, while that of Dirac[37] is Einsteinian.

The transformation of Newton's Laws between reference frames has other important implications. In many texts, one finds a statement to the effect that Newton's Second Law can be applied only in inertial reference frames. This assertion has one defect: It is false. Newton's Second Law can be applied in any reference frame, provided one is willing to pay the price. Consider a general transformation of form (I.39), with $a_0 \neq 0$, so that at least one of the unprimed and primed reference frames is noninertial. Given $\mathbf{F} = m\mathbf{a}$ in the unprimed

---

[36] Erwin Schrödinger, 1887–1961
[37] Paul Adrien Maurice Dirac, 1902–1984

frame, and with the same force laws in play, the primed variables satisfy

$$\mathbf{F}' = m\left(\mathbf{a}' + \mathbf{a}_O\right) , \qquad (\text{I.81a})$$

using Eq. (I.39c). This is equivalent to

$$\mathbf{F}' - m\mathbf{a}_O = m\mathbf{a}' . \qquad (\text{I.81b})$$

Observer O and her colleagues are perfectly entitled to use Newton's Second Law as long as they accept the additional force term $-m\mathbf{a}_O$. Such a term is sometimes called a *fictitious force* or a *pseudoforce*,[38] but such terminology is misleading. If the reader is carrying a cup of hot coffee while driving around a sharp curve, when the coffee lands in the reader's lap, it will be no comfort to label the force that put it there "fictitious." One can be squashed to death by a "pseudoforce." It is much more illuminating and useful to refer to such a term as an *inertial force*; this term will be used here exclusively.

G Engine performance: momentum conservation

Propulsion for the model rocket is provided by a miniature version of the solid-fuel rocket engines used, for example, to augment the Space Shuttle's main engines at liftoff and in various other NASA and U.S. Air Force rockets. As sketched in Fig. I.12, a model rocket engine consists of a cardboard casing, a ceramic nozzle, a plug of fuel/oxidizer mixture,[39] smoke powder, and a "separation charge" that deploys the rocket's recovery system. The engines are made to provide various thrust levels and durations, classified alphabetically from 1/2-A to L. (Sizes larger than G require a license; they are not for typical hobbyists.) We can relate the performance of such a rocket engine to its design parameters by applying principles of Newtonian mechanics.

Linear momentum

The quantity we need to analyze to understand rocket propulsion is *linear momentum*. This is a descriptor or measure of motion. It is a vector quantity, defined for a single point body or particle as

$$\mathbf{p} \equiv m\mathbf{v} , \qquad (\text{I.82a})$$

with m the inertial mass and $\mathbf{v}$ the velocity of the body. The momentum of a system of particles or an extended body is the vector sum of the momenta of the individual parts:

$$\mathbf{P} \equiv \sum_i \mathbf{p}_i . \qquad (\text{I.82b})$$

---

[38] The term *artificial gravity* is used in some circumstances.

[39] Hence, like its larger counterparts, the model rocket engine is self-contained. It would work in a vacuum, e.g., if one wished to launch model rockets on the Moon.

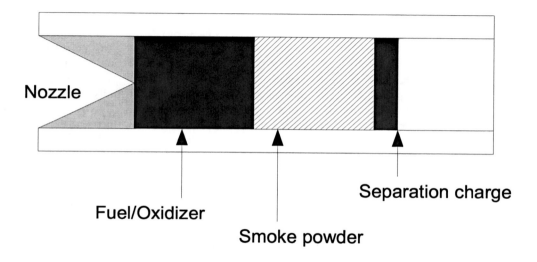

Figure I.12: Cutaway diagram of a typical model rocket engine. Burning of the fuel/oxidizer mixture provides thrust; the smoke powder burns to produce a visible trail during the coasting portion of the flight; the separation charge separates the rocket's nose cone to deploy a small parachute.

If the mass of a particle does not vary with time, the dynamics of its momentum follow from form (I.43) of Newton's Second Law:

$$\frac{d\mathbf{p}}{dt} = \sum \mathbf{F} \, , \qquad (I.83a)$$

where the right-hand side is the net force on the particle. This, in fact, is Newton's original form of the Second Law. The rate of change of the momentum of a system, then, is the vector sum of all the forces on all its parts. But the forces various parts exert on each other, called *internal forces,* come in action/reaction pairs. The sum of all forces, then, reduces to a sum of the *external forces* exerted on the system from outside:

$$\frac{d\mathbf{P}}{dt} = \sum \mathbf{F}_{\text{ext}} \, , \qquad (I.83b)$$

the pairs of internal forces summing to zero by Newton's Third Law.

This last has a profound consequence. For any system for which the external forces sum to zero, or for any isolated system upon which no external forces act, the total momentum of the system remains constant throughout its motion. This is the *Law of Conservation of (Linear) Momentum.* The importance of this law springs from its generality. We encounter it in many circumstances. Remarkably, it is found to be a more general law than the Newtonian Laws of Motion from which we obtained it here: It applies to galaxies and atoms and elementary particles, in regimes where Newton's Laws are superseded by more precise descriptions. Here, we can use it to analyze rocket engine performance.

First and Second Rocket Equations

An highly idealized analysis of rocket propulsion is illustrated in Fig. I.13, which can be considered frames of a "movie" of a rocket under power in free space. In the first frame, a rocket of mass M has velocity V, in the x direction, say, relative to the camera. In the next frame, the rocket has emitted a burst of exhaust of mass $-dM$ at nozzle speed U *relative to the rocket.* As a result, the rocket, now with diminished mass $M + dM$, has acquired additional velocity dV relative to the camera. As the rocket/exhaust system is subject to no external forces, the total momentum in the second frame must equal that in the first:

$$(M + dM)(V + dV) + (-dM)(V + dV - U) = MV \, . \qquad (I.84)$$

These are the momentum components in the direction of motion; the other components are identically zero. Rearranging this, and "dividing" by the time increment dt between the two snapshots, we obtain the *First Rocket Equation in free space*:

$$M \frac{dV}{dt} = -U \frac{dM}{dt} \, . \qquad (I.85)$$

The right-hand expression is the thrust force provided by the rocket engine. The left-hand expression is positive because the engine expels mass, implying $dM/dt < 0$.

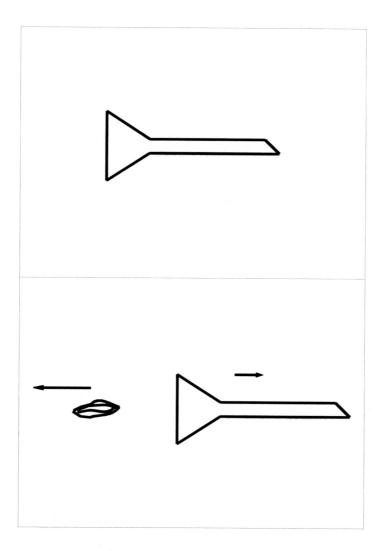

Figure I.13: Successive frames of a "motion picture" of a rocket in free space. The rocket emits exhaust mass $-dM$, and it acquires additional velocity dV to the right.

Suppose an engine burn is executed so that the rocket begins with initial mass M_i and finishes with mass $M_f < M_i$. We can rearrange and integrate the First Rocket Equation to calculate the velocity increment ΔV achieved by the burn:

$$\int_V^{V+\Delta V} dV' = \int_{M_i}^{M_f} (-U)\,\frac{dM}{M}$$

$$\Delta V = -U \ln\left(\frac{M_f}{M_i}\right)$$

$$= +U \ln\left(\frac{M_i}{M_f}\right) \ . \tag{I.86}$$

This result is the *Second Rocket Equation in free space.* We note that if a sufficiently large fraction of the rocket's mass is burned and exhausted, the rocket can acquire a speed greater than the nozzle speed U.

A typical small (B) model-rocket engine might provide thrust $F_e = 5.80$ N for $\Delta t = 1.72$ s, consuming $\Delta M = 12.5$ g of fuel/oxidizer mixture in the process.[40] Assuming a constant burn rate, we obtain from the First Rocket Equation

$$U = \frac{F_e}{-\Delta M/\Delta t}$$

$$= \frac{5.80 \text{ N}}{(0.0125 \text{ kg})/(1.72 \text{ s})}$$

$$= 798. \text{ m/s} \tag{I.87}$$

for the exhaust speed of the engine. The speed of sound in air is about 340 m/s; clearly, the model-rocket engine is not a mere toy. In comparison, the highest nozzle speeds for chemical rockets currently in use are approximately 4000 m/s.

H   The Einsteinian rocket

The analysis of rocket propulsion offers us an opportunity to examine the revision of classical mechanics formulated by Einstein in his 1905 Special Theory of Relativity. Although a thorough exposition of this theory must await another course and another book, we are in a position to apply Einstein's dynamics to this problem and to see how his dynamics differs from Newton's.

Terminology can lead to confusion here. It is common to refer to the mechanics of Newton and the electromagnetic theory of Maxwell,[41] i.e., pre-1900 physics, as *classical,* and the physics of the twentieth century, including Einstein's theories and quantum mechanics, as *modern.* In courses titled *Modern Physics,* the distinction is usually drawn this way. Physicists, however, regard both Einstein's Special and General Theories of Relativity as classical, because the framework within which questions are asked and answers sought is the same

---

[40]Data supplied by manufacturer.
[41]James Clerk Maxwell, 1831–1879

for Einstein's mechanics as for Newton's; the distinction is drawn between *classical* and *quantum* physics. We shall keep to the physicists' usage in this text.[42]

Lorentzian kinematics

The need for a new mechanics arose from the realization that inertial reference frames cannot be connected by Galilean transformations, except approximately. This is because the speed of light waves in vacuum is set by laws of nature. It must be the same in all inertial reference frames if they are to be physically equivalent, and that equivalence is the *principle of relativity* common to both Galileo and Einstein. Why that speed is fixed will be explored further in Ch. XI. But the requirement is inconsistent with Galilean transformations [Eq. (I.39b)], which leave no finite speed unchangeable. The necessary[43] transformations of position and time measurements between inertial reference frames are called Lorentz[44] transformations—Lorentz derived them, but Einstein recognized their ultimate significance. For example, if observer O and her colleagues are moving in our reference frame with velocity v_O in the $+x$ direction, she passes through our origin when her clock and ours there both read zero, and her coordinate axes are set up parallel to ours, then the Lorentz transformation between her (primed) position and time measurements and our (unprimed) measurements is:

$$ct' = \frac{ct - \dfrac{v_O}{c}\,x}{\left(1 - \dfrac{v_O^2}{c^2}\right)^{1/2}} \tag{I.88a}$$

$$x' = \frac{x - \dfrac{v_O}{c}\,ct}{\left(1 - \dfrac{v_O^2}{c^2}\right)^{1/2}} \tag{I.88b}$$

$$y' = y \tag{I.88c}$$

$$z' = z\,, \tag{I.88d}$$

with $c = 299,792,458$ m/s (exactly) the speed of light in vacuum. The time measurements are multiplied by c to give all the equations the same units and to make them more symmetric. This transformation approaches a Galilean transformation in the limit $|v_O| \ll c$, i.e., Galilean kinematics is the low-speed (compared with c) limit of Lorentzian kinematics. Certainly, any speed familiar to Galileo, or that we encounter in common experience, is well within this limit.

---

[42]Strictly, *classical* physics should refer to that of the ancient Greeks and Romans; Newtonian physics should be *neoclassical* or even *Baroque*; Maxwellian electromagnetism should be *Romantic* physics; Einstein's theories and quantum physics should be *modern*; and *postmodern* physics has yet to be determined. This author has not gotten much traction with this proposal, however.

[43]Einstein did not pull the transformations "out of a hat"; they follow from an inexorable chain of logic.

[44]Hendrik Antoon Lorentz, 1853–1928

Lorentz transformations encompass the kinematic phenomena usually emphasized in descriptions of the Special Theory: *Time Dilation*—Clocks in motion with respect to our reference frame tick at a slowed rate, as measured in our frame. Our clocks are, symmetrically, observed to tick slowly in the moving reference frame. *Length Contraction* or *Lorentz-FitzGerald*[45] *Contraction*—Lengths of objects in motion with respect to our frame, measured parallel to the direction of motion, are contracted as measured in our frame and symmetrically. *Relativity of Simultaneity*—Clocks in motion with respect to our reference frame, separated parallel to the direction of motion and synchronized in their own frame, are not synchronized according to clocks in our frame and symmetrically; i.e., "the same time" in different locations is reference-frame dependent. It is customary to emphasize the counterintuitive nature of these phenomena, the "strangeness" of the Special Theory. Of greater interest here, however, is the dynamics that emerges from the Special Theory and the conceptual continuity between that dynamics and Newton's.

The Lorentz transformations imply that velocities measured in different inertial frames are not related in the Galilean manner,[46] i.e., as per Eq. (I.39b). Instead, Eqs. (I.88a) and (I.88b) imply that a velocity v'_x measured in O's frame, in the direction of the frames' relative motion, is given by

$$
\begin{aligned}
v'_x &= \frac{dx'}{dt'} \\
&= \frac{dx - v_O\,dt}{dt - \dfrac{v_O}{c^2}\,dx} \\
&= \frac{v_x - v_O}{1 - \dfrac{v_O v_x}{c^2}} \ ,
\end{aligned}
\tag{I.89a}
$$

with $v_x = dx/dt$ the velocity of the same object measured in our frame. (*Caution*: The relations for the other velocity components are different!) The inverse relation is easily obtained from this, or from the inverse of transformation (I.88):

$$
v_x = \frac{v'_x + v_0}{1 + \dfrac{v_0 v'_x}{c^2}}
\tag{I.89b}
$$

For example, if O is traveling at $v_O = \frac{3}{4}c$ in our frame, and she fires a projectile forward at speed $v'_x = \frac{3}{4}c$ as she measures it, then the speed of the projectile in our frame is

$$
\begin{aligned}
v_x &= \frac{v'_x + v_0}{1 + \dfrac{v_0 v'_x}{c^2}} \\
&= \frac{\frac{3}{4} + \frac{3}{4}}{1 + \left(\frac{3}{4}\right)^2}\,c \\
&= \tfrac{24}{25}\,c \ ,
\end{aligned}
\tag{I.90}
$$

---

[45] George Francis FitzGerald, 1851–1901
[46] Many have found this a huge obstacle to understanding the Special Theory.

not $\frac{3}{2}c$. In fact, it is easy to show that any combination of sublight velocities in this fashion will always produce another sublight velocity.

There is a cute trick for dealing with this velocity-combination law, especially useful for one-(spatial-)dimensional problems. Rules (I.89) look almost like the angle difference/sum identities for the tangent function, except the sign in the denominator is wrong. In fact, they match the difference/sum identities for the *hyperbolic* tangent function. We can define parameter θ via

$$v_x = c \tanh \theta \ , \tag{I.91}$$

and similarly for v_O and v'_x. This parameter is variously termed the *velocity parameter, rapidity,* or *fugacity*; we shall use the first here. Velocity-combination law (I.89a) then takes the form

$$\theta' = \theta - \theta_O \ , \tag{I.92}$$

with the obvious inverse. That is, velocity parameters combine as Galilean velocities do, even though actual velocities do not.

Four-scalars and four-vectors

A new kinematics calls for a new dynamics. Relations (I.89) show that the three-dimensional vectors of Galilean kinematics are no longer suitable for expressing dynamical laws, because their transformations between inertial frames are too messy. We seek quantities that transform simply under Lorentz transformations. A relationship among such quantites then *takes the same form*—even if the values of the quantities change—in both reference frames, since both sides of the relationship transform the same way. To identify such quantities, we need to look at the notion of a vector from a different viewpoint.

A quantity consisting of a single value that does not change from inertial frame to inertial frame we shall call a *scalar*. One example of a Lorentz scalar is the *proper-time increment $d\tau$,* given by:

$$c\,d\tau = [c^2\,dt^2 - (dx^2 + dy^2 + dz^2)]^{1/2} \ . \tag{I.93}$$

If we calculate this using primed increments dt', dx', dy', and dz', as given by Eqs. (I.88), it comes out exactly the same. This is the time interval measured on a specific clock: the one that travels from time and position (t, x, y, z) to $(t + dt, x + dx, y + dy, z + dz)$ at constant speed.

A set of four components that transform from frame to frame exactly as the coordinates (ct, x, y, z) or their increments $(c\,dt, dx, dy, dz)$ do we shall now call a *vector*.[47] An obvious way to construct such an object is to take the coordinate increments themselves, which transform in the desired way, and "divide" them by the proper-time increment. This, being a scalar, does not alter the

---

[47] If one wishes to impress people at parties, one calls such a set of components a *contravariant tensor of rank one,* and then observes that such objects form a vector space.

transformation behavior. The result is called a *four-velocity*:

$$u^0 \equiv c\,\frac{dt}{d\tau} = c\gamma$$

$$u^0 \equiv \frac{dx}{d\tau} = v_x\gamma$$

$$u^0 \equiv \frac{dy}{d\tau} = v_y\gamma \qquad\qquad (\text{I.94a})$$

$$u^0 \equiv \frac{dz}{d\tau} = v_z\gamma \; ,$$

with

$$\gamma \equiv \left(1 - \frac{v^2}{c^2}\right)^{-1/2} \; , \qquad\qquad (\text{I.94b})$$

and v_x, v_y, and v_z the (ordinary) velocity components and v the speed of the moving object in question. The components of the four-velocity are numbered 0, 1, 2, and 3, and these are written as superscripts—those are *not* exponents. This is the standard notation; introducing it here saves the effort later. The four-velocity of an object is its velocity (through time and space) rescaled by the time-dilation factor γ, but its components transform from frame to frame in a manner analogous to Eqs. (I.88).

These two types of quantities are often called *four-scalars* and *four-vectors*, because the time and spatial components together number four. Minkowski[48] described the features of such objects in terms of the geometry of four-dimensional *spacetime*. The older terms *world scalars* and *world vectors* are also found in some texts.

Einsteinian momentum and energy (four-momentum)

Following Einstein, we are now in a position to define dynamical quantities with which to describe dynamical laws consistent with Lorentz transformations. Multiplying the four-velocity u^α (with $\alpha = 0, 1, 2, 3$) of a body by its mass m gives a four-vector, the *four-momentum*:

$$p^0 = mc\gamma = \frac{1}{c}(mc^2 + \tfrac{1}{2}mv^2 + \cdots) \equiv \frac{E}{c}$$

$$p^1 = mv_x\gamma = mv_x + \cdots \qquad\qquad (\text{I.95})$$

$$p^2 = mv_y\gamma = mv_y + \cdots$$

$$p^3 = mv_z\gamma = mv_z + \cdots \; ,$$

the expressions on the right giving the leading term(s) in expansions in powers of v^2/c^2; the factor γ is that of Eq. (I.94b). The first of these components defines the quantity

$$E = mc^2\gamma \; , \qquad\qquad (\text{I.96})$$

---

[48]Hermann Minkowski, 1864–1909

which consists of a large *rest energy mc^2*, the Newtonian kinetic energy (more about this in Ch. II) $\frac{1}{2}mv^2$, plus Einsteinian correction terms. Relation (I.96) is the complete form of the famous $E = mc^2$, one of the most widely known and least understood equations in science. It is difficult to find an introductory treatment that correctly identifies the logical status of this equation: *It is simply the definition of the quantity E.* The other three components consist of Newtonian momentum mv_x, et cetera, plus correction terms.

Einstein's Special Theory asserts that it is to *these* quantites, rather than their Newtonian counterparts, that physical laws such as the Law of Conservation of Energy (see Ch. II) and the Law of Conservation of Momentum apply. These replacements allow for the description of new dynamical phenomena, which Newtonian mechanics does not admit; ultimately, nature determines which version is correct.[49] Newtonian mechanics is not *disproved*; it is *superseded,* because Einstein's mechanics approaches Newton's as a limit when speeds are small compared to c. In situations where Newtonian mechanics gets the right answers, Einsteinian mechanics gets the same answers.

A word is in order about "relativistic mass": Don't. Many older treatments define a "relativistic mass" $m_{\text{rel}} \equiv m\gamma$, so that the spatial components of Eq. (I.95) still look like Newtonian momentum components, and label the mass m used here "rest mass." This is not current usage and will not be used here for several reasons: The relationship between momentum and velocity [Eqs. (I.95)] is nonlinear; there is no point in trying to conceal this. "Relativistic mass" is only energy E divided by the constant c^2; it does not represent an independent dynamical quantity. Nor does it represent any measured mass, because the mass of an object is normally measured when it is at rest. A measurement of the mass of an object in motion is actually a measurement of its momentum or energy.

Einsteinian rocket equations

We can see Einsteinian mechanics in action by considering a rocket under power in free space, as in Fig. I.13, but without assuming the speeds of rocket or exhaust are small compared to c. Here, more care must be taken with reference frames. The accelerating rocket does not define an inertial reference frame; rather, it is instantaneously at rest in a sequence of such frames, in motion with respect to one another. We assume the rocket is instantaneously at rest, with mass M, in the (inertial) frame of the camera in the first snapshot. In the second snapshot, the rocket has emitted a burst of exhaust of mass δm and speed U; the rocket's mass has decreased to $M+dM$, and the rocket has acquired *velocity parameter $d\theta$*. The second snapshot is sufficiently soon after the first that $d\theta$ can be taken as infinitesimal ($d\theta \ll 1$), although the speed U of the exhaust cannot be assumed to be small.

The governing dynamical law for this process is the conservation of four-momentum. Because the components of four-momentum transform from refer-

---

[49]Spoiler alert: Einstein wins.

ence frame to reference frame via relations akin to Eqs. (I.88), momentum in one frame depends on both momentum and energy in another, and vice versa. Hence, if momentum is to be conserved in every reference frame, energy must be as well. That also means that *conservation of mass,* familiar from Newtonian mechanics, cannot be imposed as an independent requirement. Unlike Newtonian mechanics, Einsteinian mechanics allows for tradeoffs between mass and energy—and nature does indeed include such processes. Here, this means we may *not* assume equality between the mass increments dM and $-\delta m$, as we did in Sec. G.

Four-momentum has two nontrivial components here, the "0" component (energy) and the "1" component (momentum). The conservation of energy between the two snapshots implies the relation

$$(M + dM)c^2 + \delta m\, c^2 \gamma_U = Mc^2 \ , \tag{I.97a}$$

and the conservation of momentum implies

$$Mc\, d\theta - \delta m\, U \gamma_U = 0 \ . \tag{I.97b}$$

Here, γ_U is the Lorentz factor (I.94b) associated with speed U. The corresponding factor for speed $c\tanh(d\theta)$ is unity to first order in the infinitesimal $d\theta$, and the speed itself is $c\, d\theta$ to that order (that is, dropping terms of order $d\theta^2$ and higher as too small to consider). Eliminating $\delta m\, \gamma_U$ from these two equations yields

$$d\theta = -\frac{U}{c}\frac{dM}{M} \tag{I.97c}$$

as the *Einsteinian first rocket equation* in free space.

This equation is couched in terms of $d\theta$ rather than dV, because velocity parameter increments can be added as the rocket accelerates from inertial frame to inertial frame, in the manner of Eq. (I.92). Velocity increments cannot. The first rocket equation can be integrated, starting from rest for example, to give final velocity parameter

$$\theta = \frac{U}{c}\ln\left(\frac{M_i}{M_f}\right) \tag{I.98a}$$

and velocity

$$V = c\tanh\left[\frac{U}{c}\ln\left(\frac{M_i}{M_f}\right)\right] \ , \tag{I.98b}$$

in terms of the initial and final masses M_i and M_f of the rocket, as the *Einsteinian second rocket equation* in free space. If the value of θ is small, this result accords with the Newtonian version (I.86). If it is not, the velocity V asymptotically approaches c for large values of θ; the rocket cannot accelerate beyond or even to the speed of light.

For example, suppose an advanced rocket used a laser beam (with $U = c$) for propulsion—the most efficient way to use its energy. For a final speed $V =$

$0.500\,c$, the second rocket equation implies final-to-initial mass ratio

$$\frac{M_f}{M_i} = \exp\left[-\tanh^{-1}\left(\frac{V}{c}\right)\right]$$
$$= \exp[-\tanh^{-1}(0.500)]$$
$$= 0.577 \;. \tag{I.99}$$

This is not the same as a Newtonian rocket expelling 42.3% of its mass as exhaust. Rather, the rocket must convert 42.3% of its mass to energy and expel that in the laser beam. The energies associated with an Einsteinian rocket would be enormously greater than those associated with a Newtonian rocket.

One still encounters statements, even in modern texts, to the effect that Einstein's Special Theory deals only with unaccelerated motion. As this examination of the rocket displays, such statements lack understanding: The dynamics of the Special Theory treats accelerated motion quite handily.

Einsteinian Second Law of Motion

It is relatively (!) straightforward to formulate an Einsteinian counterpart to Newton's Second Law of Motion. Some authors attempt to relate a three-dimensional force vector to the rate of change, with time t, of the three momentum components of the four-momentum (I.95). The result is a mess, because the behavior of such an object under Lorentz transformations would be very unwieldy; such a construction violates the fundamental harmony between Einsteinian dynamics and Lorentzian kinematics. A more suitable "Second Law" relates a force four-vector F^α to the rate of change, with *proper-time* increment (I.93), of the four-momentum p^α:

$$F^\alpha = \frac{dp^\alpha}{d\tau}$$
$$= m\,\frac{du^\alpha}{d\tau} \;, \tag{I.100}$$

this last for a particle with mass m and four-velocity u^α; once again the superscript α takes values $0, 1, 2, 3$.

The challenge here is to find suitable force laws for F^α. In Newtonian mechanics, the force law can be any function of position, velocity, time, et cetera, but there are mathematical constraints that an Einsteinian four-force must satisfy. Electromagnetic forces will work; in fact, the *simplest* force law that can be used in Eq. (I.100) is that for the electric and magnetic forces to be examined in Chs. VII–XI. A constant gravitational force, such as that used in the preceding sections, will *not* work. Einstein's 1915 *General* Theory of Relativity reveals that for gravitation, the force F^α is zero, but the derivative $d/d\tau$ is different: The geometry of four-dimensional spacetime is curved!

References

Constant-acceleration, drag-free motion in one, two, and three dimensions is covered in all introductory physics texts. For example, Resnick, Halliday, and Krane [RHK02], Fishbane, Gasiorowicz, and Thornton [FGT96], and Bauer and Westfall [BW14] all contain detailed treatments. These texts also treat frictional and drag forces and terminal speeds; Resnick, Halliday, and Krane's discussion of linear drag, i.e., Stokes'-law drag, is especially thorough.

Detailed analyses of motion with drag are readily found in intermediate-level mechanics texts. For example, Thornton and Marion [TM04] (the same Thornton) treat linear drag both analytically and numerically, and quadratic drag, i.e., Newton's-law drag, numerically. Taylor [Tay05] gives an analytic treatment of both forms. Barger and Olsson [BO95] confine their analysis to one-dimensional, i.e., vertical, motion.

Galilean relativity is introduced in all introductory texts, and Einsteinian relativity in all texts (or extended versions of texts) with sections on modern physics. Intermediate mechanics texts treat both versions. Specialized texts on the subject are also useful: Mermin's [Mer68] discussion of Lorentzian kinematics is especially accessible, and Taylor and Wheeler [TW92] (not the same Taylor) offer a thorough exposition at the undergraduate level. Resnick [Res72] (the same Resnick) emphasizes the experimental support for Einstein's theory. Rindler [Rin06] offers a detailed treatment at a more advanced level, as does Hartle [Har03]; these texts are introductions to the General Theory of Relativity.

The Newtonian rocket is treated in all introductory and intermediate texts. The Einsteinian version appears in more specialized texts; Taylor and Wheeler [TW92] give a thorough analysis.

Aristotle's *Physics,* which helped launch the whole adventure, is available in translation [McK41]. Galileo's discoveries in kinematics, *inter alia,* are detailed in his *Dialogues Concerning Two New Sciences.* This too is available in translation [Gal54], although the lack of modern symbolic notation for mathematics makes this a challenging read. Newton's *Philosophiae Naturalis Principia Mathematica* is also available in modern translation [New62]. First editions of the *Principia,* of course, are museum treasures. There is one in the Huntingdon Library in San Marino, California, for example. The Rockefeller Library at the University of Cambridge not only has first editions of the *Principia,* but also has Newton's manuscript notes. Many of these are in English, and after more than three hundred years, are amazingly legible.

Problems

1. A body moves in one dimension on trajectory

$$x(t) = x_0 \left(1 + \frac{t^2}{t_0^2}\right)^{1/2} ,$$

where x_0 and t_0 are constants with suitable dimensions. Calculate the velocity $v_x(t)$, acceleration $a_x(t)$, and jerk $j_x(t)$ for this motion. *A challenge*: Find out what motion this trajectory represents.

2. You are driving on a straight, level road in Kenya at constant speed 30.0 m/s. An elephant steps into the road in front of you. Reaction time 200. ms elapses before you step on the brake. After that, the car decelerates at constant rate $a_x = -0.40\ g$, where g is the usual gravitation-acceleration value. Calculate the total stopping distance from the time you see the elephant.

3. A tennis ball is fired vertically upward from the ground at initial speed v_0. At the instant it reaches the peak of its trajectory, a second ball is fired vertically, at the same inital speed, from the same location. Calculate the *fraction* of the maximum height reached by the first ball at which the two balls collide. Assume aerodynamic effects are negligible.

4. A body free-falls from rest with constant downward acceleration g; aerodynamic effects are negligible. Calculate the ratios of the distances it covers in the second, third, fourth, and nth intervals Δt of its fall (not the *total* distance—the individual distances covered in each interval) to the distance covered in the first interval Δt. (This pattern was discovered by Galileo.)

5. The Japanese term *nawabari* refers to the design elements of a castle relevant to its defense against attack; the *nawabari* of a castle determines its fate. You are responsible for the *nawabari* of a castle in which stone curtain walls rise straight up from the edge of a moat 150.0 meters wide. The moat is filled with carnivorous swans and cannot be crossed if the drawbridges are up. *Defense*: What is the *minimum* height to which you must build the walls so that archers firing bows capable of launching arrows at any angle, at speed 40.0 m/s off the bowstring, from ground level on the far side of the moat, cannot pick off the defenders atop the walls? Assume aerodynamic effects on the arrows are negligible, and neglect also the heights of the archers and defenders. *Offense*: At what angle should the archers aim to hit as high up on the walls as possible? When the arrows, thus shot, reach their targets, are they ascending, level, or descending?

6. Archery from horseback has been a formidable military technology in many parts of the world for millennia. Mounted archers in the thirteenth-century armies of Genghis Khan[50] were able to shoot enormous distances by firing forward from the back of a galloping horse. Suppose that the speed of the arrow off the bowstring (relative to the bow) is $v_B = 40.0$ m/s and the horizontal speed of the horse is $v_H = 18.0$ m/s. (a) Calculate the angle θ_B above the horizontal at which the archer should aim to achieve maximum range, *measured on level ground from the point of the shot to*

---

[50]Genghis Khan (Temujin), 1162–1227

the point where the arrow falls. Neglect all aerodynamic effects and the height of the horse and rider; assume that the arrow leaves the bow at the angle θ_B at which it is pointed, in the reference frame of the archer. (b) Calculate the maximum range thus achieved.

Many centuries earlier, Parthian (ancient Persian) archers were able to discourage pursuers by firing *backward* from the back of a galloping horse— the famous "Parthian shot." Suppose the Parthian has the same bow and the same horse as the Mongol. (He didn't.) Use the same approximations as in the previous problem. (c) Calculate the angle θ_B' above the *backwards* horizontal at which the Parthian should aim to achieve maximum range for his shot, measured in the same way as for the Mongol. (d) Calculate the maximum range achieved for this shot.

(e) Calculate the angle θ_B'' for maximum range and the maximum range achieved by an archer with the same bow standing on the ground, and compare these with results (a)–(d). Again, neglect aerodynamic effects and the height of the archer.

7. Aristotle is sometimes quoted as claiming that a body falls at a speed proportional to its mass. Galileo refuted this, both by logical argument and by dropping different masses from the Leaning Tower of Pisa (if, indeed, he actually performed the experiment). Suppose, however, that Aristotle's assertion was based on his experience with objects falling in air, with turbulent-flow or quadratic drag, essentially at their terminal speeds.

   (a) If Aristotle considered spherical objects, say, all of the same size but of different materials (hence different densities), what would the actual relation between terminal fall speed and mass be? That is, if the terminal speed is written $v_T = K_1 f_1(m)$, where $f_1(m)$ is a function only of mass and K_1 contains all the factors independent of mass, what would the function $f_1(m)$ be?

   (b) Contrariwise, if Aristotle considered spheres all of the same material and density but of different radii, so that mass varied with radius, what would the mass dependence of v_T be? If in this case it is written $v_T = K_2 f_2(m)$, where $f_2(m)$ depends only on mass and K_2 contains all the mass-independent factors, what would the new function $f_2(m)$ be?

8. A bullet is fired vertically upward from ground level at initial speed $v_0 = 300.$ m/s. *DO NOT perform this experiment; this is a MONU-MENTALLY bad idea!* Assume that the bullet is a solid lead (Pb) sphere of diameter 1.00 cm and drag coefficient $C_D = 0.470$—the usual value for a sphere.

   (a) Calculate the ratio of (quadratic) drag to gravitational force on the bullet at the instant of firing. Given this result, is the neglect of drag a good approximation here?

(b) Calculate the maximum height reached by the bullet, its speed when it returns to the ground, and its total flight time, first neglecting drag, then including quadratic drag. Compare these results and consider again the question of part (a).

9. A body is fired vertically upward from ground level with initial speed v_0. It moves under the influence of gravitation and quadratic drag; let v_T represent the terminal speed of the body in descent. The body reaches maximum height and falls back to ground level, striking at final speed v_f. Assume that the variation in air density over the body's trajectory is negligible. Then these three speeds satisfy the simple relationship

$$\frac{1}{v_f^2} = \frac{1}{v_0^2} + \frac{1}{v_T^2} .$$

Prove this. The student should be cognizant of the meaning of "prove" here.

10. Solve Problem 3 preceding, but this time assume the tennis balls are subject to quadratic drag as well as gravity. Each ball has mass m, radius r, and drag coefficient C_D. Find the fraction of the (revised) maximum height reached by the first ball at which the balls collide in terms of these quantities and initial speed v_0.

11. A body of mass m moves in the $+x$ direction, under the influence of a resistive force

$$F_x = -b\,v_x^\alpha ,$$

where b is a positive constant, v_x is the body's velocity, and α is some power (not necessarily an integer) with $\alpha > 0$. At time $t = 0$ the body is at position $x = 0$ and has velocity $v_x = v_0$, where v_0 is some positive value.

(a) Determine all values of α for which the body comes to rest in a finite *time*. *CAUTION*: Different values of α may call for solutions of different form.

(b) Determine all values of α for which the body comes to rest in a finite *distance*.

12. In the manner of Sec. E, analyze the flight of a projectile moving in two dimensions under the influence of gravity ($m\mathbf{g}$) and *linear* (Stokes'-law) drag $\mathbf{f}_D = -b\mathbf{v}$, with b constant. Assume the projectile is launched from the origin at speed v_1 and angle θ_1 above the horizontal, i.e., disregard "Phase I." Find the trajectory of the projectile, and calculate as many of these features as you can: down-range distance to apogee; height of apogee; ground-to-ground flight time; ground-to-ground range. For any calculations that cannot be completed analytically in closed form, describe what the calculation requires. *Caution*: Do not assume that the trajectory is symmetric about the apogee point as a drag-free trajectory would be.

13. *Specific Impulse* is a performance specification of rocket engines. It is defined as the ratio of thrust force produced to fuel (and oxidizer) consumption rate by weight (not by mass, for historical reasons), weight being measured at the Earth's surface. (a) Calculate the specific impulse for a rocket engine with nozzle speed (exhaust speed relative to the engine nozzle) U. (b) Evaluate and compare the specific impulses of the model rocket engine described by Eq. (I.87) and a more advanced chemical rocket with $U = 4.00$ km/s.

14. In Lorentzian kinematics and Einsteinian dynamics, *motion with constant acceleration* is not taken to mean dx/dt has constant, nonzero value in an inertial reference frame. No means of propulsion could maintain that, even in principle. Rather, it is interpreted thus: The instantaneous acceleration in the inertial frame in which the body is momentarily at rest—the acceleration "felt" by observers riding with the body—has a fixed value as the body accelerates from reference frame to reference frame. In terms of the variables used in Sec. H preceding, this condition takes the form

$$c \frac{d\theta}{d\tau} = a_0 \ ,$$

with a_0 the constant acceleration value and $d\tau$ the proper-time increment of Eq. (I.93), i.e., the time measured by a clock moving with the body. Suppose a spaceship used an Einsteinian rocket engine, with nozzle speed U, to maintain such acceleration. Calculate the ship's velocity (in the reference frame in which it was originally at rest, at $\tau = 0$) and mass as functions of shipboard time τ.

Chapter II

Nonstop Service to Anywhere

- *Determine the fuel requirements for a large jet airliner capable of nonstop flight over half the Earth's circumference.*

- *Examine further physical constraints on flight performance.*

Figure II.1: This Boeing 747-400 jumbo jet carries cargo on long, intercontinental flights. Can similar technology be used for flights between any two points on Earth? (Photograph by Ara Shakaryan.)

It is possible to fly nonstop from London to New Delhi, from New York to Tokyo, from Los Angeles to Sydney. If it were possible to fly a distance equal to half the Earth's circumference, then one could fly a nonstop, great-circle route between any two points on Earth: London to Auckland, or Hong Kong to Rio de Janiero. But such a distance is beyond the range of even today's longest-range airliners, such as the Boeing 747-400 shown in Fig. II.1. The experimental aircraft *Voyager* flew nonstop around the world in 1986. Neither the aircraft nor the circumnavigation, however, is of commercial use. The U.S. Air Force has for more than half a century used a fleet of tanker planes, and midair refueling, to send its aircraft all over the world. The costs and hazards of such a program—always major concerns in any engineering project—doubtless make it impractical for commercial aviation. Our purpose in this chapter is to examine, in a very basic way, the physical constraints that limit the range of an airliner, because limits cannot be overcome until their origins are understood. Is a jet airliner capable of nonstop service to anywhere approximately within

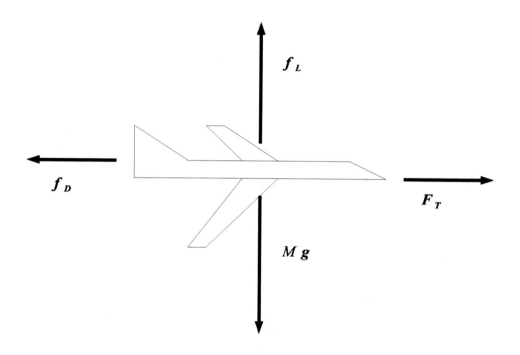

Figure II.2: A *free-body diagram* for an airplane in straight, level flight, showing the four forces in play.

reach of current technology, or must some alternative—suborbital spacecraft, exotic fuels, novel propulsion systems, for example—be sought?

A Dynamics of straight, level flight

As an introduction to the problem, we consider the fuel requirements for an airplane in straight, level flight at constant speed. Of course, the plane needs additional fuel for takeoff and climb, landing, and maneuvering, as well as a safety margin, but this is a start. The governing dynamics is Newton's Second Law: In straight, level flight at constant speed, the plane has constant velocity and zero acceleration. Hence, it is in *equilibrium*; the net force on it is zero.

An airplane in flight is subject to four forces: thrust $\mathbf{F}_T$ provided by its engine(s); aerodynamic drag $\mathbf{f}_D$; aerodynamic lift $\mathbf{f}_L$; and its weight $M\mathbf{g}$. These are arrayed as in Fig. II.2. The horizontal and vertical forces must separately sum to zero. The horizontal forces are related via:

$$\mathbf{F}_T = -\mathbf{f}_D$$
$$= +\tfrac{1}{2}C_D\rho Sv^2\,\hat{\boldsymbol{x}}\ . \tag{II.1a}$$

The vertical forces are related via:

$$\mathbf{f}_L = +\tfrac{1}{2}C_L\rho Sv^2\,\hat{\boldsymbol{y}}$$
$$= -M\mathbf{g}$$
$$= +Mg\,\hat{\boldsymbol{y}}\ . \tag{II.1b}$$

Here C_D and C_L are the drag and lift coefficients, respectively, of the airplane, ρ is the density of the air, S is the *total* area of the airplane's wings, v is its velocity, M its mass, and g the gravitational acceleration. Unit vector $\hat{\boldsymbol{x}}$ is taken to be horizontal, in the direction of flight, and $\hat{\boldsymbol{y}}$ is vertically upward.

The force laws in Eqs. (II.1) for drag and lift merit some scrutiny. They both resemble the Prandtl expression eq0155c for drag, but with some modifications. By historic tradition, the area factor S—the *cross-sectional* area in Eq. (I.55c)— is taken to be the *total* area when these laws are applied to wings. In such case the two areas, say $S_\times$ for the cross section and S_tot for the total, are related by $S_\times = S_\text{tot}\sin\alpha$, with α the *angle of attack* the wings make with the direction of flight. By tradition, the $\sin\alpha$ factor is absorbed into the coefficients C_D and C_L.

In introductory treatments, it is customary to say that the *airfoil* shape of the wing cross section forces the air to flow faster over the top of the wing than under the bottom, and the resulting reduction in pressure, in accord with the Bernoulli[1] Principle of fluid mechanics, causes lift. That can be only part of the story. For example, Newton's Third Law requires that if the plane is lifted upwards, something—the air—must be driven downwards; it cannot flow undisturbed past the wings. Both lift and drag forces are created by the air-pressure distribution over the wings. This is determined by several features of the air flow past the moving wings, sketched in Fig. II.3. Differences in flow speed create *Bernoulli lift,* while the downward deflection of air by the wings, inclined at angle α to the flow, creates *impact lift.*[2] The net force produced is perpendicular to the wing, as shown. The vertical component constitutes the lift force $\mathbf{f}_L$; the backward-directed horizontal component is called *induced drag* and is an inevitable cost of producing lift. It combines with the drag associated simply with moving the entire airplane through the air to yield the total drag force $\mathbf{f}_D$. We note that, as in Eq. (I.55c), the coefficients C_D and C_L depend only on the shape of the airplane. They can be determined, for any aircraft design, by measuring the lift and drag forces on scale models in a wind tunnel.

The engine thrust required for straight, level flight at constant speed is found

---

[1] Daniel Bernoulli, 1700–1782

[2] This is why the F-104 *Starfighter,* the "missile with a man in it," could fly with stubby, razor-thin wings. Few *Starfighters* still fly: NASA retired theirs in 1994, and the Italian air force grounded the last commissioned F-104's in 2004. Only a few, operated by civilian companies, remain in use. (Professor Jean Potvin of the Saint Louis University Department of Physics provided this information.)

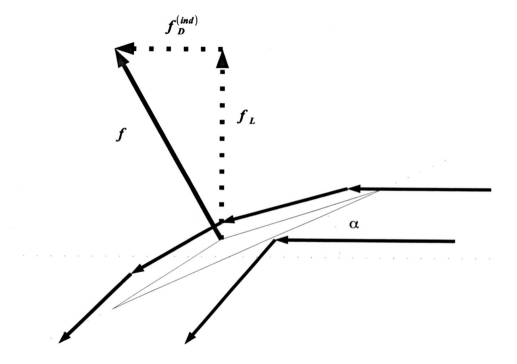

Figure II.3: Air flow past a wing, inclined at angle of attack α to the flow, creates both lift $\mathbf{f}_L$ and induced drag $\mathbf{f}_D^{(\mathrm{ind})}$.

by combining Eqs. (II.1a) and (II.1b). It has magnitude

$$F_T = \frac{C_D}{C_L}\,Mg\;. \tag{II.2}$$

The higher the lift-to-drag ratio C_L/C_D, the less thrust required for a given weight. But this relation gives no indication of how *far* thrust can be maintained. We need new descriptors of motion to address this question.

B   Work and energy

The terms used to describe the relevant aspects of motion here have many meanings and connotations in ordinary usage. In scientific usage, the goal is to strip away all meanings but one for each term to communicate unambiguously.

Work

The quantity needed to relate force and the distance through which it acts is *work*. The work done by force $\mathbf{F}$ acting on a body that undergoes displacement $d\mathbf{s}$ is the differential

$$\begin{aligned} dW &\equiv F_{\parallel}\,ds \\ &= |\mathbf{F}|\,|d\mathbf{s}|\cos\theta\;, \end{aligned} \tag{II.3}$$

where $F_{\parallel}$ denotes the component of vector $\mathbf{F}$ in, or opposite to, the direction of vector $d\mathbf{s}$, and θ is the angle between the two vectors placed tail-to-tail. We note that work is a *scalar* quantity, which can be positive, negative, or zero. No work is done if there is no motion[3] or if the force $\mathbf{F}$ is perpendicular to the displacement $d\mathbf{s}$.

It is useful at this point to introduce some mathematical terminology. The combination of vectors encountered in Eq. (II.3) is common enough, and has enough nice properties, to be given a name: the *scalar product*, or *dot product*, or *inner product* of two vectors. It is one of several ways to "multiply" vectors, i.e., one of several operations with enough features in common with the multiplication of numbers to be called a product. (One of them, the product of a vector and a scalar, was described in Ch. I Sec. A. And yes, there is an *outer product*, though it will not be needed here.) We define the dot product of vectors $\mathbf{A}$ and $\mathbf{B}$ as the scalar

$$\mathbf{A}\cdot\mathbf{B} \equiv |\mathbf{A}|\,|\mathbf{B}|\cos\theta\;, \tag{II.4}$$

with θ the angle between $\mathbf{A}$ and $\mathbf{B}$ placed tail-to-tail. It is the magnitude of $\mathbf{A}$ times the component of $\mathbf{B}$ in the direction of $\mathbf{A}$, or vice versa. Like ordinary multiplication, this product is *commutative* ($\mathbf{A}\cdot\mathbf{B} = \mathbf{B}\cdot\mathbf{A}$) and *distributive*:

---

[3]Yet a person gets tired pushing on an unmoving wall or sitting before a physics test writing no solutions. In those cases, work is done contracting and relaxing muscle fibers or moving neurotransmitter molecules, i.e., internally.

$\mathbf{A} \cdot (\mathbf{B} + \mathbf{C}) = \mathbf{A} \cdot \mathbf{B} + \mathbf{A} \cdot \mathbf{C}$. This last implies the dot product also obeys a Leibniz rule for derivatives:

$$\frac{d}{dt} \mathbf{A} \cdot \mathbf{B} = \frac{d\mathbf{A}}{dt} \cdot \mathbf{B} + \mathbf{A} \cdot \frac{d\mathbf{B}}{dt} , \qquad \text{(II.5)}$$

with $\mathbf{A}$ and $\mathbf{B}$ vector functions of t In terms of the components of the two vectors, this takes the form

$$\begin{aligned}
\mathbf{A} \cdot \mathbf{B} &= (A_x \,\hat{\boldsymbol{x}} + A_y \,\hat{\boldsymbol{y}} + A_z \,\hat{\boldsymbol{z}}) \cdot (B_x \,\hat{\boldsymbol{x}} + B_y \,\hat{\boldsymbol{y}} + B_z \,\hat{\boldsymbol{z}}) \\
&= A_x B_x \,\hat{\boldsymbol{x}} \cdot \hat{\boldsymbol{x}} + A_x B_y \,\hat{\boldsymbol{x}} \cdot \hat{\boldsymbol{y}} + \cdots + A_z B_z \,\hat{\boldsymbol{z}} \cdot \hat{\boldsymbol{z}} \\
&= A_x B_x + A_y B_y + A_z B_z ,
\end{aligned} \qquad \text{(II.6)}$$

this last obtained by applying definition (II.5) to the dot products of the unit vectors $\hat{\boldsymbol{x}}$, $\hat{\boldsymbol{y}}$, and $\hat{\boldsymbol{z}}$. As with the treatment of vectors in Ch. I Sec. A, this description differs from that mathematicians might use. They could *define* the dot product[4] as expression (II.6), then use it to define vector magnitudes and angles via expression (II.5).

The work increment (II.3) can be written in the compact form

$$dW = \mathbf{F} \cdot d\mathbf{s} . \qquad \text{(II.7a)}$$

The net work done by force $\mathbf{F}$ on an object that moves from position $\mathbf{r}_0$ to position $\mathbf{r}_1$ is obtained by breaking up the motion into segments as small as necessary and adding up the work increment on each segment. That is, the net work is the integral

$$\Delta W = \int_{\mathbf{r}_0}^{\mathbf{r}_1} \mathbf{F} \cdot d\mathbf{s} . \qquad \text{(II.7b)}$$

This is a *line integral,* not an ordinary Riemann[5] integral. In general, it depends on the path of the motion, as well as the end points $\mathbf{r}_0$ and $\mathbf{r}_1$.

The units of work follow from its definition. The SI unit of work is

$$\begin{aligned}
1 \text{ N m} &= 1 \,\frac{\text{kg m}^2}{\text{s}^2} \\
&\equiv 1 \text{ J} \quad \text{(joule)} ,
\end{aligned} \qquad \text{(II.8)}$$

named for James Joule.[6]

The notion of work and its units can be illustrated via a simple demonstration: lifting an elephant over my head. I exert an upward force F only infinitesimally greater than the elephant's weight, as tossing an elephant is ill-advised. The elephant is an adult male African elephant of mass 5.0 metric tons, i.e., 5000 kg, and the lift raises the elephant 2.0 m. The force is constant,

---

[4]If one wishes to impress people at parties, one calls this the *Euclidean inner product.*
[5]Georg Friedrich Bernhard Riemann, 1826–1866
[6]James Prescott Joule, 1818–1889

so integral (II.8) reduces to

$$\Delta W = F\,\Delta s$$
$$= Mg\,\Delta s$$
$$= (5000\ \text{kg})(9.81\ \text{m/s}^2)(2.0\ \text{m})$$
$$= 98.\ \text{kJ}\ , \tag{II.9}$$

to two significant figures.

One of the most important discoveries in science is Joule's discovery that mechanical work and heat energy can be interconverted at a "fixed exchange rate"; the student may have seen this in a laboratory-course experiment. That rate is 4.184 joules per calorie, or 4.184 kJ/kcal. The kilocalorie is the Calorie, or Large Calorie, of the dietician. (In many countries, the caloric content of packaged foods is given in kilojoules.) Hence, one lift of the elephant corresponds to 23. kcal, or ten lifts to 230 kcal. That's one doughnut: Eating one doughnut—assuming I can metabolize it with perfect efficiency—supplies sufficient caloric content for me to lift an elephant over my head ten times.[7]

Energy

Perhaps no term in the lexicon of physics is more widely used: Everyone is aware of energy issues; the U.S. government has a cabinet-level Department of Energy. In physics, energy is a measure of motion. Its first appearance is as *vis viva* (the stuff of life):

$$K \equiv \tfrac{1}{2}mv^2\ , \tag{II.10}$$

for a body of mass m moving with speed v. This is now known as *kinetic energy*. It is a scalar quantity. Like work, it is measured in SI in joules.

We have already encountered other measures of motion, e.g., velocity and momentum. Which is the "correct" measure? In Newton's time and afterward, this was a matter of some controversy; we now appreciate that each of these measures has its uses, depending on the questions to be answered. Our current understanding of the significance of energy was first expressed by Thomas Young[8] around 1801: Energy is *the capacity to do work*. This seems remarkably simple, but throughout the rest of this text we shall see the extraordinary sweep of Young's notion.

The Work-Energy Theorem

The connection between work and energy is made explicit in a simple relationship. The net work ΔW_{net} done on an object of mass m, moving from initial position $\mathbf{r}_i$ to final position $\mathbf{r}_f$ under the influence of force or forces $\mathbf{F}_{\text{net}}$, is

---

[7]The student should not attempt this demonstration. The author is a trained professional.
[8]Thomas Young, 1773–1829

given by

$$\Delta W_{\text{net}} = \int_{\mathbf{r}_i}^{\mathbf{r}_f} \mathbf{F}_{\text{net}} \cdot d\mathbf{r}$$

$$= \int_{t_i}^{t_f} m\frac{d\mathbf{v}}{dt} \cdot \mathbf{v}\, dt$$

$$= \int_{t_i}^{t_f} \tfrac{1}{2}m\frac{d}{dt}(\mathbf{v} \cdot \mathbf{v})\, dt$$

$$= \int_{t_i}^{t_f} \frac{d}{dt}\left(\tfrac{1}{2}mv^2\right) dt$$

$$= K_f - K_i$$

$$\Delta W_{\text{net}} = \Delta K\;, \tag{II.11}$$

where in the second line we have employed Newton's Second Law and the definition of velocity, as well as introduced initial and final times t_i and t_f; in the fourth line we used $\mathbf{v} \cdot \mathbf{v} = v^2$. The initial and final values of the kinetic energy are K_i and K_f, respectively, and ΔK denotes the change in this quantity. As a consequence of Newton's Second Law, the net work done on a body (treated as a point) is equal to the change in its kinetic energy. This change can be positive, negative, or zero.

Power

Another term with a plethora of colloquial meanings, *power* in physics is the *rate,* in time, at which work is done. Force $\mathbf{F}$ acting on a body moving with velocity $\mathbf{v}$ expends or transfers power

$$P \equiv \frac{dW}{dt}$$

$$= \mathbf{F} \cdot \mathbf{v}\;, \tag{II.12}$$

from expression (II.7). This definition implies the SI unit of power:

$$1\,\frac{\text{J}}{\text{s}} = 1\,\frac{\text{kg m}^2}{\text{s}^3}$$

$$\equiv 1\,\text{W} \qquad (\text{watt})\;, \tag{II.13}$$

named for James Watt.[9] In many engineering disciplines, however, it remains common to use the unit *horsepower,* a term coined by Watt and Matthew Boulton.[10] The original definition of one horsepower was 33,000 foot-pounds per minute, or 550 foot-pounds per second.[11] But now the horsepower is defined in terms of the SI unit for power:

$$1.000\,\text{hp} \equiv 745.7\,\text{W}\;, \tag{II.14}$$

---

[9]James Watt, 1736–1819

[10]Matthew Boulton, 1728–1809

[11]Apparently, Watt and Boulton were working their horse rather hard, or they had an exceptionally strong horse.

here to four significant figures.

C   Flight range

The range of an airplane is determined by the *work* that must be done by the engines over the distance flown. This work must be supplied from the chemical energy content of the fuel burned. With thrust of magnitude F_T, the plane covers distance dx—here, in level flight at constant speed—by burning fuel mass $-dM$. The fuel is characterized by energy of combustion per unit mass (*specific* energy of combustion) E_{sp}. Because of thermal and mechanical losses, the energy of combustion is made available for work with efficiency $\eta < 1$. The work/energy "budget" takes the form

$$F_T \, dx = -\eta E_{sp} \, dM \ . \tag{II.15}$$

The negative sign appears because dM is negative, i.e., the mass M of the plane decreases as it burns fuel and expels exhaust.

At this point, it is necessary to invoke some information about the performance of aircraft engines.[12] The piston engines of older, propeller-driven planes are characterized by *horsepower-specific fuel consumption* (HSFC), defined historically as

$$c_h \equiv -\frac{1}{F_T v} \frac{d(Mg)}{dt} \ , \tag{II.16a}$$

which is approximately constant over the range of operation of the engine. This implies that the efficiency factor in Eq. (II.15) is given by

$$\eta = \frac{g}{c_h E_{sp}} \ , \tag{II.16b}$$

and is constant for a given engine and a given fuel. But jet engines—turbojets and turbofans, in which the jet exhaust turns large fans that force huge volumes of air through the engine—are characterized by *thrust-specific fuel consumption* (TSFC)

$$c_t \equiv -\frac{1}{F_T} \frac{d(Mg)}{dt} \ , \tag{II.17a}$$

which is approximately constant over the operating range of the engine. The efficiency factor for these engines is given by

$$\eta = \frac{gv}{c_t E_{sp}} \ , \tag{II.17b}$$

increasing with increasing speed v. Jet planes fly faster (and higher) than propeller-driven planes not only because they can, but because they must in order to be efficient enough to be economical.

---

[12]The author teaches at Saint Louis University, whose Parks College of Engineering, Aviation, and Technology grew from the first federally chartered flight school in the U.S. Hence, such knowledge is not far to seek.

Our proposed world-spanning airliner will certainly be powered by turbofan engines, the best current technology for commerical aircraft. The work/energy requirement (II.15), with efficiency (II.17b), implies instantaneous range or "mileage"

$$-\frac{dx}{dM} = \frac{gv}{c_t F_T}$$
$$= \frac{v}{c_t} \frac{C_L}{C_D} \frac{1}{M} \ , \qquad (II.18)$$

this last obtained by substituting F_T from Eq. (II.2). The total range—the figure we seek—for an aircraft with initial mass M_i and final mass M_f is given by the integral of this:

$$R \equiv \int_{M_i}^{M_f} \frac{dx}{dM} \, dM$$
$$= \frac{1}{c_t} \int_{M_f}^{M_i} v \frac{C_L}{C_D} \frac{dM}{M} \ , \qquad (II.19)$$

where the negative sign in Eq. (II.18) has been absorbed by reversing the limits of integration, which satisfy $M_f < M_i$. This result will allow us to determine the fuel mass $M_i - M_f$ required for the desired range, and to determine whether or not the project is feasible: A large airliner might carry hundreds of metric tons of fuel, but a million metric tons (10^9 kg), say, would be prohibitive.

The value of this integral actually depends on how the flight is carried out, because the parameters v, C_L, and C_D can all be adjusted in flight. Three possible flight programs for our proposed long-range flight are, *in order of increasing range*:

- Constant altitude, constant v;

- Constant altitude, constant C_L;

- Constant v, constant C_L (cruise-climb).

The first of these is easiest from the point of view of air-traffic control. It is achieved by reducing C_L—by changing the configuration of the wings, using flaps, et cetera—as the mass of the plane decreases and less lift is required. The second is achieved by leaving the wing configuration fixed and simply reducing speed as the mass decreases. For the last, the "cruise-climb" program, the speed and wing configuration are held fixed. The plane gains altitude as it loses mass[13]; the decreasing air density with altitude reduces the lift to that required for level flight, per Eq. (II.1b). This tactic allows the factor vC_L/C_D to be held at its maximum, range-maximizing value. In that case, the integral is readily evaluated, yielding

$$R_{\max} = \frac{1}{c_t} \left(v \frac{C_L}{C_D} \right)_{\mathrm{br}} \ln \left(\frac{M_i}{M_f} \right) \ , \qquad (II.20)$$

---

[13] If one observes the flight data displayed on the video system of an airliner on a transoceanic flight, one finds that current aircraft actually do this.

where the subscript br signifies "best range." This result is known as the *Breguet*[14] *range equation,* originally derived for piston-engine aircraft. Apparently, precedence in its discovery is still in dispute.

To determine the best-range parameters and calculate the maximum range, we need to use the fact that the lift and drag coefficients are related, in part because the production of lift also involves the production of induced drag. (See Fig. II.3.) The relationship between C_D and C_L is called the *drag polar* for the aircraft; historically, it was obtained from wind-tunnel measurements plotted in polar coordinates. This relationship can be complicated and sensitive to features of the aircraft design. For our purposes a simple model will suffice:

$$C_D = C_{D0} + \frac{C_L^2}{\pi e(b^2/S)} \ .$$

(II.21)

The notation here is traditional. The drag coefficient at zero lift is C_{D0}. The factor e is *not* the base of the natural logarithm; it is called the *Oswald*[15] *span efficiency*. It measures the smoothness of the airflow off the wings; modern aircraft sometimes sport "winglets" at their wing tips to increase its value. The wingspan is denoted b, total wing area is S, and the ratio b^2/S is the "aspect ratio" of the wings. Speed v is also related to C_L via Eq. (II.1b). Hence, the ratio vC_L/C_D can be written as a function of C_L and parameters which are held fixed:

$$v\frac{C_L}{C_D} = \left(\frac{2Mg}{\rho S}\right)^{1/2} \left(C_{D0}C_L^{-1/2} + \frac{C_L^{3/2}}{\pi e(b^2/S)}\right)^{-1} \ .$$

(II.22)

The second factor in large parentheses is large for small values of C_L (the first term) and large for large values of C_L (the second term). Hence, it must have a minimum value; here the overall expression is maximized. A straightforward calculus exercise yields best-range values

$$C_{L,\text{br}} = \left(\frac{\pi e b^2 C_{D0}}{3S}\right)^{1/2}$$

(II.23a)

and from drag polar Eq. (II.21)

$$C_{D,\text{br}} = \tfrac{4}{3}C_{D0} \ .$$

(II.23b)

It appears that the first factor in parentheses in Eq. (II.22) depends on M and hence varies throughout the flight. But this is not the case: This factor is held fixed, and as shall be seen, this requirement determines the altitudes at which the airplane must fly.

To evaluate our analysis, we shall apply our results to a proposed aircraft with the following design parameters:

---

[14]Louis-Charles Breguet, 1881–1955
[15]W. Bailey Oswald, 1907–1998

- "Empty" mass $M_f = 2.50 \times 10^5$ kg;

- Wing area $S = 525.$ m^2;

- Wingspan $b = 75.0$ m;

- Oswald span efficiency $e = 0.990$;

- Zero-lift drag coefficient $C_{D0} = 0.0150$;

- Best-range airspeed $v = 270.$ m/s;

- Thrust-specific fuel consumption (TSFC)

$$c_t = 0.600 \text{ hr}^{-1} = 1.67 \times 10^{-4} \text{ s}^{-1} .$$

Here, "empty" mass refers to the mass of the aircraft, passengers, cargo, and suitable fuel reserve. These parameters are modest improvements on current jetliners but do not represent radical advances in technology.

The Breguet range equation (II.20) determines the fuel required for the desired range $R_{\max} = 2.20 \times 10^4$ km, i.e., half the Earth's circumference plus a 10% "cushion," say. The initial-to-final mass ratio is

$$\frac{M_i}{M_f} = \exp\left[c_t \left(\frac{C_D}{v C_L} \right)_{\text{br}} R_{\max} \right]$$

$$= \exp\left[\frac{4 c_t}{v_{\text{br}}} \left(\frac{C_{D0} S}{3\pi e b^2} \right)^{1/2} R_{\max} \right]$$

$$= \exp\left[\frac{4(1.67 \times 10^{-4} \text{ s}^{-1})}{270. \text{ m/s})} \left(\frac{0.0150(525. \text{ m}^2)}{3\pi(0.990)(75.0 \text{ m})^2} \right)^{1/2} (2.20 \times 10^7 \text{ m}) \right]$$

$$\frac{M_i}{M_f} = 1.95 ,$$

(II.24a)

where some care has been taken with units and orders of magnitude. This implies initial mass $M_i = 4.86 \times 10^5$ kg, or fuel mass

$$\Delta M = M_i - M_f$$

$$= 2.36 \times 10^5 \text{ kg} ,$$

(II.24b)

i.e., 236. metric tons of fuel. This is 48.6% of the initial mass M_i, which is not an outrageously large value for a large transport aircraft. "Nonstop service to anywhere" begins to look feasible.

There is much more to be considered, of course. Does the proposed design have enough lift to carry that much fuel? Relation (II.1b) demands air density

$$\rho = \frac{2Mg}{(C_L v^2)_{\text{br}} S} .$$

(II.25a)

For mass $M_i = 4.86 \times 10^5$ kg, this must not be greater than the density of air at sea level, or the plane will not be able to take off.[16] The best-range lift coefficient has value

$$C_{L,\mathrm{br}} = \left(\frac{\pi e b^2 C_{D0}}{3S} \right)^{1/2}$$
$$= \left(\frac{\pi (0.990)(75.0 \text{ m})^2 (0.0150)}{2(525. \text{ m}^2)} \right)^{1/2}$$
$$= 0.408 . \tag{II.25b}$$

The required air density ρ_i at the start of the cruise-climb portion of the flight is

$$\rho_i = \frac{2M_i g}{(C_L v^2)_{\mathrm{br}} S}$$
$$= \frac{2(4.86 \times 10^5 \text{ kg})(9.81 \text{ m/s}^2)}{0.408(270. \text{ m/s})^2 (525. \text{ m}^2)}$$
$$= 0.611 \text{ kg/m}^3 , \tag{II.25c}$$

which is indeed less than the typical density $\rho_{\mathrm{air}} = 1.20$ kg/m^3 at sea level. The aircraft will have enough lift to cruise at a suitable altitude. The density ρ_f at the end of the cruise-climb portion is

$$\rho_f = \frac{2M_f g}{(C_L v^2)_{\mathrm{br}} S}$$
$$= \frac{2(2.50 \times 10^5 \text{ kg})(9.81 \text{ m/s}^2)}{0.408(270. \text{ m/s})^2 (525. \text{ m}^2)}$$
$$= 0.314 \text{ kg/m}^3 . \tag{II.25d}$$

Pilots use a *standard atmosphere* table to relate such air density values to altitudes. Typical values corresponding to ρ_i and ρ_f are

$$h_i \approx 6700 \text{ m} \tag{II.26a}$$

and

$$h_f \approx 12000 \text{ m} \tag{II.26b}$$

for the starting and finishing altitudes of the cruise-climb flight. These are 22000 ft and 39000 ft, respectively—entirely reasonable altitudes for a jet airliner. The plane will climb 5300 m while covering 20000 km, making the approximation of level flight quite accurate. We note also that the scaling of density ρ with mass M justifies the treatment of the first factor on the right-hand side of Eq. (II.22) as constant.

---

[16]The author is not making this up. Current airliners sometimes have difficulty taking off from high-altitude airports, e.g., Denver, in very hot weather because of insufficient air density and lift.

The thrust required in the cruise-climb portion of the flight is given by Eq. (II.2). At the start of this portion, this has value

$$
\begin{aligned}
F_T^{(i)} &= \left(\frac{C_D}{C_L}\right)_{\text{br}} M_i g \\
&= \frac{4C_{D0}}{3C_{L,\text{br}}} M_i g \\
&= \frac{4(0.0150)}{3(0.408)} \left(4.86 \times 10^5 \text{ kg}\right)\left(9.81 \text{ m/s}^2\right) \\
F_T^{(i)} &= 234. \text{ kN} , \tag{II.27}
\end{aligned}
$$

using results (II.23b) and (II.25b). Let us suppose, for the purposes of our model, that turbofan engines rated at 240. kN thrust (a reasonable value) are available. That rating is for operation at sea level, at air density $\rho = 1.20 \text{ kg/m}^3$, say. But like the rocket engines of Ch. I Sec. G, the turbofan engine produces thrust proportional to the rate at which it throws *mass,* hence, proportional to the air density. The thrust of one engine at starting altitude (II.26a), i.e., density (II.25c), would be 122. kN. Requirement (II.27) can be met with two engines. Of course, any sensible engineer would incorporate at least three engines (one under each wing, one at the tail)—or, more likely, four engines (two under each wing, or two on each side of the tail)—into the design as a safety margin against the possibility of engine failure. The engine thrust decreases as ρ throughout the cruise-climb flight, but the required thrust decreases in exactly the same way, since by Eq. (II.25a) ρ and M are proportional. The pilots need not make any changes to the engine control settings through the long cruise-climb other than minor adjustments for local conditions!

These results suggest that nonstop flight between any two places on Earth not only is physically possible, but might also be within reach of current technology. But this section, like the previous chapter, shows that tackling any substantial physics—or engineering—problem requires considerable information and work. In truth, we have barely scratched the surface of the complexity involved in designing a new aircraft. Designers must consider and balance an enormous variety of issues and requirements, right down to such details as the space taken up by passenger seats, the area of the windows, and the aesthetic merits of the design.

D Other flight-performance characteristics

Energy physics also bears on other aspects of flight. Both the speeds and altitudes at which an airplane can fly are determined by considerations of *power.* The power available from the engines is given by

$$
P_A = F_T v , \tag{II.28a}
$$

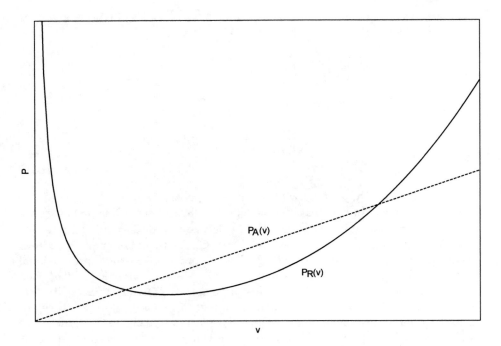

Figure II.4: Power available (P_A) and power required (P_R) as functions of speed v. Level flight can be maintained, or the aircraft has power to climb, only at speeds between the intersection points of the $P_A(v)$ line and the $P_R(v)$ curve.

where available thrust F_T is determined by engine design and air density, decreasing with altitude. The power required for level flight is given by

$$P_R = f_D v$$
$$= \tfrac{1}{2} C_D \rho S v^3$$
$$= \tfrac{1}{2} \rho S \left(C_{D0} + \frac{C_L^2}{\pi e (b^2/S)} \right) v^3$$
$$P_R = \tfrac{1}{2} \rho S C_{D0} v^3 + \frac{2 M^2 g^2}{\pi e b^2 \rho} \frac{1}{v} ,$$

(II.28b)

where drag polar (II.21) and lift requirement (II.1b) have been used. Graphed as functions of speed, $P_A(v)$ is a straight line through the origin, while $P_R(v)$ is a curve that diverges to $+\infty$ both for large v (the first term) and for v near zero (the second term). If the line and the curve intersect in two points, as illustrated in Fig. II.4, those two points identify the minimum and maximum speeds between which the aircraft has sufficient power to maintain level flight. The excess power, between the P_A line and the P_R curve, determines the rate at which the aircraft can climb.

Both line and curve shift with altitude, i.e., with air density ρ. In particular, the slope of the P_A line decreases with increasing altitude. The altitude at which the maximum climb rate is 300 ft/min is termed the *cruise ceiling* of the plane; the altitude at which the maximum climb rate is 100 ft/min is its *service ceiling*. (These are conventional designations derived from many years of pilots' experience.) The altitude at which the P_A line is only tangent to the P_R curve at a single point is the *absolute ceiling* of the airplane, the greatest altitude at which it can maintain level flight. This isn't the highest the plane can go: It can climb higher, temporarily, by pulling into a climb while maintaining thrust against drag. It then climbs as a projectile, trading kinetic energy for additional altitude. This is called "zooming." It is rarely done in a large jet airliner such as we have been considering, however.[17]

E Potential energy and energy conservation

The notion of energy can be expanded. Kinetic energy is complemented by the measure termed *potential energy*. In many introductory texts, this is defined by a phrase such as "the energy possessed by a body because of its position or condition." This is too vague to be useful. A handier definition might be: *Potential energy is a bookkeeping device for keeping track of the work done by certain types of forces.* The forces for which it is possible to define a potential energy are called *conservative forces*.

Conservative forces

A force $\mathbf{F}$ acting on a body moving in some region of space is termed *conservative* if and only if the work done by that force between *any* two points,

$$\Delta W = \int_{\mathbf{r}_0}^{\mathbf{r}_1} \mathbf{F} \cdot d\mathbf{s} , \tag{II.29a}$$

is *path-independent*, i.e., is the same for any path between points $\mathbf{r}_0$ and $\mathbf{r}_1$, depending only on the points themselves. An equivalent condition is that $\mathbf{F}$ is conservative if and only if the work done by the force on a body moving on any *closed* path—beginning and ending at the same point—is zero:

$$\Delta W_{\text{loop}} = \oint \mathbf{F} \cdot d\mathbf{s}$$
$$\equiv 0 , \tag{II.29b}$$

where the loop on the integral sign denotes integration over a closed path. These are equivalent, i.e., each implies the other[18]: Any two paths between points $\mathbf{r}_0$

---

[17]The aerobatic capabilities of large airliners are limited. The author once asked a Saint Louis University professional-pilot student to do a "loop-the-loop" in a Boeing 747. The student was able to execute the maneuver, but the plane broke up in flight. Fortunately, all of this was done on a simulator.

[18]There is a third mathematical condition on $\mathbf{F}$, equivalent to these, identifying it as conservative. But this condition is couched in the language of vector analysis, at a level too

and $\mathbf{r}_1$, taken together, constitute a closed loop, so condition (II.29b) implies the work on the two paths must be equal. Conversely, any closed loop consists of two paths between any two points on the loop, one traced forward, the other back. If the work done between the two points is path-independent, the combination—the work done around the loop—must be zero. More slickly, if the work done between any two points is path-independent, then the work done around any loop equals the work done simply remaining at the starting/finishing point, i.e, zero.

These criteria appear to be extremely restrictive. One might think conservative forces would rarely be encountered. However, several important classes of force laws meet the criteria:

- Constant forces. If $\mathbf{F}$ is constant the work integral over any path reduces to the dot product of $\mathbf{F}$ and the net displacement between the initial and final points.

- Forces $F(x)$ in one dimension, dependent only on position. In this case the work integral (II.29a) reduces to an ordinary Riemann integral, depending only on the form of $F(x)$ and the limit points.

- Central forces $F(r)\,\hat{r}$, i.e, forces directed radially toward or away from a force center, dependent only on distance from that center. Newtonian gravitation and electrostatic forces are of this type. For central forces, the work integral takes the same path-independent form as in the preceding case.

- Fundamental forces. The fundamental interactions—gravitation, weak nuclear forces, electromagnetic forces (combined, *electroweak* forces), and strong nuclear forces (i.e., *quantum chromodynamic* forces)—on which our best current understanding of matter is based are all conservative.

Hence, conservative forces are encountered in many physical situations.

Indeed, if the fundamental forces are conservative, whence *non*conservative forces? Why draw the distinction if every force behaves conservatively? Nonconservative or dissipative forces, such as friction and drag, arise from interactions with large numbers of bodies or large numbers of dynamical variables: The work done by these interactions—which, individually, are conservative—is shared out among a large number of "degrees of freedom" in such a way that it is astronomically unlikely that it can ever be gathered up again. For example, as a book slides across a table, molecular bonds are continuously formed and broken between the septillions of molecules in the book and the table top. These bonds are electromagnetic in nature; the forces are conservative. But the kinetic energy of the book is shared among the molecules in the form of heat. There is no way to orchestrate the motions of the molecules to recover that energy. It is dissipated. The work done by frictional forces (I.53) and (I.54), or drag forces (I.55), on a body moving between two points depends on the *length* of

---

advanced to be introduced at this point.

the path between them, not just on the end points. Hence, the work is not path-independent, and these "aggregate" forces are nonconservative.

Potential energy

A *potential energy function* is defined as the work done by a conservative force between an origin $\mathbf{r}_0$ (which we are free to choose) and any desired point $\mathbf{r}$:

$$U(\mathbf{r}) \equiv - \int_{\mathbf{r}_0}^{\mathbf{r}} \mathbf{F} \cdot d\mathbf{s} \ , \tag{II.30}$$

where the negative sign is introduced for later convenience. Because the integral is path-independent, this is simply a function of position $\mathbf{r}$. Path independence allows us to calculate the work done by $\mathbf{F}$ along any path between any two points $\mathbf{r}_1$ and $\mathbf{r}_2$ as the difference

$$\begin{aligned}
\Delta W &= \int_{\mathbf{r}_1}^{\mathbf{r}_2} \mathbf{F} \cdot d\mathbf{s} \\
&= -[U(\mathbf{r}_2) - U(\mathbf{r}_1)] \\
&= -\Delta U
\end{aligned} \tag{II.31}$$

in U values.

The potential energy function is not uniquely defined for a given force. Changing the choice of origin $\mathbf{r}_0$ changes the function $U(\mathbf{r})$ by adding a constant to its value at every point, viz., the work done by $\mathbf{F}$ on a body moving from the old origin to the new one. This corresponds to the freedom to set the zero of potential energy at any desired location. Work calculations like Eq. (II.31), i.e, differences in U, are unaffected by any such change.

For example, the constant Newtonian gravitational force in a small region near the surface of the Earth ($M\mathbf{g}$) is a conservative force. The potential energy corresponding to this force is given by

$$\begin{aligned}
U_g(\mathbf{r}) &= - \int_{\mathbf{r}_0}^{\mathbf{r}} M\mathbf{g} \cdot d\mathbf{s} \\
&= -M\mathbf{g} \cdot (\mathbf{r} - \mathbf{r}_0) \\
&= Mgh \ ,
\end{aligned} \tag{II.32}$$

the familiar form, with h (height) the component of displacement $\mathbf{r} - \mathbf{r}_0$ above the chosen origin in the vertically upward ($-\mathbf{g}$) direction. We shall encounter the potential energy for the *general* Newtonian gravitational force in Ch. IV. Another important example is the Hooke[19]-Law Law restoring force exerted by a linear spring in one dimension:

$$F(x) = -kx \ , \tag{II.33a}$$

---

[19] Robert Hooke, 1635–1703

with x the displacement of the spring from its unstretched, uncompressed length and k its *spring constant*. This is a one-dimensional force dependent only on position, hence, a conservative force. The corresponding potential energy is

$$U(x) = -\int_0^x (-kx')\,dx' + U_0$$
$$= \tfrac{1}{2}kx^2 + U_0\;;\tag{II.33b}$$

here the freedom to add an arbitrary constant is displayed explicitly. This potential energy will reappear in Ch. V.

For nonconservative forces, work integrals depend on paths as well as end points. They cannot be represented by functions of position. Potential energy is not defined for such forces.

Conservation of energy

The payoff: If only conservative forces act on a body in motion, then the Work-Energy Theorem (II.11) implies that over any time interval, the change in the body's kinetic energy is given by

$$\Delta K = \Delta W$$
$$= -\Delta U\;,\tag{II.34a}$$

the last from result (II.31). This implies

$$\Delta(K + U) \equiv \Delta E$$
$$= 0\;,\tag{II.34b}$$

i.e., the *total energy* $E \equiv K + U$ is unchanging, is conserved throughout the motion.

For the general case, with nonconservative forces in play, this result takes the form

$$\Delta E = \Delta W_{\mathrm{nc}}\;,\tag{II.35}$$

where the right-hand side denotes the work done by the nonconservative forces. What elevates this above a simple theorem in Newtonian mechanics is a generalization first observed by Joule and Helmholtz[20]: *In every situation in which dissipative forces act, it has always proved possible to identify the work done by those forces with another entry in the "energy ledger," so that the energy "books" again balance, i.e., the new total energy is conserved.* Joule and Helmholtz found the energy equivalent of heat produced by friction or drag; other examples include energy associated with the production of sound (pressure) waves or light (electromagnetic) waves.

It is not possible to overstate the significance of this Law of Conservation of Energy. The reader will learn few other ideas of comparable profundity and sweep—and that is not hyperbole. Although derived here from the

---

[20]Hermann Ludwig Ferdinand von Helmholtz, 1821–1894

Work-Energy Theorem, a consequence of Newton's Second Law of Motion, the Law of Conservation of Energy is found to apply far more broadly than Newton's Laws. It is the First Law of Thermodynamics. It applies in Einstein's Special and General Theories of Relativity, to bodies moving near the speed of light, and to black holes and galaxy clusters. It applies to Newtonian (i.e., Schrödinger/Heisenberg[21]) quantum mechanics and to Einsteinian (Dirac) quantum mechanics and quantum field theory. It isn't just that the speed of a car on a roller coaster can be determined from its height, or that perpetual-motion machines don't work. The fundamental structure of existence itself is linked to energy conservation. We shall encounter some of its many aspects throughout the remainder of this text.

Some examples

All introductory physics texts devote one or more chapters to the use of energy and energy conservation in solving physics problems. Because energy is a scalar, energy calculations are often simpler than vector force analyses. Here, we shall only touch upon some flight-related applications.

Energy height is an aircraft performance parameter. It is defined as the total mechanical energy of the aircraft divided by its weight, viz.,

$$
\begin{aligned}
H_e &\equiv \frac{K + U_g}{Mg} \\
&= \frac{v^2}{2g} + h \ ,
\end{aligned}
\tag{II.36}
$$

for an aircraft flying at speed v at altitude h. This quantity represents the maximum height the aircraft can reach via the zooming maneuver described in Sec. D. In a dogfight—aerial combat between two aircraft—superior energy height can be a decisive advantage.

In coastal areas around the world, it is often possible to observe seagulls floating in air, all but motionless in the wind. Remarkably, the height at which the birds float is not a matter of choice. Detailed analysis of avian flight is beyond the scope of this text, but energy methods can "rough in" the picture. In the reference frame of the air—which is moving with wind speed v_w with respect to the ground—a standing bird is moving with speed v_w. The floating bird is almost at rest, only occasionally flapping its wings to maintain lift. It has traded its kinetic energy in the air frame for gravitational potential energy. Hence, its height h and the wind speed are related by

$$
\Delta U \approx -\Delta K \ , \qquad \text{i.e.,}
$$
$$
mgh \approx \tfrac{1}{2}mv_w^2 \ .
\tag{II.37a}
$$

This yields

$$
h \approx \frac{v_w^2}{2g}
\tag{II.37b}
$$

---

[21] Werner Heisenberg, 1901–1976

for the height. For example, in a stiff wind of 40. km/hr, a bird would float at height

$$h \approx \frac{v_w^2}{2g}$$

$$\approx [4.0 \times 10^3 \text{ m/hr})(1 \text{ hr}/3600 \text{ s})]^2 2(9.81 \text{ m/s}^2)$$

$$\approx 6.3 \text{ m} , \tag{II.38}$$

a reasonable-looking value. We note that the mass of the bird drops out of the calculation—in a given wind, a common tern and a great albatross would float at the same height.[22]

In many physics texts the Law of Conservation of Energy is stated with a phrase such as, "Energy can be transferred or transformed, but *the total energy of the universe* remains constant." Alas, that is the one example we cannot use: The "total energy of the universe" is not meaningful. There is nowhere to stand to measure the total energy of the universe, no way to "weigh" it even in principle. It is not that the universe *violates* energy conservation; the notion simply has no meaning applied to the universe as a whole. In intermediate and advanced mechanics courses, it is shown that energy conservation is intimately connected with a feature called "time-translation invariance." Colloquially put: If an experiment is performed on some system at noon and an identical experiment performed on an identically prepared system at 1 p.m., the second experiment will duplicate the first one hour later. It is shown in analytical mechanics that a system with time-translation invariance has a conserved total energy.[23] But there is no meaning to the notion of "starting the universe one hour later," since, by definition, the universe contains all the clocks; no meaning to time-translation invariance or energy conservation. We can profitably apply these concepts to any subsystem of the universe, but not to the universe itself.

Forces and motions from potential energy

In chemistry and atomic and molecular physics, it is common to describe dynamics in terms of potential energies rather than forces. The two descriptions must be logically equivalent: Since potential energy is obtained from force by integration, force can be derived from potential energy by differentiation. The Fundamental Theorem of Calculus implies that if the integral in Eq. (II.30) is differentiated with respect to coordinate x, leaving y and z fixed [equivalently, if we calculate ΔU between points $(x + \Delta x, y, z)$ and (x, y, z), divide by $-\Delta x$, and take the limit $\Delta x \to 0$], the result is force component

$$F_x = -\frac{\partial U}{\partial x} , \tag{II.39a}$$

---

[22]Of course, one is not likely to see this. One would expect to see two common terns floating side by side; as is well known, one good tern deserves another.

[23]One of the payoffs of the development of physics is our ability to appreciate connections such as this.

with similar expressions for components F_y and F_z. The derivative here is a *partial derivative,* a derivative with respect to one of several independent variables, treating the others as constants. It is usually encountered in the second or third term of the introductory calculus sequence. Fortunately, the *mechanics* of differentiation is no different for these than for ordinary derivatives df/dx of functions of a single independent variable. The vector force **F**, then, can be written

$$\mathbf{F} = -\left(\frac{\partial U}{\partial x}\, \hat{\boldsymbol{x}} + \frac{\partial U}{\partial y}\, \hat{\boldsymbol{y}} + \frac{\partial U}{\partial z}\, \hat{\boldsymbol{z}} \right)$$
$$\equiv -\nabla U \ . \qquad\qquad\qquad (\text{II.39b})$$

The vector ∇U, defined by the previous expression, is called the *gradient* of U; it is read "grad U" or "del U," although the ∇ symbol is called a "nabla." The gradient of any function is a vector, the direction of which is the direction that the function increases most rapidly with distance, and the magnitude of which is the rate of change of the function with distance in that direction. Hence, because of the negative sign in the original definition of U, the force vector is in the direction in which potential energy *decreases* most rapidly with distance (it points "downhill") and has magnitude equal to that maximum rate of *decrease* with distance. *CAUTION*: The gradient is given in terms of partial derivatives by the simple expression of Eq. (II.39b) only in Cartesian (rectangular) coordinates. The expression is more complicated in other (curvilinear) coordinate systems; the reader will encounter these later.

Many features of the motion of a body subject to a conservative force can be determined from a graph of its potential-energy function. This approach is particularly useful for motion in one or two dimensions. Consider, for example, one-dimensional motion with potential energy as in Fig. II.5. Such a potential-energy function might be used to model interacting atoms, ions, or nuclear particles. A *classical* particle must move only in those regions in which the potential-energy curve lies below the line corresponding to its conserved total energy; the difference between that line and the curve is its kinetic energy.[24] Hence, a particle with energy E_0, as indicated in the figure, must either oscillate between *turning points* a and c or move in from infinity, come to rest momentarily at *turning point* e, and accelerate away again. At points b and d, the slope of the graph is zero, i.e., the force is zero at these *equilibrium points*. Moreover, if a particle at rest at point b is disturbed slightly, it will oscillate about that point. This local minimum of the potential energy is a *stable* equilibrium point. A particle at rest at point d, slightly disturbed, will accelerate away from that point. A local maximum of potential energy, point d is an *unstable* equilibrium point. If the graph were flat over some interval, those points would be *neutral* equilibrium points—a particle at rest at such a point, slightly disturbed, neither returns nor accelerates away.

---

[24]Quantum mechanics changes the rules: A quantum particle can *tunnel* through barriers it lacks the energy to surmount.

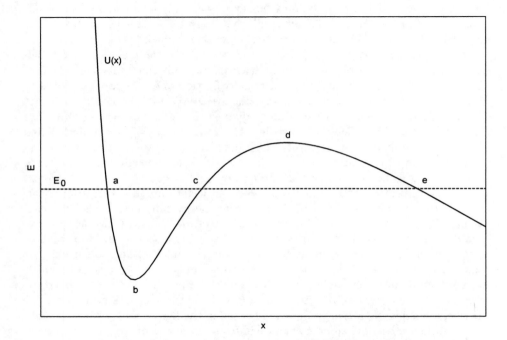

Figure II.5: A graph of potential energy in one dimension. Many features of motion under the influence of this interaction can be deduced from this graph.

References

Energy and its applications are extensively discussed in all introductory physics texts and all intermediate and advanced mechanics texts. The example of the floating seagull in Sec. E is discussed in Barger and Olsson [BO95].

The dynamics of flight is presented in specialized texts on the subject. For example, Anderson [And85] and Hale [Hal84] provide accessible treatments much more detailed than that given here; the former includes a wealth of historical detail. Von Mises [vM59] is a classic reference.

The inapplicability of energy conservation to the universe is discussed in Misner, Thorne, and Wheeler [MTW73], pp. 457–459. After forty years and only a single edition, this is still the gold-standard advanced text on Einstein's General Theory of Relativity.

Problems

1. Lacking thrust, a glider cannot maintain level flight at constant speed in still air. It can, however, glide at a constant speed at a fixed downward angle. Find this angle in terms of the lift and drag coefficients C_L and C_D. Recall that lift is perpendicular to the glider's velocity, and drag is antiparallel to its velocity. (Force laws (II.1) are appropriate here.)

2. A car experiences tractive (rolling) friction with coefficient of friction $\mu_t = 0.0200$ and turbulent-flow (quadratic) drag with drag coefficient $C_D = 0.350$. The car has mass (including fuel, which is a negligible fraction of the total) $M = 900.$ kg and frontal cross-sectional area $S = 2.00$ m^2. The density of air is approximately 1.20 kg/m^3. The heat of combustion of gasoline is approximately 36.0 MJ per *liter,* but when thermal and mechanical losses are considered, only 20.0% of this is available for the work of propulsion. Calculate the gas mileage (km/liter will do) of this car traveling on a straight, level road at a constant speed of 30.0 m/s. WARNING: Formula (II.18) for the "mileage" of the jet plane does *not* apply here; only the principles on which it is based do.

3. Analyze the *energy* requirements for a rocket in free space. Although the momentum of the rocket and its exhaust is conserved (constant), their energy is not; energy must be supplied by the combustion of fuel. WARNING: This is not the same as the energy requirement of the jet plane.

   (a) Using the same approach used in Ch. I Sec. G to derive the First and Second Rocket Equations—sequential "snapshots" of the rocket—obtain an expression for the total kinetic energy gained by rocket and exhaust during the engine burn in terms of the initial and final rocket masses M_i and M_f, the nozzle speed U of the exhaust, et cetera. This is the energy that must be supplied by combustion. You may use the fact that squares and products of differentials are negligible compared to the differentials themselves.

(b) Define an "energy efficiency" η for the rocket as the ratio of the final kinetic energy of the *rocket*—assuming it starts from rest—to the required energy calculated in part (a). (This is *not* the same as the efficiency of the jet engine in Sec. C.) Find an expression for this efficiency as a function of the ratio of fuel mass to final rocket mass, $x \equiv (M_i - M_f)/M_f$.

(c) Find the maximum value of the efficiency η and the value of x at which it occurs. Good numerical estimates will do.

4. Using the numerical parameters of the long-range jet airliner described in Sec. C, evaluate the engine *power* supplied for level flight at the start and at the end of the cruise-climb portion of its journey.

5. Complete the power analysis described in Sec. D.

(a) Determine the minimum and maximum speeds for level flight in terms of engine thrust F_T and any other relevant parameters.

(b) Determine the condition on these parameters corresponding to the absolute ceiling of the aircraft.

Chapter III

"Trouble in River City"

- *A cue ball, rolling without slipping, strikes an identical ball at rest head-on. Find the final velocities of both balls.*

- *Analyze collisions in one dimension, neglecting rotation.*

- *Examine collisions in two and three dimensions.*

- *Understand the kinematics and dynamics of rotation and apply these to the motion of the billiard balls.*

- *Examine rotational motion in three dimensions and gyroscopic motion.*

Figure III.1: Collision theory and rotational dynamics in action on the green baize. (Photograph by Krzysztof Grzymajlo.)

Pool may mean "trouble in River City" [from "The Music Man" (1957), music and lyrics by Meredith Willson,[1] book by Meredith Willson and Franklin Lacey[2]], but the game and its relatives—billiards, snooker, et cetera—are excellent demonstrations of important physical phenomena. Pool balls *collide* and *scatter*; they also *spin* and *roll*. Here, we shall use some simple examples from the pool table as starting points to explore these varied and crucial aspects of mechanics.

A   Collisions: interactions of short duration

A *collision* is an interaction between or among two or more bodies, of "short" duration, for which it is possible to identify well-defined "before" and "after" states. Collisions are found on every scale in physics: Elementary particles

---

[1] Robert Meredith Willson, 1902–1984
[2] Franklin Lacey, 1917–1988

collide, nuclei and atoms and molecules collide, billiard balls collide, automobiles collide, planetary bodies collide, galaxies collide. On the face of it, "short" duration makes no sense. Duration is a quantity with units and cannot be deemed large or small except in comparison to another quantity with the same units. Hence, a collision must involve two timescales, one much shorter than the other. One would not describe the gravitational interaction between the Earth and the Sun as a collision—the interaction goes on and on, defining only one timescale (the year), featuring neither "before" nor "after." On the other hand, a rogue comet might enter the solar system from interstellar space, make a close pass by the Sun, and return to the depths of space. That would be a collision— actual "contact" is not required. The comet would roam the galaxy for billions of years, while its passage through the inner solar system might take only a few months, a "short" duration by comparison. Billiard balls move about the table on a timescale of seconds, but they interact via elastic forces, in contact with each other, for a few *milli*seconds at a time. Elementary particles move through accelerators and detectors on a timescale of nanoseconds but might interact over intervals of order 10^{-23} s. Despite the enormous range of sizes and timescales, all of these can be described using the language of collisions.

Newtonian collision in one dimension

We can begin our exploration of collision phenomena with the simplest of examples: a two-body collision in one spatial dimension in which the bodies retain their identities. As depicted in Fig. III.2, point bodies of masses m_1 and m_2 move along a common line with velocity components v_1 and v_2 (with $v_1 > v_2$), collide, and separate. They retain their masses but acquire new velocity components V_1 and V_2, respectively. We shall regard the masses m_i and initial velocities v_i as known quantities and the final velocities V_i as unknowns to be determined.

The forces the bodies exert on each other during their brief interaction might be exceedingly complicated, so that solving Newton's Second Law for each body, instant by instant, might be hopeless. We can, however, attack the problem by comparing the "before" and "after" states of the two bodies. Here, conservation laws, asserting that certain features of the system are the same before and after, are especially useful. For example, we might assume that the forces exerted by the bodies on each other during the collision are much larger than any external forces on the system. The effects of the external forces can be neglected, in comparison with those of the forces between the bodies, *over the duration of the collision.* But those internal forces are an action/reaction pair. Hence, as described in Ch. I Sec. G, the total momentum of the system is the same just before and just after the collision. This yields the condition

$$m_1v_1 + m_2v_2 = m_1V_1 + m_2V_2 \tag{III.1}$$

on the two unknown velocities.

To obtain a second condition and solve the problem, we resort to an approach similar to that used to describe friction and drag interactions in Ch. I Sec. E, a

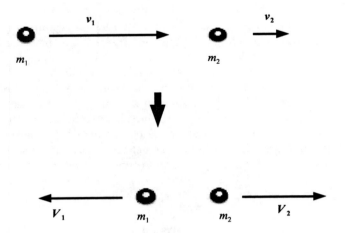

Figure III.2: A two-body collision in one dimension.

rather nineteenth-century approach. We define a parameter to represent a very complicated effect, such as the elastic forces that arise between two colliding solid bodies, relying on experiment to determine the parameter. In this case, we introduce the *coefficient of restitution*, the ratio r of the speed of recession of the two bodies to their speed of approach, via the relation

$$V_2 - V_1 = -r(v_2 - v_1) \; . \tag{III.2}$$

The negative sign is included because $V_2 - V_1$ is positive (the bodies separate after colliding), while $v_2 - v_1$ is negative (they approach before); this makes r a non-negative quantity. The coefficient is dimensionless. For ordinary collisions involving elastic forces, it is constrained to lie in the interval $0 \leq r \leq 1$. The usefulness of this definition hinges on the observation that for many bodies and materials, r is constant over a wide range of collision speeds, determined only by the materials involved. The coefficient of restitution of many bodies can be estimated by dropping them to a floor from height h_d, say, and measuring the height h_r to which they rebound. Using the constant-acceleration kinematics of

Ch. I Sec. B yields the relation

$$r \equiv \frac{v_r}{v_d}$$

$$= \frac{(2gh_r)^{1/2}}{(2gh_d)^{1/2}}$$

$$= \left(\frac{h_r}{h_d}\right)^{1/2} , \tag{III.3}$$

with v_d the speed of impact with the floor and v_r the speed of rebound, and aerodynamic drag neglected. Engineering laboratories use more precise variants of this experiment to determine r values.

Relations (III.1) and (III.2) give two linear equations for the two unknown velocities. They can be solved by any of the standard algebraic methods, with results

$$V_1 = \frac{m_1 - rm_2}{m_1 + m_2} v_1 + \frac{(1+r)\,m_2}{m_1 + m_2} v_2 \tag{III.4a}$$

and

$$V_2 = \frac{(1+r)\,m_1}{m_1 + m_2} v_1 + \frac{m_2 - rm_1}{m_1 + m_2} v_2 . \tag{III.4b}$$

These are a complete and general solution of the one-dimensional, two-body collision problem as stated.

Several special cases of this analysis are worth noting.

- The *perfectly inelastic collision* corresponds to $r = 0$; the two bodies adhere, couple, or fuse in the collision. Their common final velocity is

$$V_1 = V_2 = \frac{m_1 v_1 + m_2 v_2}{m_1 + m_2} , \tag{III.5}$$

  the total momentum of the system divided by the total mass.

- *Elastic collisions* correspond to $r = 1$; this is algebraically equivalent to the condition that the total kinetic energy of the bodies is conserved in the collision. Some particular examples of such collisions are illuminating:

  – If the bodies are of equal mass ($m_1 = m_2$), and the second body is initially at rest ($v_2 = 0$; the reference frame in which this obtains is called the LAB frame), then the first body comes to rest ($V_1 = 0$), and its velocity is simply transferred to the second: $V_2 = v_1$. This is readily observed on the pool table, the air-hockey table, or *Newton's Cradle,* the familiar desk toy with steel balls suspended from a wooden frame.

  – If the first body is much more massive than the second ($m_1 \gg m_2$), and the second is again initially at rest—e.g., bowling ball strikes ping-pong ball—then the velocity of the first body is essentially unaltered ($V_1 \cong v_1$), and the second body flies off with twice that speed: $V_2 \cong 2v_1$.

— If the first body is much more massive than the second ($m_1 \gg m_2$), and the two bodies converge with equal speeds and opposite velocities ($v_2 = -v_1$), then again the velocity of the first body is essentially unchanged ($V_1 \cong v_1$), while the second body flies off with *three times* the original speed: $V_2 \cong 3v_1$. Such a collision can be approximated by holding a tennis ball, say, slightly above a basketball and dropping them together so that the balls collide just after the basketball has rebounded from the floor. *CAUTION*: Any reader who intends to attempt this demonstration is urged to be *very* careful!

Of course, solutions (III.4) also cover all cases between the two extremes of perfectly inelastic and elastic collisions.

Einsteinian collisions

This problem is ideal for illustrating the changes in mechanics introduced by Einstein in 1905. Indeed, Einsteinian mechanics is most frequently invoked in analyzing collisions, since bodies moving at speeds near that of light are almost always encountered on Earth in nuclear or elementary-particle collision experiments.

Here, it is clear that the choice of transformations connecting physically equivalent reference frames (inertial frames) strongly constrains the physical laws that can be invoked. In all the collisions analyzed previously, momentum was conserved, and mass was also conserved. This was explicit in the assumption that the two bodies retained their masses m_1 and m_2, but it was not optional. Consider a very general sort of interaction in which some particles with masses m_i and velocities $\mathbf{v}_i$ come together, *something* happens—collisions, chemical reactions, nuclear reactions, whatever—and the same or different particles, with masses $\tilde{m}_j$ and velocities $\tilde{\mathbf{v}}_j$, emerge. (The subscript is different because the number of particles might change.) Momentum conservation, with Newtonian momentum, requires

$$\sum_i m_i \mathbf{v}_i = \sum_j \tilde{m}_j \tilde{\mathbf{v}}_j \ . \tag{III.6a}$$

If observer O and her colleagues, in uniform motion with respect to our reference frame, are equally entitled to use the law of conservation of momentum, then the masses and (primed) velocities they assign to all the particles obey

$$\sum_i m_i \mathbf{v}_i' = \sum_j \tilde{m}_j \tilde{\mathbf{v}}_j' \ , \tag{III.6b}$$

the masses the same as in our frame because mass is a scalar quantity. But if O's velocities are related to ours via Galilean transformation (I.39b), with constant velocity $\mathbf{v}_O$, then this implies

$$\sum_i m_i \mathbf{v}_i - \left(\sum_i m_i \right) \mathbf{v}_O = \sum_j \tilde{m}_j \tilde{\mathbf{v}}_j - \left(\sum_j m_j \right) \mathbf{v}_O \ , \tag{III.6c}$$

or, given Eq. (III.6a),

$$\sum_i m_i = \sum_j \tilde{m}_j \; . \tag{III.6d}$$

That is, *conservation of mass is a logical consequence of Newtonian momentum conservation and Galilean covariance.*

By contrast, Einsteinian energies and momenta are related in different inertial frames by Lorentz transformations (I.88), e.g.,

$$E' = \gamma_O(E - v_O p^1) \tag{III.7a}$$

and

$$p^{1\prime} = \gamma_0 \left(p^1 - \frac{v_O}{c^2} E \right) \; , \tag{III.7b}$$

with

$$\gamma_O = \left(1 - \frac{v_O^2}{c^2} \right)^{-1/2} \; , \tag{III.7c}$$

for observer O and friends moving in the common $+x$ direction with speed v_O. These imply that Einsteinian momentum **p** can be conserved in all inertial frames only if energy E is also, and vice versa. Since E and **p** determine mass m via

$$E^2 - [(p^1)^2 + (p^2)^2 + (p^3)^2] c^2 = m^2 c^4 \; , \tag{III.8}$$

as follows from definitions (I.95), mass must fall out where it will; its conservation cannot be imposed independently.

An highly simplified example illustrates this clearly. Suppose the first body in our two-body collision is a proton of mass m_p, moving at a speed such that its energy is $E_1 = 2.000 \, m_p c^2$. The second body is another proton, initially at rest. Suppose further that the collision is perfectly inelastic—the two protons fuse to become a single new particle after the collision. (This is *passing* unlikely, because protons are messy, composite particles,[3] but it will serve as an example.) The first proton has momentum

$$\begin{aligned} p_1 &= (E_1^2 - m_p^2 c^4)^{1/2}/c \\ &= [(2.000 \, m_p c^2)^2 - m_p^2 c^4]^{1/2} \\ &= 1.732 \, m_p c \; , \end{aligned} \tag{III.9a}$$

a single component in the one spatial dimension of the problem, and speed

$$\begin{aligned} v_1 &= \frac{p_1 c^2}{E_1} \\ &= \frac{1.732 \, m_p c^3}{2.000 \, m_p c^2} \\ &= 0.8660 \, c \; . \end{aligned} \tag{III.9b}$$

---

[3]The author invites his experimentalist colleagues to do this with actual protons.

By energy and momentum conservation, the new particle has energy $E = 3.000\, m_p c^2$ and momentum $P = 1.732\, m_p c$. Hence, it has speed

$$
\begin{aligned}
V &= \frac{Pc^2}{E} \\
&= \frac{1.732\, m_p c^3}{3.000\, m_p c^2} \\
&= 0.5774\, c \ ,
\end{aligned}
\tag{III.9c}
$$

and mass

$$
\begin{aligned}
M &= (E^2 - P^2 c^2)^{1/2}/c^2 \\
&= [(3.000\, m_p c^2)^2 - (1.732\, m_p c)^2 c^2]^{1/2}/c^2 \\
&= 2.449\, m_p \ .
\end{aligned}
\tag{III.9d}
$$

Speed V is *two-thirds* the initial speed v_1 of the first proton, in contrast to the Newtonian result (III.5), which would be one-half (of a different initial speed, as well). And mass M clearly contrasts with the Newtonian, mass-conserving value $2m_p$. In this collision, kinetic energy has been traded for additional mass, a process that Einsteinian dynamics allows—demands—but Newtonian dynamics does not. The fact that this tradeoff is observed in nuclear and particle collisions more elaborate than this one affirms the validity of Einstein's supersession of Newtonian mechanics.

B   Collisions in two and three dimensions

In two or three spatial dimensions,[4] even the simple two-body collision, in which the bodies retain their masses, presents a greater challenge. In the Newtonian limit, momentum conservation takes the vector form

$$
m_1 \mathbf{v}_1 + m_2 \mathbf{v}_2 = m_1 \mathbf{V}_1 + m_2 \mathbf{V}_2 \ .
\tag{III.10}
$$

In two (three) dimensions, this represents two (three) equations for four (six) unknown final-velocity components. The perfectly inelastic collision is still soluble: If the bodies adhere, their common final velocity is

$$
\mathbf{V}_1 = \mathbf{V}_2 = \frac{m_1 \mathbf{v}_1 + m_2 \mathbf{v}_2}{m_1 + m_2} \ ,
\tag{III.11}
$$

the total momentum divided by the total mass. But more general collisions are problematic. The meaning of the coefficient of restitution is unclear. In some cases, it can be applied to particular components of the bodies' relative velocities, but its interpretation is sensitive to the details of the geometry of each collision. In general, its introduction would supply one additional relationship to the problem, since the coefficient of restitution is a ratio of speeds. Likewise,

---

[4]If one wishes to explore the higher-dimensional spaces of certain modern elementary-particle models, there could be more.

for elastic collisions, kinetic-energy conservation can supply only one additional relationship (in place of that supplied by a coefficient of restitution, not in addition to it), because energy is a scalar. Hence, we have three (four) equations for four (six) unknowns. The problem is indeterminate: One (two) piece(s) of information is (are) still missing.

For example, if the two bodies have equal mass ($m_1 = m_2$), and the second body is initially at rest ($\mathbf{v}_2 = \mathbf{0}$), then momentum conservation, Eq. (III.10), implies

$$\mathbf{V}_1 + \mathbf{V}_2 = \mathbf{v}_1 \ . \tag{III.12a}$$

If the collision is elastic, then energy conservation implies

$$
\begin{aligned}
V_1^2 + V_2^2 &= v_1^2 \\
&= \mathbf{v}_1 \cdot \mathbf{v}_1 \\
&= (\mathbf{V}_1 + \mathbf{V}_2) \cdot (\mathbf{V}_1 + \mathbf{V}_2) \\
V_1^2 + V_2^2 &= V_1^2 + V_2^2 + 2\mathbf{V}_1 \cdot \mathbf{V}_2 \ ,
\end{aligned}
\tag{III.12b}
$$

which in turn implies

$$\mathbf{V}_1 \cdot \mathbf{V}_2 = V_1 V_2 \cos \vartheta = 0 \ , \tag{III.12c}$$

with ϑ the angle between the final velocity vectors $\mathbf{V}_1$ and $\mathbf{V}_2$. Unless one of the bodies is left at rest after the collision, *the final trajectories of the two bodies, under the circumstances assumed here, must be perpendicular.* This is illustrated in Fig. III.3. This analysis does not give the bodies' final directions, only that the angle between them must be $\pi/2$ radians, or $90°$.

If the direction of one of the bodies after the collision had been specified, the problem could have been solved completely. Most introductory physics texts contain a number of problems of this sort. In general, the number of missing pieces of information is equal to the number needed to specify one direction in two (three) dimensions.

If the interaction between the bodies is known, e.g., contact, electrostatic, gravitational, et cetera, then it may be possible to solve Newton's Second Law for the encounter and obtain a complete solution. Conversely, by measuring the outcomes of various collisions, we might deduce features of the interaction. This is the general approach of the broad class of observations known as *scattering experiments.* For example, we might formulate a model of the interaction between atomic nuclei, calculate the behavior of nuclei colliding under the influence of this interaction, fit the results of these calculations to data on actual collisions by adjusting parameters of the model, then use the parameter values obtained to enhance our understanding of the nuclear interaction. This may seem a rather esoteric exercise, but it is not: Much of what we know about the world beyond the reach of our senses—especially at the level of the elementary particle, the nucleus, the atom, and the molecule—has been obtained from scattering experiments. Indeed, our vision itself is a process of scattering analysis of "photons" or particles of light.

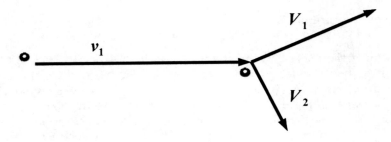

Figure III.3: Newtonian, elastic collision between two bodies of equal mass in the LAB frame (i.e., with the second body initially at rest).

The description of scattering experiments is couched in the language of probabilities and *cross sections*. For example, suppose a beam of particles of intensity $\mathcal{I}$ particles per unit area per unit time is directed at a target, and dN particles per unit time are collected by a detector at a particular direction, occupying solid angle $d\Omega$ (the area of the detector perpendicular to the incoming, scattered particles divided by the square of the distance from the scattering center, measured in *steradians*[5]). The *differential cross section* for scattering in that direction is given by

$$\frac{d\sigma}{d\Omega} \equiv \frac{1}{\mathcal{I}} \frac{dN}{d\Omega} \; . \qquad \text{(III.13)}$$

The integral of this quantity over all scattering directions, i.e., all solid angles, is the *total cross section* for the scattering process involved. These are quantities with units of area, hence the names.[6] In atomic, nuclear, and particle physics, these are often measured in "barns" (10^{-24} cm^2), millibarns, microbarns, et cetera, and even "sheds" (10^{-24} barns).[7]

C   Center of mass and the CM frame

Center of mass—definition

An important feature of systems of bodies and extended bodies—as opposed to bodies approximated as points—is the *center of mass*. In the Newtonian limit, this is defined as the *mass-weighted average position* of the system or body:

$$\mathbf{r}_{\text{cm}} \equiv \frac{1}{M_{\text{tot}}} \sum_i m_i \, \mathbf{r}_i \; , \qquad \text{(III.14a)}$$

for a collection of discrete point bodies labeled by index i with total mass M_{tot}. For an extended body regarded as a continuous distribution of matter, we imagine the body broken into suitable pieces and take the limit in which the number of pieces is very large, i.e.,

$$\begin{aligned}
\mathbf{r}_{\text{cm}} &\equiv \frac{1}{M_{\text{tot}}} \int_{\text{body}} \mathbf{r} \, dm \\
&= \frac{1}{M_{\text{tot}}} \int_{\text{body}} \mathbf{r} \, \rho(\mathbf{r}) \, d\mathcal{V} \; , \qquad \text{(III.14b)}
\end{aligned}$$

where $\rho(\mathbf{r})$ is the mass density of the body and $d\mathcal{V}$ a suitable increment of volume. In the mathematical language familiar to statisticians, the center of

---

[5] Hence, the sphere of all possible directions surrounding the scattering center occupies 4π steradians of solid angle.

[6] The choice of terminology is related to the fact that, e.g., if you hang targets on a barn wall and shoot at them, the probability of a hit is proportional to the cross-sectional area of the targets.

[7] These names may relate to the ability to hit "the broad side of a barn" as a measure of shooting skill, or lack thereof.

mass is the *first moment of the mass distribution*. The zeroth moment is simply the total mass; we shall encounter the second moment later.

For example, an 8.00 kg mass at the origin ($\mathbf{r}_1 = \mathbf{0}$) and a 2.00 kg mass at position $\mathbf{r}_2 = (10.0 \text{ m})\,\hat{\boldsymbol{x}}$ is a system of total mass 10.00 kg and center of mass

$$
\begin{aligned}
\mathbf{r}_{\text{cm}} &= \frac{1}{M_{\text{tot}}}\,(m_1 \mathbf{r}_1 + m_2 \mathbf{r}_2) \\
&= \frac{1}{10.00 \text{ kg}}\,[(8.00 \text{ kg})\,\mathbf{0} + (2.00 \text{ kg})\,(10.0 \text{ m})\,\hat{\boldsymbol{x}}] \\
&= (2.00 \text{ m})\,\hat{\boldsymbol{x}} \ .
\end{aligned}
\tag{III.15}
$$

This example illustrates the fact that there need not be any mass at the center of mass. As another example, consider a right circular cone of uniform mass density, height h, and base radius R. For convenience, let us take the center of the base of the cone as the origin and the axis of the cone as the $+z$ axis. By symmetry, the center of mass of the cone must lie on that axis: $x_{\text{cm}} = y_{\text{cm}} = 0$. How high up? If we imagine the cone sliced into a stack of disks, as sketched in Fig. III.4, we obtain the height of the center of mass from the integral

$$
\begin{aligned}
z_{\text{cm}} &= \frac{1}{M_{\text{tot}}} \int_0^h z\,\rho\,\pi \left(R\,\frac{h-z}{h} \right)^2 dz \\
&= \frac{1}{\frac{1}{3}\pi R^2 h \rho} \int_0^h \frac{\pi R^2 \rho}{h^2}\,(h^2 z - 2hz^2 + z^3)\,dz \\
&= \frac{3}{h^3}\,\left(\tfrac{1}{2}h^2 z^2 - \tfrac{2}{3}hz^3 + \tfrac{1}{4}z^4 \right)\Big|_0^h \\
z_{\text{cm}} &= \tfrac{1}{4}h \ ,
\end{aligned}
\tag{III.16}
$$

where the standard formula for the volume of a cone is used in the second line. Most introductory calculus texts contain exercises like this, although the integrals can be evaluated analytically only for a few simple geometries. Hence, such calculations will not be emphasized here.

Kinematics

The motion of the center of mass (CM) point displays useful simplicity. The velocity of the CM, obtained directly by differentiating definition (III.14), is

$$
\begin{aligned}
\mathbf{v}_{\text{cm}} &= \frac{1}{M_{\text{tot}}} \sum_i m_i \mathbf{v}_i \ , \\
&= \frac{\mathbf{P}_{\text{tot}}}{M_{\text{tot}}} \ ,
\end{aligned}
\tag{III.17}
$$

and similarly for the integral form. The last expression—valid in the Newtonian limit, of course—shows that *the total momentum of the system is the momentum it would have if its entire mass were a point located at the center of mass*. This even though no part of the system need actually be located at the CM.

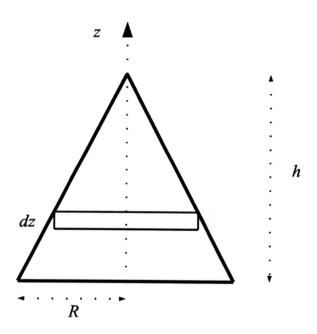

Figure III.4: A right circular cone of height h and base radius R, seen side-on, showing a disk slice of thickness dz.

The acceleration of the CM is obtained by a second differentiation. It is the mass-weighted average of the accelerations of the parts of the system:

$$\mathbf{a}_{cm} \equiv \frac{1}{M_{tot}} \sum_i m_i \, \mathbf{a}_i \; , \tag{III.18}$$

and similarly for the integral over an extended body.

Dynamics

Applying Newton's Laws to Eq. (III.18) reveals further simplification. The total mass of the system times the acceleration of the CM is, by Newton's Second Law, the sum of all the forces acting on all the parts of the system:

$$\begin{aligned} M_{tot}\mathbf{a}_{cm} &= \sum_i \left(\sum \mathbf{F}_i \right) \\ &= \sum \mathbf{F}_{ext} \; . \end{aligned} \tag{III.19}$$

The double sum over all the forces collapses to the sum of the *external* forces $\mathbf{F}_{ext}$ acting on the system from outside. By Newton's Third Law, the internal forces— the forces exerted by parts of the system, or pieces of the body, on each other— form action/reaction pairs and add to zero. This means *the CM moves as a point body of mass M_{tot} would under the net external force on the system.* For example, if the drag-free rocket of Ch. I Sec. C had exploded after Burnout into a million pieces, moving in a million different directions, the CM of the fragments would have continued to move on the parabolic trajectory of a projectile until the first piece struck the ground, adding a new external force to the system. The forces of the explosion itself, being internal forces, would have no effect on the motion of the CM.

We have already seen that for a system for which the external forces sum to zero, or for an isolated system subject to no external forces, the total momentum of the system is constant, i.e., conserved. Results (III.17) and (III.19) show that this is equivalent to the statement that for any system for which $\sum \mathbf{F}_{ext} = \mathbf{0}$ obtains, the velocity $\mathbf{v}_{cm}$ of the center of mass is constant.

Energies

The center of mass is also useful in describing the energies of a system or extended body. The velocity $\mathbf{v}_i$ of any part of the system can be written as the vector sum of the CM velocity $\mathbf{v}_{cm}$ and the velocity $\mathbf{v}_i^{(cm)}$ of the part relative to the CM:

$$\mathbf{v_i} = \mathbf{v}_{cm} + \mathbf{v}_i^{(cm)} \; , \tag{III.20}$$

whether this represents a Galilean transformation or not. The kinetic energy of a system then takes the form

$$
\begin{aligned}
K &= \tfrac{1}{2} \sum_i m_i v_i^2 \\
&= \tfrac{1}{2} \sum_i m_i (\mathbf{v}_{\mathrm{cm}} + \mathbf{v}_i^{(\mathrm{cm})}) \cdot (\mathbf{v}_{\mathrm{cm}} + \mathbf{v}_i^{(\mathrm{cm})}) \\
&= \tfrac{1}{2} \sum_i m_i (v_{\mathrm{cm}}^2 + 2\mathbf{v}_{\mathrm{cm}} \cdot \mathbf{v}_i + v_i^{(\mathrm{cm})2}) \\
&= \tfrac{1}{2} M_{\mathrm{tot}} v_{\mathrm{cm}}^2 + M_{\mathrm{tot}} \mathbf{v}_{\mathrm{cm}} \cdot \mathbf{v}_{\mathrm{cm}}^{(\mathrm{cm})} + \sum_i K_i^{(\mathrm{cm})}
\end{aligned}
$$

$$
K = \tfrac{1}{2} M_{\mathrm{tot}} v_{\mathrm{cm}}^2 + \sum_i K_i^{(\mathrm{cm})} , \tag{III.21}
$$

with a similar integral expression for an extended body. Here, the middle term in the third and fourth lines vanishes because $\mathbf{v}_{\mathrm{cm}}^{(\mathrm{cm})} = (1/M_{\mathrm{tot}}) \sum_i m_i \mathbf{v}_i^{(\mathrm{cm})}$, the velocity of the CM relative to the CM, must be zero. The kinetic energy of each part as seen from the CM is defined via $K_i \equiv \tfrac{1}{2} m_i v_i^{(\mathrm{cm})2}$ for the last expression. This is an important and general result: *The kinetic energy of a system or body decomposes cleanly into the kinetic energy of a point mass of the same total mass, moving as the CM, plus a sum of kinetic energies relative to the CM, i.e., "internal" energies.*[8]

The Work-Energy Theorem of Ch. II Sec. B was derived for a point body. Its application to systems and extended bodies is more complicated. The quantity of work

$$
\begin{aligned}
\int \left(\sum \mathbf{F}_{\mathrm{ext}} \right) \cdot d\mathbf{r}_{\mathrm{cm}} &= \int M_{\mathrm{tot}} \frac{\mathbf{v}_{\mathrm{cm}}}{dt} \cdot (\mathbf{v}_{\mathrm{cm}} \, dt) \\
&= \Delta \left(\tfrac{1}{2} M_{\mathrm{tot}} v_{\mathrm{cm}}^2 \right)
\end{aligned} \tag{III.22}
$$

is equal to the change in the kinetic energy associated with the center-of-mass motion. But this is not necessarily the work done by any particular force or combination of forces. The rest of the work done by both external and internal forces—the internal forces must sum to zero in pairs, but the work they do on different parts of the system need not—must equal the change in the total "internal" energy of the system.

The potential energy of a system or body in the local gravitational field of the Earth—*not* the potential energy associated with the system's *own* gravitation— is simply related to its center of mass. As per example (II.32), that external

---

[8]Result (III.17) shows that the total momentum of the system decomposes similarly, except that the sum of "internal" momenta is zero.

potential energy is given by

$$
\begin{aligned}
U_g &= \sum_i (-m_i \mathbf{g} \cdot \mathbf{r}_i) \\
&= -\mathbf{g} \cdot \sum_i m_i \mathbf{r}_i \\
&= -M_{\text{tot}} \mathbf{g} \cdot \mathbf{r}_{\text{cm}} \\
&= Mgh_{\text{cm}} ,
\end{aligned}
\tag{III.23}
$$

with h_{cm} the vertical height of the CM, *provided* the spatial extent of the system is small enough that $\mathbf{g}$ can be treated as a constant. The potential energy is the same as that of a point mass of the same total mass at the CM. This is one of two reasons (we shall encounter the other later in this chapter) that the center of mass (CM) is often called the *center of gravity* (CG).[9]

This result appears in many contexts. For example, it explains the utility of the maneuver known as the "Fosbury flop." At the 1968 Mexico City Olympic Games, Dick Fosbury[10] startled the athletic world with a new technique for the high jump: At the end of his run, he leapt *backwards* at the bar, arching his upper body over the bar with his legs trailing behind, bringing his legs over only after his upper body had cleared the bar. These acrobatics allowed the CM of Fosbury's body—which was not *in* his body—to pass under the bar. As result (III.22) indicates, the height of his CM was limited by the work Fosbury could get out of his leg muscles. By passing his CM under the bar, he could clear a higher bar than was possible with traditional technique. Fosbury won, and the Fosbury flop has since become standard procedure for high jumpers.

Center-of-mass (CM) frame

For systems with zero net external force, i.e., for which total momentum is conserved, the CM defines an inertial frame of reference, the *center-of-mass* frame. The analysis of collisions and scattering events is simplified in this frame because the total momentum of the system is zero. For example, for a two-body collision like those analyzed in the preceding sections, with initial velocities $\mathbf{u}_1$ and $\mathbf{u}_2$ and final velocities $\mathbf{U}_1$ and $\mathbf{U}_2$ (to identify them as CM-frame velocities), momentum conservation takes the form

$$
\begin{aligned}
m_1 \mathbf{u}_1 + m_2 \mathbf{u}_2 &= m_1 \mathbf{U}_1 + m_2 \mathbf{U}_2 \\
&= \mathbf{0} .
\end{aligned}
\tag{III.24a}
$$

This implies that the initial and final velocities are not independent, but satisfy

$$
\mathbf{u}_2 = -\frac{m_1}{m_2} \mathbf{u}_1
\tag{III.24b}
$$

---

[9]This usage is common but somewhat careless, as the CM and CG are conceptually different, and there are situations—extended structures in space, for example—in which the two points would not coincide.

[10]Richard Douglas Fosbury, 1947–

and

$$\mathbf{U}_2 = -\frac{m_1}{m_2}\,\mathbf{U}_1\;. \tag{III.24c}$$

If the collision is characterized by a coefficient of restitution, thus:

$$|\mathbf{U}_2 - \mathbf{U}_1| = r\,|\mathbf{u}_2 - \mathbf{u}_1|\;, \tag{III.25}$$

then the final *speeds* are given by

$$|\mathbf{U}_1| = r\,|\mathbf{u}_1| \tag{III.26a}$$

and

$$|\mathbf{U}_2| = r\,|\mathbf{u}_2|\;. \tag{III.26b}$$

The corresponding initial and final total kinetic energies of the system are related via

$$K_f^{(\mathrm{cm})} = r^2\,K_i^{(\mathrm{cm})}\;, \tag{III.26c}$$

a simple result that does *not* apply, in general, in other frames of reference. In a one-dimensional collision, the velocities of the bodies simply reverse directions, with magnitudes scaled by r. (The perfectly inelastic collision is rather dull in the CM frame: The bodies collide, adhere, and sit.) In two or three dimensions, the angle θ_{cm} or angles $(\theta_{\mathrm{cm}}, \phi_{\mathrm{cm}})$ through which the line of the two velocities is rotated must still be determined by other considerations.

The modeling of scattering processes described in Sec. B preceding is usually carried out in the CM frame. Kinematic relationships can be used to connect the scattering angle θ_{cm}, say, in the CM frame to scattering angles ψ and ζ measured in the LAB frame, in which the target body is initially at rest. The geometry relating these angles is illustrated in Fig. III.5.

Einsteinian center-of-*momentum* frame

The center of mass defined by Eqs. (III.14) is not as useful beyond the Newtonian limit as it is within. When Einsteinian dynamics is in play, i.e., for bodies moving at speeds not small compared to the speed of light, it is more useful to define a *center-of-momentum* frame for isolated systems. (This is also identified by the initials CM; it is usually clear from context which is meant.) The center-of-momentum frame for a system is the reference frame in which the total Einsteinian momentum of the system is zero. Form (III.14b) of the Lorentz transformation for momentum indicates that the velocity of this frame, relative to any given inertial frame, is given by

$$\mathbf{v}_{\mathrm{CM}} = c^2\,\frac{\mathbf{P}_{\mathrm{tot}}}{E_{\mathrm{tot}}}\;, \tag{III.27}$$

momentum $\mathbf{P}_{\mathrm{tot}}$ and energy E_{tot}, in the given frame, obtained from definitions (I.95).

The total energy of a system in its CM frame is the energy available for reactions, particle production, and so on—in any other frame, some energy

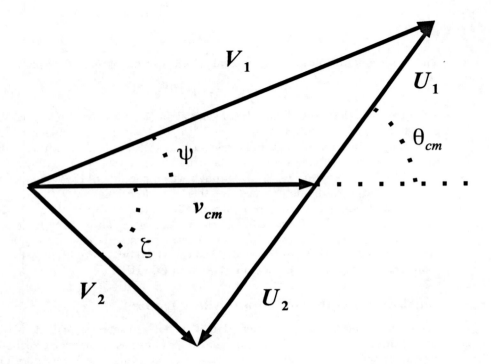

Figure III.5: Final velocities in a two-body collision: $\mathbf{U}_1$ and $\mathbf{U}_2$ in the CM frame, $\mathbf{V}_1$ and $\mathbf{V}_2$ in the LAB frame, with $\mathbf{v}_{cm}$ the velocity of the system center of mass in the LAB frame. Scattering angles are θ_{cm} in the CM frame, ψ and ζ in the LAB frame.

is reserved simply to conserve momentum. The relation between the CM-frame energy and that in, say, the LAB frame can be obtained by noting that Lorentz transformations leave unchanged—invariant—the quantity $E^2 - P^2c^2$, as is easily checked. Suppose in the LAB frame a particle with mass m_1 and energy $E_1^{(\text{lab})}$ collides with a particle of mass m_2 at rest. With total momentum in the CM frame zero by definition, the total CM energy is given by

$$
\begin{aligned}
E_{\text{tot}}^{(\text{CM})} &= \left[\left(E_1^{(\text{lab})} + m_2c^2 \right)^2 - P_1^{(\text{lab})2}c^2 \right]^{1/2} \\
&= \left[2m_2c^2 E_1^{(\text{lab})} + (m_1c^2)^2 + (m_2c^2)^2 \right]^{1/2} \\
&\approx \left(2m_2c^2 E_1^{(\text{lab})} \right)^{1/2} \qquad \text{for} \qquad E_1^{(\text{lab})} \gg m_1c^2 , \ m_2c^2 , \qquad \text{(III.28)}
\end{aligned}
$$

where the identity $E_1^{(\text{lab})2} - P_1^{(\text{lab})2}c^2 = m_1^2c^4$ is used in the second line. The approximate result in the last line is of great importance for very high-energy physics experiments. In that limit, a tenfold increase in CM energy requires a hundredfold gain in LAB-frame energy. To evade this, the highest-energy particle accelerators—such as the Tevatron at Fermilab in Batavia, Illinois, or the Large Hadron Collider in (under) Geneva—feature collider designs: Beams of high-energy particles are directed at each other, rather than at stationary targets, putting the laboratory itself into the CM frame.[11]

The somewhat fanciful inelastic collision of two protons in Sec. A preceding provides an example. The LAB-frame velocity of the center-of-momentum frame is given by Eq. (III.9c):

$$
v_{\text{CM}} = 0.5774 \, c . \qquad \text{(III.29)}
$$

In the CM frame, the two protons approach each other with equal speeds and opposite velocities. Their CM-frame initial-velocity components are

$$
u_1^{(\text{CM})} = -u_2^{(\text{CM})} = 0.5774 \, c , \qquad \text{(III.30)}
$$

since the second proton is initially at rest in the LAB frame. The total energy in the CM frame is the rest energy of the final particle, viz.,

$$
E_{\text{tot}}^{(\text{CM})} = 2.449 \, m_p c^2 , \qquad \text{(III.31)}
$$

as per Eq. (III.9d).

D   Rotation about a fixed direction

If pool, billiards, and snooker were played on air-hockey tables, they would be very different games. The rotation of the balls plays a major role in their

---

[11]The price to be paid for this more-efficient use of beam energy is that even the highest-intensity beams are very fine vacua compared with a thin metal target, so the number of particles that interact is much reduced in comparison to a traditional design.

dynamics. With the possible exception of fluid mechanics, no subfield of classical mechanics has been more elaborately developed than the mechanics of rotation. Rotation is a pattern of motion of systems or extended bodies—a point mass can rotate, i.e., revolve, about another point, but the intrinsic rotation of a point is not meaningful.[12] The term describes a wide variety of phenomena. A general definition for the purposes of physics[13] is: *Rotation is a pattern of motion in which all parts of the system or body travel in circles on parallel planes, with the centers of all the circles lying on a common line, the axis of rotation.*

Rotational motion, therefore, is inherently at least two-dimensional. But the special case of rotation about a fixed axis, or about a moving axis with a fixed direction, shares many features with one-dimensional motion such as that described in Ch. I Sec. B. This case will occupy this and the following section. The more general case, in which the direction of the rotation axis can move, is treated in Sec. F to follow.

Kinematics

Since rotation is defined by circular motion, a general description of circular motion is needed. A circular trajectory in the x-y plane, say, is given by

$$\mathbf{r}(t) = r \cos\theta(t)\,\hat{\boldsymbol{x}} + r \sin\theta(t)\,\hat{\boldsymbol{y}} \ . \qquad (III.32)$$

Here, r is the (constant) radius of the circle and $\theta(t)$ the time-dependent angular position measured, say, from the $+x$ axis as is customary. The velocity of a body on this trajectory is

$$\begin{aligned}
\mathbf{v}(t) &= \frac{d\mathbf{r}}{dt} \\
&= r\,\frac{d\theta}{dt}\,[-\sin\theta(t)\,\hat{\boldsymbol{x}} + \cos\theta(t)\,\hat{\boldsymbol{y}}] \\
&= r\omega\,\hat{\boldsymbol{\theta}} \ , \qquad (III.33)
\end{aligned}$$

the last expression introducing the *angular velocity* $\omega = d\theta/dt$ and the unit vector $\hat{\boldsymbol{\theta}}$ in the tangential direction around the circle. The acceleration of the body—circular motion is *a fortiori* (necessarily) accelerated motion—is

$$\begin{aligned}
\mathbf{a}(t) &= \frac{d\mathbf{v}}{dt} \\
&= -r\left(\frac{d\theta}{dt}\right)^2 [\cos\theta(t)\,\hat{\boldsymbol{x}} + \sin\theta(t)\,\hat{\boldsymbol{y}}] + r\,\frac{d^2\theta}{dt^2}\,[-\sin\theta(t)\,\hat{\boldsymbol{x}} + \cos\theta(t)\hat{\boldsymbol{y}}] \\
&= -r\omega^2\,\hat{\boldsymbol{r}} + r\alpha\,\hat{\boldsymbol{\theta}} \ ,
\end{aligned}$$
$$(III.34)$$

---

[12]It would seem to be an exception to this observation that elementary particles have intrinsic spins. The resoluton of this apparent paradox is twofold: Elementary particles are not point objects, and their spin is not classical rotary motion.

[13]Mathematicians might use a more abstract but more far-reaching definition involving orthogonal matrices.

this last expression introducing the unit vector $\hat{r}$ in the radial direction from the center of the circle and the *angular acceleration* $\alpha = d\omega/dt$. The first term in that expression is the *centripetal* (center-seeking) acceleration[14] of the body, the second term its *tangential* acceleration. The centripetal, radial acceleration component is given by

$$a_r = -r\omega^2$$
$$= -\frac{v^2}{r} , \tag{III.35}$$

with speed v the magnitude of the velocity given by Eq. (III.33).

An important special case of this general circular motion is *uniform* circular motion. This is defined by zero angular acceleration, $\alpha \equiv 0$. (This is still accelerated motion, however, as the radial acceleration remains.) The angular velocity ω is constant in this case. Uniform circular motion is periodic motion, repeating exactly after a fixed time interval. The language appropriate to periodic phenomena can be applied here: The *frequency*[15] of the motion—the number of cycles per unit time—is given by

$$\nu = \frac{\omega}{2\pi} , \tag{III.36a}$$

as ω is the rate at which the body moves through angle in radians, and 2π radians is one cycle. The *period* of the motion—the time for one cycle—is related via

$$T = \frac{1}{\nu}$$
$$= \frac{2\pi}{\omega} \tag{III.36b}$$

to the frequency and angular speed.

A word about units is in order. Angular velocity is measured in radians per second, but an angle in radians is an arc length divided by a radius, i.e., dimensionless. Hence, the unit for ω is algebraically s^{-1}. The unit of frequency, cycles per second, is also algebraically s^{-1}. In this case, the inverse second is often labeled the *hertz* (Hz), for Heinrich Hertz[16]: 1 $s^{-1} = 1$ Hz. But only for frequency—angular velocity is always measured in s^{-1}, not Hz.

The Earth itself provides examples of these quantities. The Earth rotates on its axis once a day, which is 86,400. seconds long. (Astronomers might quibble about the *exact* length of a solar day, but this will do for now.) The angular

---

[14]This is not to be confused with *centrifugal* (center-fleeing) acceleration, which is encountered in rotating—hence, noninertial—reference frames.

[15]The lower-case Greek ν used here should not be confused with the Latin v for velocity or speed. Many texts use f for frequency, but f is already widely used, e.g., for forces or general functions. Hence, the answer to the question "What's nu?" is "Frequency."

[16]Heinrich Rudolph Hertz, 1857–1894

velocity associated with this motion given by

$$\Omega = \frac{2\pi}{T_{\text{day}}}$$
$$= \frac{2\pi}{86400. \text{ s}}$$
$$= 7.2722 \times 10^{-5} \text{ s}^{-1} \ . \tag{III.37}$$

The Earth's annual revolution around the Sun is not exactly circular motion. The radius of its orbit varies by about 2% from its average value at its maximum (aphelion) and minimum (perihelion), as will be considered in Ch. IV. But approximated as circular motion, it has angular velocity

$$\omega = \frac{2\pi}{T_{\text{year}}}$$
$$= \frac{2\pi}{3.156 \times 10^7 \text{ s}}$$
$$= 1.99 \times 10^{-7} \text{ s}^{-1} \ , \tag{III.38a}$$

linear speed

$$v = r\omega$$
$$= (1.495 \times 10^{11} \text{ m})(1.991 \times 10^{-7} \text{ s}^{-1})$$
$$= 29.8 \text{ km/s} \ , \tag{III.38b}$$

and centripetal acceleration

$$a_r = -r\omega^2$$
$$= (1.495 \times 10^{11} \text{ m})(1.991 \times 10^{-7} \text{ s}^{-1})^2$$
$$= 5.93 \times 10^{-3} \text{ m/s}^2$$
$$= 6.04 \times 10^{-4} \ g \ . \tag{III.38c}$$

Here, the value $r = 1.495 \times 10^{11}$ m is introduced for the Astronomical Unit (AU), the mean radius of the Earth's orbit, and g is the familiar acceleration due to gravity near the Earth's surface. The results are rounded to three significant figures, corresponding to the accuracy of the circular approximation, although an extra significant figure is kept for intermediate values in the calculations.

The kinematics of rotation with a fixed axis direction parallels that of one-dimensional translational (linear) motion. Hence, the kinematic relationships explored in Ch. I Sec. B, say, can be used for the corresponding rotational motions simply by replacing linear quantities (Latin symbols) with the associated rotational quantities (Greek symbols). This table lists some of the parallels:

Quantity	Translation	Rotation	Relation
Displacement	Δx	$\Delta\theta$	$\Delta s = r_\perp \, \Delta\theta$
Velocity	$v = dx/dt$	$\omega = d\theta/dt$	$v = r_\perp \omega$
Accleration	$a = dv/dt$	$\alpha = d\omega/dt$	$a_\parallel = r_\perp \alpha$

Here, Δs is arc length in circular motion, $r_\perp$ the perpendicular distance from the body or body part to the axis of rotation, and $a_\|$ the component of acceleration tangential to the circle. The centripetal acceleration, always present, is not listed here.

A second special case, that of motion with *constant angular acceleration α*, is thus readily treated. Relations (I.11)–(I.13) for linear motion with constant acceleration have rotational counterparts

$$\omega(t) = \omega_0 + \alpha(t - t_0) \,, \tag{III.39}$$

$$\theta(t) = \theta_0 + \omega_0(t - t_0) + \tfrac{1}{2}\alpha(t - t_0)^2 \,, \tag{III.40}$$

and

$$\omega^2 = \omega_0^2 + 2\alpha(\theta - \theta_0) \,. \tag{III.41}$$

Hence, all problems involving these two classes of motion can be treated similarly.

All parts of a rotating system or body execute circular motion about a common axis. In general, it is not necessary that the parts cover angle on their respective circles at the same rate, i.e., the angular velocity ω_i for each part need not be the same. In the special case that ω_i *is* the same for all parts, the motion is called *rigid rotation*. This is the only form of motion other than uniform translation (in which the linear displacement is the same for all parts) consistent with the *rigid-body constraint*—the requirement that all parts of the body maintain fixed distances from each other as the body moves. In fact, it can be shown as a mathematical theorem that any motion of a body satisfying this constraint, a rigid body,[17] can be decomposed into a combination of a uniform translation and a rigid rotation. A body need not be rigid to rotate rigidly: The Earth rotates *almost* rigidly, although it is closer to a fluid body than a rigid body. The general case, in which different parts of the body have different angular velocities, is called *differential rotation*. For example, the Sun and the Milky Way galaxy rotate differentially.

Dynamics

The dynamical features of rotation about a fixed direction also parallel those of one-dimensional linear motion. However, as the reader can readily demonstrate, the tendency of force to produce or alter rotational motion depends not only on the force itself, but also on where and in what direction it is applied. The appropriate measure is *torque* or *moment of force,* defined in this context as

$$\tau \equiv r_\ell F$$
$$= r_\perp F_\| \,. \tag{III.42}$$

Here, F is the magnitude of a force in the plane of rotation, i.e., perpendicular to the rotation axis. The *lever arm r_ℓ* of the force is the perpendicular distance

---

[17]Of course, no real body is perfectly rigid in this sense; this is an approximation.

from the *line of action* of the force to the rotation axis, while $r_\perp$ is, as previously, the perpendicular distance from the point of application of the force to the rotation axis, and $F_\parallel$ is the component of the force tangential to the rotational motion. These are illustrated in Fig. III.6. The trigonometric relationship between $F_\parallel$ and F is the same as that between r_ℓ and $r_\perp$, so the two products in Eq. (III.42) are equal.[18] The SI units of torque are newton-meters (N m). Although these are algebraically equivalent to the units of work and energy, torque is *not* measured in joules.

The tendency of mass to resist changes in rotational motion also depends both on the mass and its geometric distribution. For rotation about a fixed direction, this is represented by the *moment of inertia*

$$I = \sum_i m_i r_{\perp,i}^2 \qquad\qquad \text{(III.43a)}$$

for a system of point masses, or

$$I = \int_{\text{body}} r_\perp^2\, dm$$
$$= \int_{\text{body}} r_\perp^2 \rho(\mathbf{r})\, d\mathcal{V} \qquad\qquad \text{(III.43b)}$$

for an extended body, with mass density $\rho(\mathbf{r})$. In the language of moments used in Sec. C in connection with the center of mass, the moment of inertia is a *second moment* of the mass distribution because of the second power of the distance $r_\perp$ in Eqs. (III.43).

For example, for a thin, uniform ring of mass M and radius R, rotating about an axis through the center and perpendicular to the plane of the ring, all parts of the ring are distance R from the axis. Integral (III.43b) gives moment of inertia

$$I_{\text{ring}}^\perp = MR^2 \ . \qquad\qquad \text{(III.44)}$$

In contrast, the mass of a uniform *disk* of mass M and radius R, rotating about an axis through the center and perpendicular to the plane of the disk, is distributed at various distances from the axis between zero and R. The moment of inertia can be calculated by dividing the disk into concentric rings of radius r, circumference $2\pi r$, and thickness dr. Integral (III.4b) yields

$$I_{\text{disk}}^\perp = \int_0^R r^2\, \frac{M}{\pi R^2}\, 2\pi r\, dr$$
$$= \frac{2M}{R^2} \int_0^R r^3\, dr$$
$$= \tfrac{1}{2} MR^2 \ . \qquad\qquad \text{(III.45)}$$

---

[18]The fully three-dimensional definition of torque actually simplifies this description. It will be examined in Sec. F following.

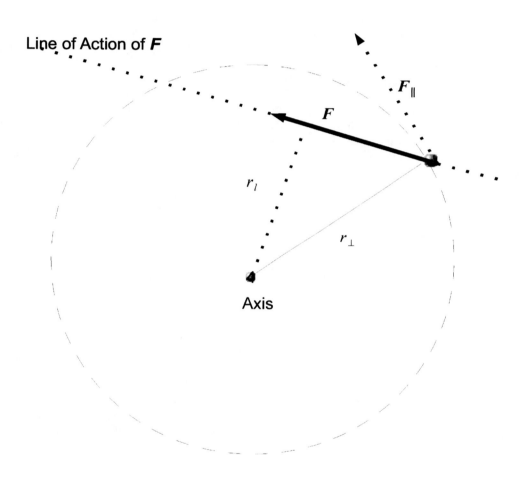

Figure III.6: Force $\mathbf{F}$ and lever arm r_ℓ, seen looking down the axis of rotation.

Calculations like these, of moments of inertia for uniform solids of simple geometries, are standard calculus exercises.

For a given object or system, the moment of inertia depends on the choice of rotation axis. Moments for axes in the same direction are related via the *Parallel-Axis Theorem*:

$$I_O = I_{cm} + Mr^2_{\perp,cm} \ , \tag{III.46}$$

where I_O is the moment of inertia about an axis through origin O, I_{cm} is the moment of inertia of the same object about an axis, parallel to the first, but passing through the center of mass of the object, and $r_{\perp,cm}$ is the perpendicular distance from the center of mass to the first axis. That is, the moment of inertia about any axis decomposes into the moment of inertia about a parallel axis through the CM, plus the moment of inertia of a point body of the same mass located at the CM. [The proof of this theorem is similar to the derivation of result (III.21); it is easier to carry out in the full three-dimensional formulation of Sec. F following.] For example, for a uniform rod of mass M and length ℓ and an axis through the center of the rod, perpendicular to the rod, integral (III.43b) gives moment of inertia

$$
\begin{aligned}
I^{(center)}_{rod} &= \int_{-\ell/2}^{+\ell/2} x^2 \frac{M}{\ell} \, dx \\
&= \tfrac{1}{12} M\ell^2 \ .
\end{aligned}
\tag{III.47a}
$$

For a parallel axis through one end of the rod, the moment is

$$
\begin{aligned}
I^{(end)}_{rod} &= \int_0^\ell x^2 \frac{M}{\ell} \, dx \\
&= \tfrac{1}{3} M\ell^2 \\
&= I^{(center)}_{rod} + M \left(\frac{\ell}{2} \right)^2 \ ,
\end{aligned}
\tag{III.47b}
$$

in accord with theorem (III.46). These results indicate why a drum major's baton is much easier to twirl about its center than about one end.

These quantities are simply related: For a system with fixed moment of inertia I, e.g., for a rigid body, torque, moment of inertia, and angular acceleration satisfy

$$\tau = I\alpha \ , \tag{III.48}$$

a rotational version of Newton's Second Law.

As an example, consider the mass hung from a rotating spool[19] illustrated in Fig. III.7. Mass m accelerates downward under the influence of its own weight and tension T in the cord. The same tension provides torque to spin up the spool, unwinding the cord as the mass descends. The acceleration of the mass is given by

$$ma = mg - T \ , \tag{III.49a}$$

---

[19]This example describes an experiment performed in physics labs in Saint Louis University's Parks College of Engineering, Aviation, and Technology.

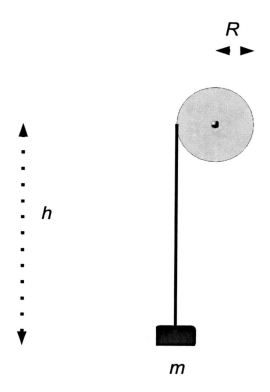

Figure III.7: A mass m suspended from a rotating spool of radius R. The descending mass spins up the spool; linear and rotational versions of Newton's Second Law govern both motions.

with downward positive, and the angular acceleration of the spool by

$$I\alpha = \mathcal{T}R \ ,\tag{III.49b}$$

since the tension has lever arm R. Since the cord remains taut at all times, the mass must descend at the rate the cord unwinds. Hence, the linear and angular accelerations obey the constraint

$$a = \alpha R \ .\tag{III.49c}$$

Combining these yields

$$m\alpha R^2 = mgR - I\alpha \ ,\tag{III.50a}$$

or angular acceleration

$$\begin{aligned}\alpha &= \frac{mgR}{mR^2 + I} \\ &= \frac{mgR}{mR^2 + \frac{1}{2}MR^2} \ ,\end{aligned}\tag{III.50b}$$

this last because the spool is a uniform disk of mass M, and moment of inertia (III.45). With sample values $M = 1.00$ kg, $R = 0.100$ m, and $m = 0.100$ kg, this gives angular acceleration

$$\begin{aligned}\alpha &= \frac{mg}{\left(m + \frac{1}{2}M\right)R} \\ &= \frac{(0.100 \text{ kg})(9.807 \text{ m/s}^2)}{\left[0.100 \text{ kg} + \frac{1}{2}(1.00 \text{ kg})\right](0.100 \text{ m})} \\ &= 16.3 \text{ s}^{-1} \ .\end{aligned}\tag{III.51}$$

The angular velocity of the spool, from Eq. (III.41), is

$$\begin{aligned}\omega &= (2\alpha\Delta\theta)^{1/2} \\ &= \left(2\alpha\frac{h}{R}\right)^{1/2} \\ &= \left(\frac{2mgh}{\left(m + \frac{1}{2}M\right)R^2}\right)^{1/2} \ ,\end{aligned}\tag{III.52}$$

with $\Delta\theta$ the angular displacement of the spool and h the distance fallen and the length of cord unwound. If a total length $h = 1.00$ m of cord unwinds off the spool, the final angular speed of the spool is

$$\begin{aligned}\omega &= \left(\frac{2mgh}{\left(m + \frac{1}{2}M\right)R^2}\right)^{1/2} \\ &= \left(\frac{2(0.100 \text{ kg})(9.807 \text{ m/s}^2)(1.00 \text{ m})}{\left[0.100 \text{ kg} + \frac{1}{2}(1.00 \text{ kg})\right](0.100 \text{ m})^2}\right)^{1/2} \\ &= 18.1 \text{ s}^{-1} \ .\end{aligned}\tag{III.53}$$

Here, the calculation is performed using the original data, rather than the calculated value (III.51) of α, as a precaution against a mistake in that calculation. The frequency of the now uniform rotation of the spool is

$$
\begin{aligned}
\nu &= \frac{\omega}{2\pi} \\
&= \frac{18.08 \text{ s}^{-1}}{2\pi} \\
&= 2.88 \text{ Hz} \\
&= 173. \text{ rpm} .
\end{aligned} \tag{III.54}
$$

Here, an extra signficant figure is again kept in the intermediate value for ω; the result is rounded to the appropriate precision. The final value is expressed in the familiar units rpm, *revolutions per minute*.

The energy of a rotating system or body can be expressed in terms of rotational quantities. The kinetic energy of the system is the sum of that of its parts, viz.,

$$
\begin{aligned}
K &= \sum_i \tfrac{1}{2} m_i v_i^2 \\
&= \tfrac{1}{2} \sum_i m_i r_{\perp,i}^2 \omega_i^2 \\
&= \tfrac{1}{2} \left(\sum_i m_i r_{\perp,i}^2 \right) \omega^2 \\
&= \tfrac{1}{2} I \omega^2 ,
\end{aligned} \tag{III.55}
$$

the last two lines appropriate *for rigid rotation only*, and with comparable integral expressions for an extended body. Rotational work can be given as

$$
\begin{aligned}
dW &= F_\parallel \, ds \\
& \quad F_\parallel r_\perp \, d\theta \\
&= \tau \, d\theta ,
\end{aligned} \tag{III.56}
$$

utilizing Eq. (III.42). Power in rotational dynamics takes the form

$$
\begin{aligned}
P &= \frac{dW}{dt} \\
&= \tau \, \omega .
\end{aligned} \tag{III.57}
$$

None of these are new quantities—these are the same energy measures described in Ch. II Sec. B expressed in terms of rotational variables. The Work-Energy Theorem of that section can also be applied to rotational dynamics, though some care must be taken. For a system or body rotating about an axis through its center of mass, contribution (III.22) is zero; all of the energy is "internal." For rigid rotation, the internal forces can do no work; all the torque in Eqs. (III.56) and (III.57) must come from external forces.

Angular momentum

Angular momentum is the rotational analog of linear momentum. Unlike the energy and work measures of Eqs. (III.55)–(III.57), however, it is not linear momentum expressed in rotational variables. It is a distinct quantity measured in different units than linear momentum. The angular momentum of a "point" mass about a given axis is its linear momentum times its lever arm—because momentum, like force, is a vector quantity with a line of action—viz.,

$$\ell = r_\ell\, p$$
$$= r_\perp\, p_\parallel$$
$$= mvr_\perp \; , \tag{III.58a}$$

the first two expressions being general and the last appropriate to circular motion. Here, $p_\parallel$ denotes the tangential component of momentum, and the radii are as in Eq. (III.42) preceding. The angular momentum of a system of masses or an extended body is the sum of that of its components, e.g.,

$$L = \sum_i \ell_i$$
$$= \sum_i m_i v_i r_{\perp,i}$$
$$= \left(\sum_i m r_{\perp,i}^2 \right) \omega$$
$$= I\omega \; , \tag{III.58b}$$

the first expression being general, the second for circular motion, and the last two for rigid rotation.

The dynamics of angular momentum follows from its definition and Newton's Second Law. Its rate of change for a single particle is given by

$$\frac{d\ell}{dt} = r_\ell \frac{dp}{dt}$$
$$= r_\ell\, F$$
$$= \tau \; , \tag{III.59a}$$

the last from Eq. (III.42). [It is not immediately obvious that the time derivative of Eq. (III.58a) should not also include a term involving dr_ℓ/dt; the absence of such a term is made clear in the three-dimensional derivation of this relation, described in Sec. F following.] The rate of change of the angular momentum of a system or body, then, is the sum of the torques on all its parts. If the *internal* torques cancel in pairs as the internal forces do, then this relation takes the form

$$\frac{dL}{dt} = \sum \tau_{\text{ext}} \; , \tag{III.59b}$$

a result more general than Eq. (III.48)—it is not assumed here that the moment of inertia of the system or body is constant. The assertion that the torques due to internal forces cancel is more restrictive than Newton's Third Law as stated in Ch. I Sec. E. That statement—the *weak form* of Newton's Third Law—did not require that action/reaction pairs of forces act along a common line. The forces, equal in magnitude and opposite in direction, could still form a *couple,* exerting a net torque but no net force.[20] If it is further asserted that action/reaction pairs of forces act, say, along the common line between interacting bodies—the *strong form* of Newton's Third Law—then the cancellation of internal torques is assured.

If the strong form of Newton's Third Law is in play, then for an isolated system subject to no external torques, or if the external torques sum to zero, the total angular momentum of the system remains constant as the system moves. This result, the *Law of Conservation of Angular Momentum,* is a conservation law of surpassing importance comparable to those for linear momentum and energy we have encountered previously. More general than Newton's Laws themselves, it holds sway where Einstein's Special or General Theories of Relativity govern and applies to molecules, atoms, nuclei, and elementary particles for which quantum mechanics must be invoked.

A spectacular example of angular-momentum conservation is found on the ice rink. A figure skater begins a spin with arms and nonsupporting leg extended as far as possible from the vertical axis through her or his standing foot. The area of contact between the tip of one skate blade and the ice is so small that the forces involved have negligible lever arms—they exert negligible torque, and angular momentum is conserved. The skater then draws her/his extended arms and legs into a tight tuck around the rotation axis, reducing her/his moment of inertia, say, from I_1 to $I_2 \ll I_1$. Angular-momentum conservation

$$I_1\omega_1 = I_2\omega_2 \qquad\qquad \text{(III.60a)}$$

implies that the skater thus acquires much-increased angular velocity

$$\omega_2 = \frac{I_1}{I_2}\,\omega_1 \qquad\qquad \text{(III.60b)}$$

and much-increased kinetic energy

$$
\begin{aligned}
K_2 &= \tfrac{1}{2}I_2\omega_2^2 \\
&= \frac{I_1}{I_2}\,K_1 \; .
\end{aligned}
\qquad\qquad \text{(III.60c)}
$$

Whence the additional energy?[21] It comes from the work the skater's muscles must do to move the body parts inward over and above the required centripetal forces. The internal forces supplied by the muscles contribute no net force nor any net torque, but they do work that need not add to zero. The skater's body during the spin-up maneuver is certainly not rigid!

---

[20]For example, one normally applies a *couple* to the steering wheel of a car. One's purpose is to turn the wheel; actually accelerating its CM anywhere is quite undesirable. This is why it is said that the most important component of any car is the nut behind the steering wheel.

[21]A question one asks only about a conserved quantity. The tacit assumption that if energy appears, it must come *from* some source is an expression of its conservation.

Rolling—translation plus rotation

Rolling is a configuration of motion consisting of translation plus rotation about an axis perpendicular to the direction of translation. A special case of this is *rolling without slipping*. This motion consists of translation of a round object with center moving at speed v_{cen}, plus rigid rotation about a perpendicular axis with angular speed $\omega = v_{\text{cen}}/r$, where r is the radius of the object. Thus, a *contact point* on the circumference of the object is instantaneously at rest. An equivalent description of rolling without slipping is instantaneous pure, rigid rotation about an axis through the contact point, with angular speed $\omega = v_{\text{cen}}/r$. Both characterizations assign the same pattern of velocities to all parts of the body.

The kinetic energy of any rolling body is usefully written using Eqs. (III.21) and (III.55):

$$K = \tfrac{1}{2}Mv_{\text{cm}}^2 + \tfrac{1}{2}I^{(\text{cm})}\omega^2 \; . \tag{III.61}$$

This form is particularly useful for a body with a symmetric mass distribution, so that its geometric center (cen) and its center of mass (cm) are the same point.

A classic example is the demonstration known as the Rolling Race. Round objects are allowed to roll without slipping, from rest, down an inclined plane of height h, as illustrated in Fig. III.8—either side-by-side or racing against the clock. What determines the order of finish of the race? For symmetric objects of mass M, radius R, and moment of inertia $I^{(\text{cm})} = \beta M R^2$, conservation of energy implies

$$
\begin{aligned}
Mgh &= \tfrac{1}{2}Mv^2 + \tfrac{1}{2}I^{(\text{cm})}\omega^2 \\
&= \tfrac{1}{2}Mv^2 + \tfrac{1}{2}(\beta M R^2)\left(\frac{v}{R}\right)^2 \\
&= \tfrac{1}{2}(1+\beta)Mv^2 \; ,
\end{aligned}
\tag{III.62a}
$$

where Mgh is the gravitational potential energy lost (kinetic energy gained) in the descent, and v and ω are the values at the bottom of the ramp. Energy is conserved because the *static* frictional force between the rolling body and the ramp does no work if the contact point does not slip. Aerodynamic drag and *tractive* friction are treated as neglgible here. The final speed of the object is

$$v = \left(\frac{2gh}{1+\beta}\right)^{1/2} \; . \tag{III.62b}$$

As this is a constant-acceleration problem, the average speed of the object is half this, and the time of descent is inversely proportional to v. With uniform-

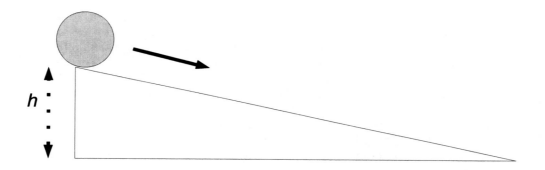

Figure III.8: The Rolling Race. Place your bets!

density contestants of simple shapes, the Racing Form looks like this:

Contestant	$\beta = I^{(\mathrm{cm})}/(MR^2)$	Finish
Solid Sphere	$\frac{2}{5}$	Win
Disk	$\frac{1}{2}$	Place
Spherical Shell	$\frac{2}{3}$	Show
Ring	1	Also Ran

Neither the mass nor the radius of each object matters; all objects of a given shape (β value) will finish in the same time and the same order.[22]

E   Rotation and billiard-ball dynamics

We have now assembled the tools needed to analyze the pool shot described at the start of this chapter. First, the cue ball is set into motion by a stroke from the cue stick, as illustrated in Fig. III.9. The cue ball need not roll without slipping initially, but it can be struck so that it does. The height above the center of the ball at which the horizontal[23] stroke should be delivered for this can be calculated using the dynamics of preceding sections. The stroke is a collision: The stick and ball make contact for a few milliseconds; both undergo elastic deformations giving rise to large, complicated forces; they return to their original shapes, and the forces vanish when they separate.

The appropriate quantity to describe such a situation is *impulse*. This is a vector quantity defined as the time integral of a force:

$$\mathbf{J} \equiv \int_{\Delta t} \mathbf{F}\, dt \; ; \qquad \text{(III.63a)}$$

in one dimension, this is the area under a graph of F vs. t over the interval Δt. In general, $\mathbf{J}$ depends on the time interval over which it is evaluated. But it is most useful for describing *impulsive* forces with definite start and stop times, especially if those forces are large and complicated. Newton's Second Law connects impulse and momentum thus:

$$\mathbf{J} = \int_{\Delta t} \frac{d\mathbf{p}}{dt}\, dt$$
$$= \Delta \mathbf{p} \; , \qquad \text{(III.63b)}$$

---

[22] It is possible to construct a very slow contestant, with $\beta > 1$, by attaching large, massive wheels that hang over the sides of the ramp to an axle of much smaller radius, which rolls on the ramp. On the other hand, the Block of Dry Ice can beat all comers by cheating—by sliding frictionlessly down the ramp without rolling ($\beta = 0$)!

[23] The cue ball need not be struck horizontally; that is a simplifying assumption used here. Expert players can use a variety of angles. In a *massé* shot, the cue ball is struck vertically or nearly vertically to make it swerve or jump around another ball. This shot can, however, get one unceremoniously evicted from some billiards establishments.

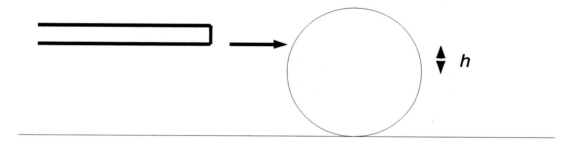

Figure III.9: The cue stick strikes the cue ball a horizontal blow at height h above ($h > 0$) or below ($h < 0$) its center.

where $\Delta\mathbf{p}$ is the *change* in momentum of the body to which the force is applied over the given time interval. [This result parallels the Work-Energy Theorem (II.11).] *Angular impulse* can be defined similiarly, viz.,

$$J_\theta = \int_{\Delta t} \tau\, dt$$
$$= \Delta L\ , \tag{III.64}$$

the last following from Eqs. (III.59).

If the cue ball is struck horizontally at height h above its center, the stick delivers both impulse and angular impulse to the ball. The lever arm of the force, about the CM of the ball,[24] is always $r_\ell = h$. So however complicated the force and torque, the linear and angular impulses are related via

$$J_\theta = hJ\ , \tag{III.65a}$$

where the positive direction for rotation is taken to correspond to forward rolling. Hence, results (III.63b) and (III.64) imply that the velocity v and angular velocity ω acquired by the ball as it comes off the stick are related by

$$I^{(\mathrm{cm})}\omega = Mvh\ , \tag{III.65b}$$

with M the mass of the cue ball and $I^{(\mathrm{cm})}$ its moment of inertia about an axis through its CM. Since the ball is a uniform-density sphere of radius R, say, with moment of inertia $I^{(\mathrm{cm})} = \frac{2}{5}MR^2$, the angular velocity is

$$\omega = \tfrac{5}{2}\left(\frac{h}{R}\right)\frac{v}{R}\ . \tag{III.65c}$$

Hence, the desired stroke height for rolling without slipping is

$$h_{\mathrm{rws}} = \tfrac{2}{5}R\ , \tag{III.66}$$

i.e., the ball must be struck horizontally $\frac{7}{10}$ of its diameter above the table to roll without slipping immediately. If the ball is struck at $h > h_{\mathrm{rws}}$, it acquires "topspin," its contact point sliding backward on the felt of the table, giving rise to a forward frictional force. If struck at $h < h_{\mathrm{rws}}$, including $h \leq 0$, the ball acquires "backspin," its contact point sliding forward and giving rise to a backward frictional force. Skilled players—unlike this author—can use these frictional forces to position the cue ball from one shot to another.[25]

---

[24]The reader may note that the CM of the ball, being accelerated, does not define an inertial reference frame. However, the inertial forces involved, as per Ch. I Sec. F, supply no torque about the CM, so the rotational analysis is unaffected. A comparable analysis of rotation about the momentarily stationary contact point between the ball and the table gives equivalent results.

[25]Tennis players and bowlers also use topspin and backspin to affect the motions of the balls they hit or roll. Skilled pool players can also use spin and friction to produce sideways motion, a feature known as "English."

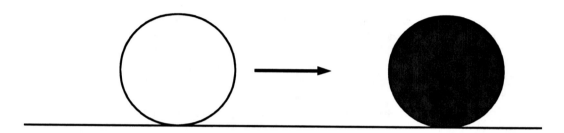

Figure III.10: The rolling cue ball strikes the target ball.

The cue ball is subject to frictional forces from the felt during its contact with the stick. But those forces are so much smaller than the elastic stick/ball forces that *over the interval of the contact,* the impulses delivered by friction are negligible in comparison to those delivered by the stroke.

Analyses along similar lines reveal the behavior of the cue ball and a target ball, initially at rest, after they collide head-on. This is illustrated in Fig. III.10. Let $v_1^{(i)}$, $v_1^{(c)}$, and $v_1^{(f)}$ denote the velocities of the cue ball before impact, immediately after impact, and after the ball attains its final, rolling-without-slipping state, respectively, and let $v_2^{(i)}$, $v_2^{(c)}$, and $v_2^{(f)}$ denote the same velocities for the target ball. Let $\omega_1^{(i)}$, $\omega_1^{(c)}$, $\omega_1^{(f)}$, $\omega_2^{(i)}$, $\omega_2^{(c)}$, and $\omega_2^{(f)}$ denote the corresponding angular velocities. Before the collision, the cue ball is rolling without slipping at speed v_0. It has velocity and angular velocity

$$v_1^{(i)} = v_0 \tag{III.67a}$$

and

$$\omega_1^{(i)} = \frac{v_0}{R} . \tag{III.67b}$$

The target is at rest; it has velocity and angular velocity

$$v_2^{(i)} = 0 \qquad\qquad\qquad (\text{III.67c})$$

and

$$\omega_2^{(i)} = 0 \ . \qquad\qquad\qquad (\text{III.67d})$$

The two balls are the same size. During the few milliseconds of impact, they experience large elastic forces directed along the line between their centers. If the balls are taken to be so smooth that *ball-ball friction is negligible,* then the *angular* impulse delivered to each ball at impact is zero—as previously, ball-felt friction delivers negligible impulse during this interval. Rotation, then, is not part of the collision: Each ball retains its angular velocity, and the balls exchange momenta as in the collisions of Sec. A preceding. Approximating the collision as *elastic,* and with balls of equal mass, we obtain for the cue ball, immediately after impact,

$$v_1^{(c)} = 0 \qquad\qquad\qquad (\text{III.67e})$$

and

$$\omega_1^{(c)} = \frac{v_0}{R} \ . \qquad\qquad\qquad (\text{III.67f})$$

The target ball leaves the collision with velocity

$$v_2^{(c)} = v_0 \qquad\qquad\qquad (\text{III.67g})$$

and angular velocity

$$\omega_2^{(c)} = 0 \ . \qquad\qquad\qquad (\text{III.67h})$$

The cue ball is momentarily at rest, still spinning. It experiences a forward frictional force that accelerates it forward while decelerating its spin. This continues until the ball is once again rolling without slipping, after which its velocity and angular velocity are constant if tractive friction and aerodyamic drag are neglected. The impulse J_f delivered by ball-felt friction during this sliding period (of the order of a second, compared to the milliseconds of the impact) is related to the angular impulse $J_{f\theta}$ delivered by

$$J_{f\theta} = -RJ_f \ , \qquad\qquad\qquad (\text{III.68})$$

as the frictional force has lever arm R, and the sign reflects the opposite effects of the two impulses. Relations (III.63b) and (III.64) then imply

$$I^{(\text{cm})} \left(\frac{v_1^{(f)}}{R} - \frac{v_0}{R} \right) = -RMv_1^{(f)} \ . \qquad\qquad\qquad (\text{III.69})$$

Hence, the cue ball attains final velocity

$$v_1^{(f)} = \tfrac{2}{7}v_0 \qquad\qquad\qquad (\text{III.70a})$$

and angular velocity

$$\omega_1^{(f)} = \tfrac{2}{7}\frac{v_0}{R} \ . \tag{III.70b}$$

The target ball emerges from the collision sliding, momentarily with no spin. It experiences a backward frictional force that decelerates it while spinning it up; this too continues until the ball attains rolling without slipping. The frictional impulse and angular impulse over the sliding period are not the same as those delivered to the cue ball, but they do obey the same relation (III.68). Here this implies

$$I^{(\mathrm{cm})}\frac{v_2^{(f)}}{R} = -R\left(M v_2^{(f)} - M v_0\right) \ . \tag{III.71}$$

The target ball, therefore, acquires final velocity

$$v_2^{(f)} = \tfrac{5}{7}v_0 \tag{III.72a}$$

and angular velocity

$$\omega_2^{(f)} = \tfrac{5}{7}\frac{v_0}{R} \ . \tag{III.72b}$$

Once again, victory! The analyses of collisional and rotational dynamics have answered the question posed. *Quod erat faciendum*—which was to be done.

This result has certain remarkable and unexpected features. For example, the total momentum of the two-ball system remains the same through the collision and the sliding motions. This means that the (linear) impulses delivered by the kinetic frictional forces—the ball-ball forces, internal to the system, cannot influence its total momentum—must be equal in magnitude, opposite in sign. If the kinetic frictional forces are given by Eq. (I.53), with the same coefficient μ_k and weight for both balls, then the forces must be equal in magnitude, opposite in direction. Hence, they must act *for the same time interval,* i.e., the two balls take the same amount of time to attain rolling without slipping after the collision. Did the reader anticipate this?

F Rotation in three dimensions

Kinematics

Generalizing linear kinematics and dynamics from one to three dimensions, as in Ch. I Sec. C, entails replacing scalar quantities with corresponding vector quantities. For rotational motion, this approach is complicated by the fact that *rotational displacement $\Delta\boldsymbol{\theta}$ is not a vector.* It has magnitude—the angle through which a body or system turns—and direction: Although rotation involves motion in an infinity of different directions, it is associated with a unique direction, that of the axis of rotation. But rotations do not add as vectors. For example: If you hold this book upright with the front cover facing you and rotate it 90° counterclockwise about a vertical axis, the front cover will face to your right, and the spine will be facing you. If you then rotate it 90° counterclockwise about a horizontal axis to your right, you will be looking at the top of the book, the front cover to your right, the spine facing downward. If you start over and perform the same rotations in the opposite order, then after the first rotation the top of the book will face you, the front cover facing downward and the spine to your left. After the second rotation, the spine will be facing you, the front cover downward—a completely different final position. Each of these transformations can be represented as a single rotation about a single axis but a different rotation for the two orders. Yet vector addition (tail-to-tip addition) is *commutative*; the order should not matter. Hence, a sequence of rotational displacements cannot be represented as a sum of vectors. The correct representation is as a product of matrices; matrix multiplication need not be commutative.

This demonstration would be less obvious if the individual rotations had been much smaller than 90°. In the limit of infinitesimal rotations, the discrepancy between one sequence and the other would be second-order in the infinitesimal displacements, i.e., negligible. That is, finite rotations cannot be treated as vectors, but infinitesimal rotations can. That means that instantaneous *angular velocities* $\boldsymbol{\omega} \equiv d\boldsymbol{\theta}/dt$ are vector quantities. The *magnitude* of $\boldsymbol{\omega}$ is the instantaneous angular speed $\omega = d\theta/dt$. Its direction is the unique direction associated with the rotation, that of the rotation axis. Actually, the axis defines two directions. By convention—in a blatant example of dextrism—the choice between the two is made via the *right-hand rule*: If the curved fingers of the right hand point in the direction of the rotational motions of the body or system, the extended thumb of that hand identifies the direction of the angular-velocity vector.

The linear velocity $\mathbf{v}$ of a point at location $\mathbf{r}$ in a body or system rotating with angular velocity $\boldsymbol{\omega}$ (where the origin of position is located on the rotation axis) is determined by the vectors $\boldsymbol{\omega}$ and $\mathbf{r}$. It has magnitude

$$\begin{aligned} v &= \omega r_\perp \\ &= \omega r \sin\theta \,, \end{aligned} \qquad\qquad \text{(III.73)}$$

where θ is the angle between vectors $\boldsymbol{\omega}$ and $\mathbf{r}$ placed tail to tail. Its direction is perpendicular to $\boldsymbol{\omega}$, i.e., to the rotation axis, and to $\mathbf{r}$, as $\mathbf{v}$ is tangent to the rim of the cone traced out by $\mathbf{r}$ as it rotates. As long as $\mathbf{r}$ is not parallel to $\boldsymbol{\omega}$ (in which case $\mathbf{v}$ is zero), there are exactly two directions perpendicular to both vectors—the directions perpendicular to the plane they define. The correct choice is identified by another version of the right-hand rule: If $\boldsymbol{\omega}$ and $\mathbf{r}$ are placed tail to tail, the correct direction of $\mathbf{v}$ is given by the extended thumb of the right hand when the curled fingers of that hand are oriented to "push" vector $\boldsymbol{\omega}$ into vector $\mathbf{r}$. This is not an independent right-hand rule; it enforces consistency with the version used previously to define the direction of $\boldsymbol{\omega}$.

The preceding combination of vector features appears so often, and has sufficient properties in common with multiplication, that it is called the *cross product* or *vector product* of two vectors[26]:

$$\mathbf{v} = \boldsymbol{\omega} \times \mathbf{r} \ . \tag{III.74}$$

This is usually introduced in the second or third term of the introductory calculus course. It is a way to multiply two vectors together, complementary to the dot or scalar product of Eqs. (II.4) and (II.6): That product picks out the component of one vector *parallel* to another; this one selects the component of one vector *perpendicular* to another. The cross product does *not* have some of the more familiar properties of a product. It is *anti*commutative rather than commutative: $\mathbf{B} \times \mathbf{A} = -\mathbf{A} \times \mathbf{B}$ because of the right-hand rule. Neither is it associative: $\mathbf{A} \times (\mathbf{B} \times \mathbf{C})$ need not equal $(\mathbf{A} \times \mathbf{B}) \times \mathbf{C}$ for general vectors $\mathbf{A}$, $\mathbf{B}$, and $\mathbf{C}$. It does obey the distributive law over vector addition: $\mathbf{A} \times (\mathbf{B} + \mathbf{C}) = \mathbf{A} \times \mathbf{B} + \mathbf{A} \times \mathbf{C}$. Consequently, the cross product obeys a Leibniz rule for differentiation:

$$\frac{d}{dt} \mathbf{A} \times \mathbf{B} = \frac{d\mathbf{A}}{dt} \times \mathbf{B} + \mathbf{A} \times \frac{d\mathbf{B}}{dt} \ . \tag{III.75}$$

The component form of the cross product is

$$\begin{aligned}
\mathbf{A} \times \mathbf{B} &= (A_x\,\hat{\boldsymbol{x}} + A_y\,\hat{\boldsymbol{y}} + A_z\,\hat{\boldsymbol{z}}) \times (B_x\,\hat{\boldsymbol{x}} + B_y\,\hat{\boldsymbol{y}} + B_z\,\hat{\boldsymbol{z}}) \\
&= A_xB_x\,\hat{\boldsymbol{x}} \times \hat{\boldsymbol{x}} + A_xB_y\,\hat{\boldsymbol{x}} \times \hat{\boldsymbol{y}} + \cdots + A_zB_z\,\hat{\boldsymbol{z}} \times \hat{\boldsymbol{z}} \\
&= (A_yB_z - A_zB_y)\,\hat{\boldsymbol{x}} + (A_zB_x - A_xB_z)\,\hat{\boldsymbol{y}} + (A_xB_y - A_yB_x)\,\hat{\boldsymbol{z}} \\
&= \begin{vmatrix} \hat{\boldsymbol{x}} & \hat{\boldsymbol{y}} & \hat{\boldsymbol{z}} \\ A_x & A_y & A_z \\ B_x & B_y & B_z \end{vmatrix} \ ,
\end{aligned} \tag{III.76}$$

where the cross products of the unit vectors are worked out directly from the definition.[27] Strictly, the last expression is meaningless—the determinant is a function on the set of matrices, which are arrays of numbers. But if expanded formally by minors, it reproduces the previous expression; it can be considered a mnemonic device for the cross product. We shall make use of the cross product throughout this section and in Chs. IX and XI to follow.

The generalization of rotational mechanics to three dimensions continues along these lines. *Angular acceleration* is a vector quantity:

$$\boldsymbol{\alpha} \equiv \frac{d\boldsymbol{\omega}}{dt} \ . \tag{III.77}$$

In three dimensions this can describe both spin up/spin down and changes in the direction of the angular velocity.

---

[26] If one wishes really to impress people at parties, one calls this the *antisymmetrized tensor product* and observes that it corresponds to a vector only in three dimensions.

[27] The fact that the axes are ordered so as to satisfy $\hat{\boldsymbol{x}} \times \hat{\boldsymbol{y}} = +\hat{\boldsymbol{z}}$, and corresponding, makes the coordinate system *right-handed*. Left-handed coordinate systems are treated in some advanced texts, but not here.

Dynamics

Rotational dynamics is actually more straightforwardly described in three dimensions than in the fixed-axis case. Angular momentum is defined for a point body, *about an origin point,* via

$$\boldsymbol{\ell} \equiv \mathbf{r} \times \mathbf{p} , \qquad\qquad \text{(III.78a)}$$

where $\mathbf{r}$ is the position of the body measured from the desired origin, and $\mathbf{p}$ is the body's momentum. There is no need to define the lever arm separately; the cross product takes care of that. The angular momentum of a system or extended body, again about a point, is simply the vector sum

$$\mathbf{L} = \sum_i \boldsymbol{\ell}_i , \qquad\qquad \text{(III.78b)}$$

or a similar integral expression.

This can exhibit surprising features. For example, for the single mass-bearing arm rotating on a vertical shaft illustrated in Fig. III.11, the angular momentum of the system about the given origin—if the masses of the shaft and arm are taken to be negligible—is, as shown, *not parallel to the angular velocity.* The velocity and momentum of the mass are into the page, represented by the symbol $\otimes$. (The opposite direction, out of the page, would be represented[28] by the symbol $\odot$.) The angular momentum $\mathbf{L}$ is perpendicular to $\mathbf{r}$ and $\mathbf{p}$, as indicated in the figure.

Hence, for this system, $\mathbf{L}$ cannot be written as a scalar multiple of $\boldsymbol{\omega}$. The relationship between the two quantities is still linear—twice as much $\boldsymbol{\omega}$ still means twice as much $\mathbf{L}$. The scalar moment of inertia in Eq. (III.58b) must be replaced by an *inertia tensor*[29] $\mathcal{I}$:

$$\mathbf{L} = \mathcal{I}\boldsymbol{\omega} . \qquad\qquad \text{(III.79)}$$

The inertia tensor can be represented by a 3×3 matrix. The components of this matrix are second moments of the mass distribution, of which the moment of inertia in Eqs. (III.43) is one example.

The reader will notice that if the rotor of Fig. III.11 had an identical arm and mass on the other side, then the total angular momentum vector *would* be parallel to $\boldsymbol{\omega}$. In fact, for any body of any shape and mass distribution, it is always possible to find three mutually perpendicular axes such that if the body is rotating about any one of these axes, its angular momentum will also lie along that axis. These are called the *principal axes* of the body.[30] An axis of symmetry of a mass distribution is always a principal axis.

---

[28]This notation comes from a time when three-dimensional or perspective figures were prohibitively expensive to produce and print in textbooks. The symbols $\otimes$ and $\odot$ represent an arrow flying away from or toward the reader, respectively.

[29]The *tensor,* often used as a mathematical bogey to frighten students, is a simple notion: It is a linear function between one set of vectors and another, here angular velocities and angular momenta. Tensors have many other applications in many areas of physics.

[30]Readers who have studied linear algebra will be pleased to note that the identification of the principal axes of a body is a matrix-diagonalization problem.

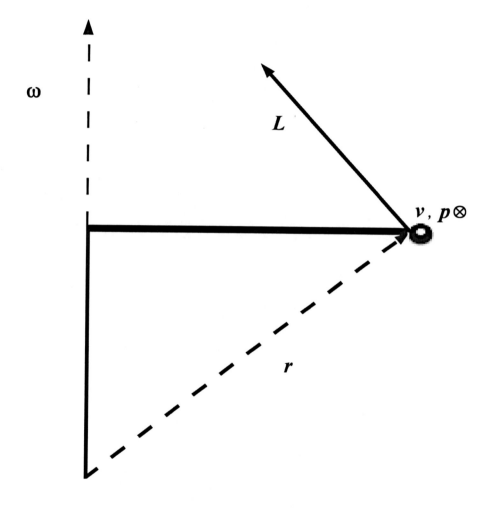

Figure III.11: Single-arm rotor. With **r** for the mass as indicated, its velocity and momentum are into the figure. The corresponding angular momentum is, as shown, *not* parallel to the angular-velocity vector.

Torque is also easily expressed in three dimensions. The torque exerted about the origin point by force $\mathbf{F}$ applied at position r is given by the cross product

$$\tau \equiv \mathbf{r} \times \mathbf{F} \; ; \qquad\qquad (\text{III.80})$$

once again, the cross product takes care of determining the lever arm. The dynamics of angular momentum (III.78a) takes the simple form

$$\frac{d\boldsymbol{\ell}}{dt} = \frac{d}{dt}\,\mathbf{r} \times \mathbf{p}$$
$$= \frac{d\mathbf{r}}{dt} \times \mathbf{p} + \mathbf{r} \times \frac{d\mathbf{p}}{dt}$$
$$= \mathbf{r} \times \mathbf{F}$$
$$\frac{d\boldsymbol{\ell}}{dt} = \tau \; . \qquad\qquad (\text{III.81a})$$

Here the first term in the second line vanishes because $\mathbf{v} = d\mathbf{r}/dt$ and $\mathbf{p}$ are parallel (the angle between them is zero), and the second term takes the given form via Newton's Second Law. The dynamics of angular momentum (III.78b) for a system or body is a sum of such contributions:

$$\frac{d\mathbf{L}}{dt} = \sum \tau \; , \qquad\qquad (\text{III.81b})$$

where the sum of torques becomes the sum of *external* torques acting on the system or body if the strong form of Newton's Third Law is in play. This last result implies the Law of Conservation of Angular Momentum in three dimensions: For an isolated system or a system subject to zero net torque, its vector angular momentum $\mathbf{L}$ is constant—conserved. Now this means both the magnitude and direction of the angular momentum are conserved, i.e., this is three conservation laws, one for each component of $\mathbf{L}$. As described in Sec. D preceding, this law applies far beyond the boundaries of the Newtonian limit.

Precession

Rotation in three dimensions gives rise to a remarkable variety of phenomena. An important class of these is *precession,* in which the angular-momentum vector of a system rotates in response to a torque perpendicular to that angular momentum. This is easily demonstrated with a "toy" gyroscope—although this is actually a simple version of a device widely used for navigation and guidance systems. The gyroscope illustrated in Fig. III.12 consists of a wheel, containing most of the gyroscope's mass, spinning on an axle through its center. It rests on a fixed point, about which the axle is free to pivot in any direction. The weight of the gyroscope exerts a torque τ about the pivot directed into the page at the instant depicted in the figure. Were the gyroscope not spinning, this torque would produce angular acceleration causing the gyroscope to topple over. But since it is spinning with angular speed ω_s, say, it has nonzero angular momentum of magnitude $L = I\omega_s$, where I is the moment of inertia of the wheel about

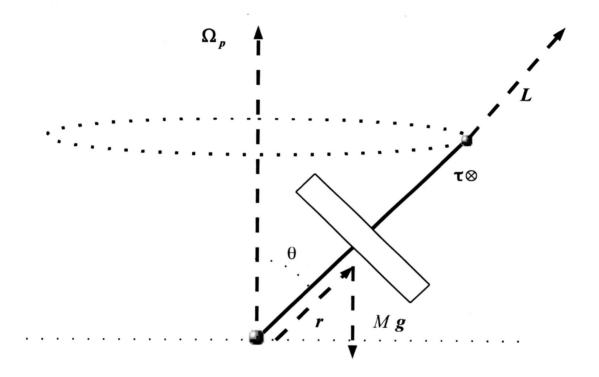

Figure III.12: A demonstration gyroscope seen side-on. The weight $M\mathbf{g}$ of the gyroscope exerts a torque $\boldsymbol{\tau}$ about the fixed pivot point directed into the page at this instant. This causes the spin angular momentum $\mathbf{L}$ to precess in a cone, with precession angular velocity $\boldsymbol{\Omega}_p$.

an axis through the axle. The torque thus causes the vector angular momentum **L**, directed along the axle, to move in the direction of $\boldsymbol{\tau}$, i.e., to trace out the cone shown in the figure, in accord with Eq. (III.81b). In the absence of frictional or drag forces, the gyroscope does not topple over at all.

Anticipating a result from the next section, we can treat the weight $M\mathbf{g}$ of the gyroscope as acting through its center of mass, i.e., the center of the wheel. It produces torque

$$\boldsymbol{\tau} = \mathbf{r} \times (M\mathbf{g})$$

$$= Mgr \sin\theta \, \hat{\otimes} \, , \tag{III.82a}$$

with the notation that of Fig. III.12 and $\hat{\otimes}$ a unit vector directed into the figure. The rotation, i.e., precession of the angular momentum **L**, is given by

$$\frac{d\mathbf{L}}{dt} = \boldsymbol{\Omega}_p \times \mathbf{L}$$

$$= \Omega_p I \omega_s \sin\theta \, \hat{\otimes} \, , \tag{III.82b}$$

a result akin to Eq. (III.74). It is assumed here that *the angular momentum associated with the precession is negligible in comparison to that associated with the gyroscope's spin.* Since $\boldsymbol{\tau}$ and $d\mathbf{L}/dt$ must be equal, the precession angular speed must be

$$\Omega_p = \frac{Mgr}{I\omega_s} \tag{III.82c}$$

in this approximation for any value of the inclination θ. The author's demonstration gyroscope has mass $M = 0.10$ kg concentrated in the rim of a wheel of radius $r_w = 3.0$ cm, i.e., with moment of inertia approximately

$$I = Mr_w^2$$

$$= (0.10 \text{ kg})(3.0 \times 10^{-2} \text{ m})^2$$

$$= 9.0 \times 10^{-5} \text{ kg m}^2 \, . \tag{III.83a}$$

It is spun up to roughly 10. revolutions per second, i.e., $\omega_s = 20.\,\pi \text{ s}^{-1} = 63.\text{ s}^{-1}$. The length of the axle from the pivot to the center of the wheel is also $r = 3.0$ cm. The resulting precession angular speed is

$$\Omega_p = \frac{Mgr}{I\omega_s}$$

$$= \frac{(0.10 \text{ kg})(9.8 \text{ m/s}^2)(3.0 \times 10^{-2} \text{ m})}{(9.0 \times 10^{-5} \text{ kg m})^2(63.\text{ s}^{-1})}$$

$$= 5.2 \text{ s}^{-1} \, . \tag{III.83b}$$

This corresponds to precession period

$$T_p = \frac{2\pi}{\Omega_p}$$

$$= \frac{2\pi}{5.2 \text{ s}^{-1}}$$

$$= 1.2 \text{ s} \, , \tag{III.83c}$$

which can readily be observed.

The actual motion of the gyroscope depends sensitively on how it is started. A more detailed analysis would show that it can precess at two different angular speeds—results (III.83b) and (III.83c) correspond to "slow precession." It can also oscillate between minimum and maximum values of angle θ, a motion known as *nutation*.

Precession is encountered in many contexts. Perhaps the most famous is the *Precession of the Equinoxes,* the precession of the spinning Earth about the perpendicular to the plane of its orbit about the Sun, with a period of 25,800 years. This motion is driven by torques produced by tidal forces from the Sun and Moon acting on the equatorial bulge of the Earth. It was discovered by the Babylonian astronomer Cidenas[31] in 343 BCE and the Greek astronomer Hipparchus[32] in 130 BCE.[33] (The ancient astronomers observed changes in the relations between the constellations and the seasons. They did not attribute the phenomenon to motions of the Earth.) The turning of a bicycle or motorcycle is a precession: The rider leans the vehicle, producing a torque that causes the angular momenta of the wheels to precess in the desired direction. Atomic nuclei possess spin angular momentum and can be made to precess. This is the basis of an important technology, which will be described in more detail in Ch. IX.

The rotational dynamics presented here and in Secs. D and E preceding is Newtonian. Einsteinian rotational dynamics displays subtleties best left to a more advanced text.

G A few words about static equilibrium

The translational (linear) and rotational versions of Newton's Second Law provide a foundation for the science of *statics,* the third—and oldest—branch of mechanics. A rigid body in *static equilibrium* must experience no acceleration. In accord with Eq. (III.19), the forces $\mathbf{F}$ applied to the body must satisfy the condition

$$\sum \mathbf{F} = \mathbf{0} \ . \tag{III.84a}$$

The body must also experience no angular acceleration. In accord with Eq. (III.81b), the torques $\boldsymbol{\tau}$ applied to the body must satisfy

$$\sum \boldsymbol{\tau} = \mathbf{0} \tag{III.84b}$$

when calculated about any point. It is easy to show that if condition (III.84a) is satisfied, then condition (III.84b) is satisfied about every point if it is satisfied about any one point, which one is free to choose. Hence, in general, conditions (III.84) constitute six conditions that can be used to determine the forces and geometry in static-equilibrium situations.

---

[31] Cidenas or Kidinnu, 4th C. BCE

[32] Hipparchus (Hipparchos) of Nicaea, *c.* 190–120 BCE

[33] Hipparchus is usually credited with the discovery; no doubt his paper was more widely disseminated.

An important simplification applies to torques produced by the weight of a system or body. The total gravitational torque about any origin is given by

$$\boldsymbol{\tau}_g = \sum_i \mathbf{r}_i \times (m_i \mathbf{g})$$

$$= \left(\sum_i m_i \mathbf{r}_i \right) \times \mathbf{g}$$

$$= \mathbf{r}_{\text{cm}} \times (M_{\text{tot}} \mathbf{g}) , \qquad\qquad (III.85)$$

or a similar integral, since the sum in the second line is the definition of the position $\mathbf{r}_{\text{cm}}$ of the center of mass times the total mass M_{tot}. That is, the weight of the body produces the same torque as if its entire mass were located at its CM, provided that *the extent of the system or body is small enough that the gravitational acceleration* $\mathbf{g}$ *can be treated as a constant.*[34] This result was used in the analysis of the gyroscope [Eq. (III.82a)]. It is the second reason [after result (III.23)] that the center of mass (CM) is often identified with the center of gravity (CG) of a system or body.

Statics is an important discipline of wide application. Most introductory physics textbooks devote an entire chapter to static-equilibrium problems. Many engineering departments teach entire courses on statics. In this text, however, we shall move on to other physics.

References

Collisions and rotation are major topics covered in all introductory physics texts and intermediate-level mechanics texts. Yet more detailed treatments are found in advanced texts, such as those of Goldstein, Poole, and Safko [GPS01] and Landau and Lifshitz [LL76]. (This last is Volume 1 of the authors' monumental *Course of Theoretical Physics.*) There are also specialized texts devoted entirely to these topics, such as those of Routh [Rou12b, Rou12a], originally published in successive editions between 1860 and 1905.

Problems

1. A racquetball (diameter 8.00 cm, mass 60.0 g) is held a negligible distance above a basketball (diameter 30.0 cm, mass 610. g); the bottom of the basketball is 4.00 m above a hard floor. The two balls are dropped together from rest. The basketball bounces off the floor and then strikes the racquetball. All collisions have $r = 0.700$ coefficient of restitution. The demonstration takes place in a rotunda with a ceiling 12.00 m above the floor. (The balls are dropped from a mezzanine.) Will the racquetball

---

[34]This assumption can fail, e.g., for large structures in space. Then the weight of the structure can produce a net torque about its CM. This torque can be utilized to control the orientation of the structure, an effect known as "tidal stabilization."

strike the ceiling? If so, at what speed? If not, by what distance will it come up short?

2. Consider a general one-dimensional collision in the LAB frame: A body of mass m_1 and velocity component v_1 collides head-on with a body of mass m_2, initially at rest ($v_2 = 0$). The collision is characterized by coefficient of restitution r. Calculate the ratio of total Newtonian kinetic energy after the collision to total kinetic energy before, K_f/K_i. Compare your result with Eq. (III.26c), which applies in the CM frame.

3. A body of mass m traveling with velocity $\mathbf{v}_0$ collides with an identical body at rest. The two bodies move off at equal angles ϕ to each side of the original direction of the first body. Fraction $f \geq 0$ of the original kinetic energy of the first body is dissipated in the collision. Find the maximum value of ϕ and the maximum value of f (not necessarily corresponding to the same collision) for which this can occur.

4. A body collides off-center with an identical body at rest, i.e., in the LAB frame. Result (III.12c) shows that if the collision is Newtonian and elastic, the final trajectories of the two bodies are perpendicular. If instead the collision is *inelastic*, i.e., kinetic energy is dissipated, is the angle between the final trajectories acute, right, or obtuse?

5. The same question, but this time for an *Einsteinian*, elastic collision—i.e., the bodies retain their original masses. How does this result approach that of Eq. (III.12c) in the Newtonian limit?

6. Looking up the necessary data, and treating the two bodies as spherical, determine the location of the center of mass of the Earth-Moon system.

7. The Sun and Jupiter move in approximately circular orbits around their common center of mass.

   (a) Looking up the necessary data, calculate the radius of the Sun's orbit.

   (b) Calculate the *angular diameter* of the Sun's orbit seen edge-on from a distance of 4.30 light years. Such an observation is one method observers at Alpha Centauri might use to search for "exoplanets" around the Sun.

   (c) Repeat calculations (a) and (b) for the *Earth*-Sun system. This is why Earth-like exoplanets are harder to find than Jupiter-like ones.

8. A uniform round body of mass M, radius R, and moment of inertia $I = \beta M R^2$ is placed at the top of a ramp inclined at angle θ to the horizontal. Calculate the minimum coefficient of static friction μ_s between the body and the ramp that allows the body to roll without slipping from rest.

9. Consider the Sun to be a uniform-density sphere of mass $M_\odot = 1.989 \times 10^{30}$ kg and radius $R_\odot = 6.963 \times 10^8$ m, rotating rigidly with period 25.0 days. (This model is only an approximation.) If the Sun were to collapse under its own gravitational forces—which obey the strong form

of Newton's Third Law—to a uniform-density, spherical neutron star of radius 20.0 km without losing any mass in the process, what would be the rotation period of the neutron star?

10. Consider the pool shot of Sec. D. Calculate the initial and final total kinetic energy of the balls and their ratio K_f/K_i. Remember that both translational and rotational kinetic energies are in play.

11. Consider again the pool shot of Sec. D. Suppose the coefficient of *kinetic* friction between the pool balls and the felt surface, before the balls attain rolling without slipping, is μ_k. Calculate the *distance* each ball slips.

12. Suppose the pool shot analyzed in Sec. E were made with thin-shell hollow steel balls (with moments of inertia $I = \frac{2}{3}MR^2$) rather than solid resin[35] balls. The game would, of course, be noisier. How else would the results of that section—the height of the cue stroke and the final speeds of the balls—be different?

13. Consider the pool shot of Sec. E again, with solid balls as before, but suppose the collision between the balls is characterized by coefficient of restitution $r \leq 1$. Find the final velocities of the balls in this case, all else being as before.

14. Consider once again the pool shot of Sec. E, with solid, perfectly elastic balls, but with appreciable ball-ball friction. *During the impact* between the balls, such forces would impart angular impulses to each ball. If the coefficient of kinetic friction between the balls is $\mu_k^{(bb)}$, what now will be the final velocities of the two balls? Assume that $\mu_k^{(bb)}$ is sufficiently large that during the impact, ball-ball friction completely dominates (is much larger than) ball-felt friction, i.e., ball-felt friction remains negligible during this interval.

15. A projectile of mass m is launched from the ground at time $t = 0$, with speed v_0, at angle θ_0 above the horizontal. Aerodynamic drag is negligible.

    (a) Write the vector angular momentum of the projectile about the original launch point as an explicit function of time.

    (b) Write the vector torque on the projectile about the original launch point as an explicit function of time.

    (c) Verify relation (III.81a).

16. When a new tire is mounted on the wheel of a car, the assembly must be *balanced*. Small weights are attached to the rim of the wheel to achieve two conditions: First, the center of mass of the tire and wheel must lie at the center of the wheel, i.e., on the axle. Second, the axle must be a *principal axis* of the tire/wheel assembly, i.e., the angular momentum of the rotating assembly must lie along the axle. Why each of these conditions?

---

[35] We do *not* countenance the use of ivory.

17. The Earth's rotation axis is tilted 23°.4 from the perpendicular to the plane of its orbit about the Sun. It precesses about that perpendicular—the Precession of the Equinoxes—with a period of 25,800 yr. Calculate the magnitude of the torque exerted on the Earth to produce this precession. Model the Earth as a uniform, solid sphere, with moment of inertia $I = \frac{2}{5}M_\oplus R_\oplus^2$; look up any necessary data. By comparing the value of this torque with calculations of the gravitational torque[36] produced by the Sun and Moon, it is possible to measure the oblateness—departure from sphericity—of the Earth.

---

[36] These calculations must be reserved for a more advanced text.

Chapter IV

Shoot the Moon

- *Determine the velocity change and timing required for a Lunar-Orbit-Insertion rocket-engine burn to send a spacecraft from low Earth orbit to the Moon.*

- *Examine and apply Newton's Law of Universal Gravitation.*

- *Examine and apply Kepler's Laws of Planetary Motion.*

- *Consider Einsteinian gravitation and celestial mechanics.*

145

On July 20, 1969, an age-old dream was realized when Apollo 11 astronaut Neil Armstrong[1] became the first human to set foot on the surface of the Moon after a journey begun four days earlier.[2] The vision, ingenuity, perserverance, and courage of thousands of people went into this feat. The achievement is perhaps less appreciated than it should be. Nowadays, many people carry images of space travel from movies and television shows in which people *fly* through space—continuously under power and steering. In reality, nobody flies through space. The distances are too vast for any spacecraft to carry the fuel necessary for that. Spacecraft *coast* through space. Space travel is *marksmanship* on an awesome scale.[3] One purpose of this chapter is to convey to the reader a slightly more realistic picture of the scientific and technical challenges of spaceflight.

The Apollo spacecraft, its Lunar Module, and the third stage of its Saturn V rocket were launched from Cape Canaveral (then Cape Kennedy), Florida, into low Earth orbit. A second, Lunar Orbit Insertion (LOI) burn from the third-stage engine put the spacecraft on a long orbit that would take it to the vicinity of the Moon. This is illustrated in Fig. IV.2. In this chapter, we shall examine and apply some of the physics principles underlying this maneuver.

This—the understanding of the motions of bodies in space—is the area in which the modern (i.e., post-medieval) science of physics gained its first great successes. This is where the field—and, by extension, all of modern science—established its credibility, took its present place in human culture, and began the transformation of the world of the seventeenth century into that of the twenty-first. For the sake of clarity, in this text, the major principles of this subject are presented in logical rather than historical order.

A   Newton's Law of Universal Gravitation

The second part of Newton's *Principia* is devoted to his work on "The System of the World"—his Law of Universal Gravitation and its implications for the motions of the heavens and the Earth.

Newton and the apple

We are all familiar with the story of Newton and the apple: Newton, sitting under an apple tree, was struck on the head by a falling apple and thus became aware of the existence of gravity. Entertaining, but nonsense. Newton was a farm boy; he knew all about apple trees. As pointed out at the start of Ch. I Sec. E, Newton had returned home when the plague broke out in Cambridge in 1665. Apparently, watching the apples fall in his orchard led Newton to contemplate that the acceleration of the apples in their fall—which he knew

---

[1] Neil Armstrong, 1930–2012

[2] The technology advances. NASA's New Horizons spacecraft, launched in 2006 bound for Pluto, crossed the orbit of the Moon *eleven hours* after launch.

[3] Interplanetary spacecraft do make a few small course corrections *en route*. The accomplishment is no less awesome for that.

Figure IV.1: Twelve men have walked on the surface of Selene, or Cynthia (the proper names of Earth's moon in Greek and Latin, respectively). What did they have to do to get there? (Photograph by Tristan3D.)

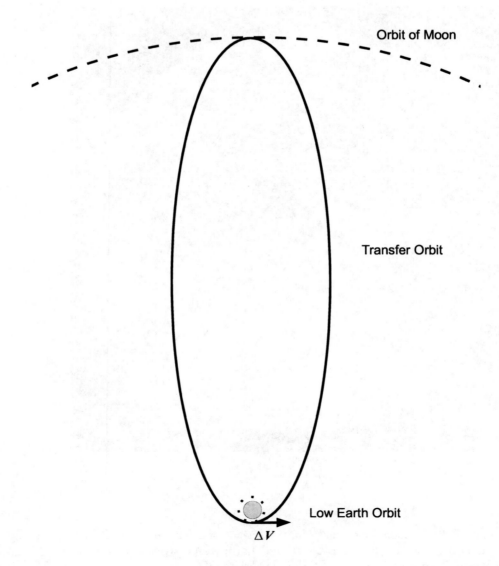

Figure IV.2: Transfer from low Earth orbit to the orbit of the Moon. Not to scale. The Lunar Orbit Insertion (LOI) burn imparts velocity change $\Delta \mathbf{V}$ to the spacecraft in low Earth orbit, putting it on the transfer orbit to the Moon.

from the earlier work of Galileo—and the acceleration of the Moon in its orbit around the Earth *might be manifestations of the same phenomenon.* That may seem trivial to the modern reader, but in the seventeenth century it was a world-rocking, revolutionary idea: that the perfect, imperishable, heavenly bodies might be subject to the same *mechanism* as the impermanent, corruptible matter of Earth. And equally bold: that that mechanism might be comprehensible to the finite human mind.

Newton was not expecting that the accelerations of the apple and the Moon took the same *value*; rather that both arose from an influence exerted by the Earth on both objects. He expected that influence to vary inversely as the square of distance from the Earth. This hypothesis was not original with Newton; it is credited to Bullialdus.[4] It's actually rather obvious: If an influence spreads out in all directions from a point source, the *intensity* of that influence—amount per unit area—will decrease inversely as the area of the spheres through which it passes, i.e., inversely as the square of distance. This is a straightforward consequence of the fact that space is three-dimensional.[5] Newton could calculate the acceleration of the Moon. He knew—because the ancient Greeks had measured—the Earth-Moon distance r_{EM}, as well as the radius of the Earth $R_\oplus$. (Here, and subsequently, the astronomers' symbol $\oplus$ is used to identify quantities associated with the Earth.) He knew the period T_M of the Moon's nearly circular orbit. Hence, he could estimate lunar acceleration

$$
\begin{aligned}
a_M &= -r_{EM}\left(\frac{2\pi}{T_M}\right)^2 \\
&= -(3.844 \times 10^8 \text{ m})\left(\frac{2\pi}{(27.3 \text{ d})(86400. \text{ s/d})}\right)^2 \\
&= 2.73 \times 10^{-3} \text{ m/s}^2 \,,
\end{aligned}
\tag{IV.1a}
$$

where modern values are supplied for the variables. The apple has acceleration $-g$; the ratio of the two accelerations is

$$
\begin{aligned}
\frac{-g}{a_M} &= \frac{9.81 \text{ m/s}^2}{2.73 \times 10^{-3} \text{ m/s}^2} \\
&= 3.60 \times 10^3 \,,
\end{aligned}
\tag{IV.1b}
$$

to three significant figures. The apple is one Earth radius from the center of the Earth; the inverse squared ratio of their distances is

$$
\begin{aligned}
\left(\frac{r_{EM}}{R_\oplus}\right)^2 &= \left(\frac{3.844 \times 10^8 \text{ m}}{6.371 \times 10^6 \text{ m}}\right)^2 \\
&= 3.640 \times 10^3 \,,
\end{aligned}
\tag{IV.1c}
$$

---

[4]Ismaël Boulliau (Bullialdus), 1605–1694

[5]If space—not space*time*—had four full-scale dimensions, then gravitation would vary as the inverse cube of distance. But in that case, planetary orbits would be unstable, and planetary systems would not exist.

here to the four significant figures of the data. The precision is somewhat misleading, as this is only a rough estimate: The motion of the Moon is much more complicated—and much more precisely known—than this, and corrections, e.g., to g for the Earth's rotation, are omitted. But it is easy to see why Newton concluded he "had found the answer pretty nearly."

It concerned Newton that while approximating the Earth as a point source for its influence on the Moon—sixty Earth radii away—was reasonable, treating it as such for the apple at the Earth's surface was not. It is, in fact, the (almost exactly) correct treatment, as shall be seen, but it would take twenty years and the invention of calculus for Newton to resolve this to his own satisfaction. Nonetheless, the lesson of the apple—that mechanics on Earth and mechanics in the heavens are one and the same—was learned.

The force law

Newton proposed a *force law* through which the influence that moves the apple and the Moon could be applied in his Second Law of Motion. We can describe this most clearly with a tool unavailable to him: modern vector notation, much of which was invented in the nineteenth century. He proposed a *universal* attraction: Every body attracts every other via a force given by

$$\mathbf{F} = -\frac{Gmm_s(\mathbf{r} - \mathbf{r}_s)}{|\mathbf{r} - \mathbf{r}_s|^3} \, , \tag{IV.2}$$

where $\mathbf{F}$ is the force acting on "receiver" mass m at location $\mathbf{r}$, produced by "source" mass m_s at location $\mathbf{r}_s$, and G is Newton's Gravitational Constant. The vector $\mathbf{r} - \mathbf{r}_s$ is the vector displacement from source to receiver, as illustrated in Fig. IV.3.

Newton's Constant, with current precision, is given by

$$G = 6.67259 \times 10^{-11} \, \frac{\mathrm{m}^3}{\mathrm{kg\,s}^2} \, . \tag{IV.3}$$

Newton lived and died without knowing the value of this constant; he worked with proportionalities. To determine its value, one must measure the force between known masses. Measuring the force between a known mass and the Earth—the weight of a known mass—is no help, because one has no independent measure of the Earth's mass. Not until 1798, i.e., seventy-one years after Newton's death, did Henry Cavendish[6] succeed in measuring the gravitational attraction between laboratory masses using a *torsion balance* or *Cavendish balance*. This is crudely sketched in Fig. IV.4: The force acting on the masses twists the suspension wire. Its deflection is measured by reflecting a light beam off the attached mirror, an example of an arrangement known as an "optical lever." A Cavendish balance for a modern teaching laboratory costs in the neighborhood of \$1500 US (and the experiment is a difficult one). The original,

---

[6]Henry Cavendish, 1731–1810

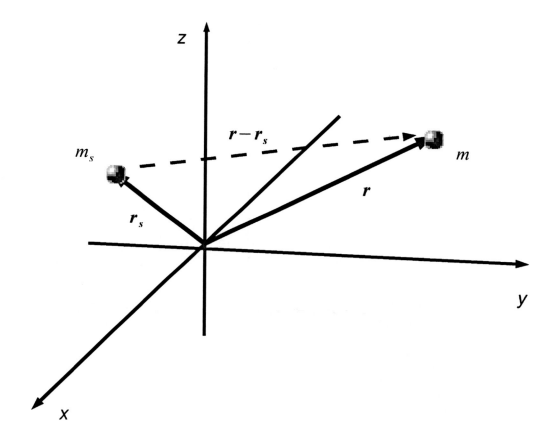

Figure IV.3: Masses and vector positions for Newton's Law of Universal Gravitation.

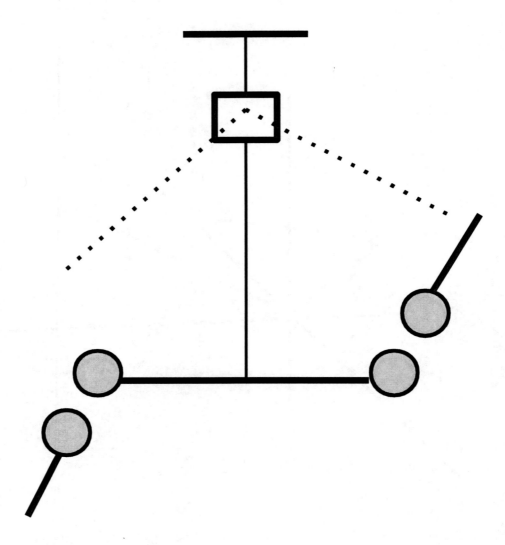

Figure IV.4: Sketch of a torsion balance or Cavendish balance for measuring the gravitational attraction between known masses. The outer masses can be moved on levers to vary the distance between them and the masses on the suspended bar.

enclosed in a bell jar to reduce disturbances from air currents, must have been a masterpiece of the metalsmith's and glassmaker's arts.[7] Once the value of G is known, one can deduce the mass of the Earth from the weights of known masses and force law (IV.2). Hence, Cavendish characterized his experiment as "weighing the Earth."

Newton's force is proportional to the masses of the "source" and "receiver" bodies. The notation we use glibly hides the fact that masses m and m_s are conceptually different notions of mass, and both are distinct from the inertial mass that appears in Newton's Second Law (I.43) and from the chemists' notion of mass as "amount of stuff." Mass m is *passive gravitational mass*; it measures the extent to which a body *receives* gravitational attraction. Mass m_s is *active gravitational mass,* a measure of the extent to which a body *produces* gravity. It is a remarkable simplicity of nature that both of these measures are equal to the inertial mass of Newton's Second Law and to the chemists' measure of mass. (Strictly, they are all proportional to each other, with proportionality constants that are the same for all objects and materials, so that two objects with the same mass by one measure have the same mass by all the other measures. Then, by judicious choice of units, we can set all the proportionality constants to unity, i.e., make all the mass measures equal.) Newton was aware of these distinctions. He was able to demonstrate, using pendula with bobs of different masses and materials, that the inertial and passive-gravitational masses he could measure were equal to within his experimental precision of parts per 10^3. At the turn of the twentieth century, Baron Eötvös,[8] using a refinement of Cavendish's techniques, was able to establish this equality to within parts per 10^9. In the 1980s the EötWash experimenters at the University of Washington, improving on the Baron's apparatus, pushed the precision down to parts per 10^{13}. From a Newtonian perspective, this remains an ever more impressive coincidence. In the alternative view of gravitation provided by Einstein's 1915 General Theory of Relativity, it becomes obviousness itself; we shall touch on this approach in Sec. D following.

The equality of passive-gravitational and active-gravitational masses ensures that force law (IV.2) satisfies Newton's Third Law: The two bodies, each of which is both source and receiver of gravitational attraction, exert forces on each other equal in magnitude, opposite in direction. The equality of the other measures of mass with that of the chemists means that all measures are conserved in chemical reactions, i.e., in the Newtonian limit.

Force law (IV.2) incorporates Bullialdus' hypothesis—it describes an inverse-square force. Three powers of the distance $|\mathbf{r} - \mathbf{r}_s|$ appear in the denominator, but the vector $\mathbf{r} - \mathbf{r}_s$ appears in the numerator. Hence, the magnitude of the

---

[7]It is easy to lose sight of the contributions to science made by the unnamed artisans who fashioned the equipment for crucial experiments. In England, at least, such people were organized and recognized to the extent of having their own guild: the *Worshipful Guild of Scientific Instrument Makers.* A plaque commemorating this organization can still be seen in London.

[8]Roland Eötvös, 1848–1919

force is

$$|\mathbf{F}| = \frac{Gmm_s}{|\mathbf{r} - \mathbf{r}_s|^2} , \qquad (IV.4)$$

varying inversely as the square of the distance between source and receiver. Twice the distance means one-fourth the force, and so on.

Force (IV.2) is opposite in direction to the vector $\mathbf{r} - \mathbf{r}_s$. That is, it acts along the line between source and receiver. Any force of this nature is called a *central force*. A central force satisfies the strong form of Newton's Third Law, as described in Ch. III Sec. D.

Although not required by the formula itself, it is a matter of observation that the masses appearing in Eq. (IV.2) are always positive. Newton's gravitational force is always attractive. A positive and a negative mass could give rise to a repulsive force, but the behavior of such a system would be odd, because the negative mass should also possess negative *inertial* mass. Hence, the negative mass would gravitate *toward* the positive mass, while the latter would gravitate away from the former. (Two negative masses would give rise to an attractive force, which would drive them away from each other!) Negative masses could give rise to other bizarre phenomena; fortunately, they have never been observed.

Newton himself proposed no *mechanism* for his universal, inverse-square, central, attractive force.[9] He simply described how bodies move—the term "gravitate" describes a motion rather than a force or influence.

Force law (IV.2) describes the gravitational attraction between two masses treated as points. The attraction created by multiple sources, or an extended source, is the vector sum of the contributions of the individual sources or pieces, with no "crosstalk":

$$\mathbf{F} = \sum_{\text{sources}} \left(-\frac{Gmm_s(\mathbf{r} - \mathbf{r}_s)}{|\mathbf{r} - \mathbf{r}_s|^3} \right)$$

$$= -Gm \int_{\text{source}} \frac{\mathbf{r} - \mathbf{r}_s}{|\mathbf{r} - \mathbf{r}_s|^3} \, \rho(\mathbf{r}_s) \, d\mathcal{V}_s , \qquad (IV.5)$$

with $\rho(\mathbf{r}_s)$ the density distribution of an extended source treated as continuous. This result is often called the *principle of superposition,* but this is a misnomer. "Superposition" only means addition, but "principle" suggests a logically necessary feature. But there can be no *logically necessary* principles in science: Such principles would have to be true regardless of the outcome of any experiment, and all scientific principles must stand the tests of observation and experiment. It is much more enlightening to recognize that result (IV.5) is a *feature* of the gravitational force, properly termed *linearity in its sources.* Newtonian gravitation and electromagnetic forces are linear in this sense; Einsteinian gravitation and the weak and strong nuclear forces are not. Most introductory physics texts contain numerous exercises requiring the reader to evaluate sums and integrals of forms (IV.5). However, because these are simply exercises in vector algebra

---

[9] *Hypotheses non fingo*—"I do not feign hypotheses."

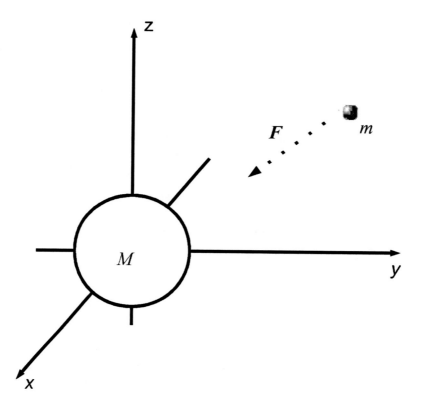

Figure IV.5: Gravitational attraction of a test mass by a spherically symmetric source.

and integral calculus, and because these sums and integrals can be evaluated exactly only for the simplest geometries, such calculations will not be emphasized here.

One example, though, is useful enough to be examined more closely: the gravitational force produced by a *spherically symmetric* source. Such a situation is illustrated in Fig. IV.5. Readers who wish to do so are invited to set up and evaluate the appropriate integrals. This author will simply present the results, leaving the calculation to Ch. VII, where a much more powerful method will be introduced:

- *Outside* a spherically symmetric source, the gravitational attraction produced is exactly the same as would be produced if the entire mass of the sphere were concentrated at its center.

- Anywhere *inside* a hollow spherical shell, the gravitational attraction produced on any test mass is exactly zero.

- Inside any sphere, the gravitational attraction is that produced by the portion of the sphere interior to (at smaller radius from the center than) the test location, which may be treated as a point mass located at the center. The portion of the sphere exterior to the test location contributes no gravitational force.

The first of these results vindicates the treatment of the Earth as a point even when dealing with the apple near its surface. That is exactly correct—not approximate—to the accuracy to which the Earth is spherically symmetric. The second result is problematic for science-fiction stories featuring Dyson[10] spheres and Edgar Rice Burroughs'[11] *Pellucidar* stories, where people live on the inside of a sphere—no gravity! The third result implies, for example, that the force on a given mass inside a sphere of uniform density varies linearly with radius from the center to the surface (beyond which it decreases inversely as radius squared). In fact, measurement of the Earth's gravity in bore holes drilled into the Earth is used to search for deviations from uniform density, e.g., mineral or oil deposits.

"Action at a distance" and gravitational *fields*

The Law of Universal Gravitation has a feature that has bothered physicists since Newton's day, a feature termed *action at a distance.* How does the Sun know where the Earth is to exert the proper force on it? Does the Earth gravitate toward the point where the Sun *is* or where it *appears to be*? (The Sun moves its own diameter across the sky every two minutes, and light from the Sun takes 8.3 minutes to reach the Earth; the actual location of the Sun is more than four solar diameters away from its apparent location.) The correct answer is that the Earth is attracted to where the Sun is—otherwise, its orbit would be unstable. But how does it know? Force law (IV.2) has no time dependence. If the Sun moves, does its gravitational influence respond instantaneously throughout the universe? Even before Einsteinian mechanics introduced finite limits on the speed at which energy or information could propagate, the notion of instantaneous change everywhere was troubling.

The resolution of the difficulty is to introduce the notion of the gravitational *field.* This is code language—this is not the field of the agronomist or of the mathematician. In physics, a *field* is an entity that exists throughout a region of space and time, in contrast to a *particle,* which exists at a point. The usage arose first in the study of fluids, which are necessarily extended—a "point fluid" makes no sense. The density and velocity distributions of a fluid, for example, are a scalar and a vector *field,* respectively. The gravitational field is a force field,[12] or more precisely, an acceleration field: It is defined as the gravitational force on a test mass per unit receiver mass. Of course, actually performing such a measurement would disturb the sources producing the field being measured. The solution is to make the measurement on a mass so small that the answer

---

[10]Freeman Dyson, 1923–
[11]Edgar Rice Burroughs, 1875–1950
[12]This is the correct use of this term.

wouldn't change if it were made smaller, i.e., the *limit*

$$\mathbf{g}(\mathbf{r}) \equiv \lim_{m \to 0} \frac{\mathbf{F}(\mathbf{r})}{m}$$

$$= -G \sum_{\text{sources}} \frac{m_s(\mathbf{r} - \mathbf{r}_s)}{|\mathbf{r} - \mathbf{r}_s|^3} \, , \tag{IV.6}$$

or a comparable integral, this last expression following from Eq. (IV.5). Here, $\mathbf{r}$ is the location at which the field $\mathbf{g}(\mathbf{r})$ is evaluated. The Sun, and every other body, exists cloaked in its gravitational field, which extends throughout space. There is no notion of a mass denuded of its gravitational field; hence, even "point" masses are to be regarded as extended objects. The Earth does not "know" where the distant Sun is—it responds to the gravitational field *at its own location*. The Sun's gravitational attraction does not *propagate* from the Sun to the Earth; it has been there since the Sun was formed. If the Sun moves, changes in its field do not appear instantaneously all over the universe. They propagate in a manner that Newton's Law, Eq. (IV.2), is not sufficient to describe.

This little subterfuge is more insidious than it appears. As will be seen in later chapters, the concept of field has gradually grown and taken over physics. Our best current understanding of matter and dynamics and existence itself is couched in the language of fields. The notion of a "point particle" is sidelined and plays no role.

The gravitational field $\mathbf{g}$ is already familiar. Near the surface of the (approximately spherical) Earth, it takes the value

$$\mathbf{g} \cong -\frac{GM_\oplus}{R_\oplus^2} \, \hat{\mathbf{r}}$$

$$\cong -\frac{[6.673 \times 10^{-11} \text{ m}^3/(\text{kg s}^2)][5.976 \times 10^{24} \text{ kg}]}{(6.371 \times 10^6 \text{ m})^2} \, \hat{\mathbf{r}}$$

$$\cong -(9.825 \text{ m/s}^2) \, \hat{\mathbf{r}} \tag{IV.7}$$

where $\hat{\mathbf{r}}$ is the unit vector radially away, i.e., vertically upward, from the center of the Earth. This is the gravitational acceleration we have used since Ch. I. The value used in previous calculations differs slightly from this because it is the *measured* value of $\mathbf{g}$ and includes corrections for the oblateness and rotation of the Earth.

Gravitational potential energy

Newton's Law of Gravitation describes a conservative force. It fits the last two categories listed in Ch. II Sec. E: It is a central force, with radial component depending only on distance between source and receiver, and it is a fundamental force. The defining integral (II.30) for gravitational potential energy is readily evaluated, yielding

$$U_g(\mathbf{r}) = -Gm \sum_{\text{sources}} \frac{m_s}{|\mathbf{r} - \mathbf{r}_s|} + C \, , \tag{IV.8}$$

or a similar integral over a continuous source. Such sums and integrals are usually slightly easier than those of Eq. (IV.5)—which is a vector sum, where this is a scalar—although they are still exactly evaluable only for a few simple geometries, and again, more powerful methods of calculation are available.

The constant C is chosen to set the zero of potential energy at a convenient location for a given class of problems. For celestial mechanics $C = 0$ is usually the simplest choice; it implies $U_g \to 0$ "at infinity," i.e., far from all sources. For terrestrial mechanics, the choice $C = +GmM_\oplus/R_\oplus$ yields potential energy, at height h above the *surface* of the Earth,

$$
\begin{aligned}
U_g &= -\frac{GmM_\oplus}{R_\oplus + h} + \frac{GmM_\oplus}{R_\oplus} \\
&= \frac{GmM_\oplus h}{R_\oplus(R_\oplus + h)} \\
&\cong m\frac{GM_\oplus}{R_\oplus^2}\, h \qquad \text{(for } h \ll R_\oplus) \\
&\cong mgh \ , \tag{IV.9}
\end{aligned}
$$

using result (IV.7). The familiar form for motion in small regions near the surface of the Earth is recovered.

Potential energy (IV.8) is a characteristic both of the sources m_s and the receiver m. A scalar field characteristic only of the sources is provided by the *gravitational potential,* i.e., potential energy per unit receiver mass. Like the gravitational field, this is defined as a limit to eliminate the perturbation of the sources arising from the test mass, viz.,

$$
\begin{aligned}
V_g(\mathbf{r}) &\equiv \lim_{m \to 0} \frac{U_g(\mathbf{r})}{m} \\
&= -\sum \frac{Gm_s}{|\mathbf{r} - \mathbf{r}_s|} + V_\infty \ , \tag{IV.10}
\end{aligned}
$$

or the corresponding integral, where result (IV.8) is used for the last expression. The constant V_∞ is the limiting value of V_g "far from all sources"; it can be freely chosen for convenience. *Potential* and *potential energy* are distinct quantities with different units: Potential has units of energy per unit mass, or velocity squared. Analyses of Newtonian gravitation more sophisticated than that presented here usually focus on the potential field, often given the symbol Φ rather than V_g.

Gravitational force and gravitational field are related to potential energy and potential, respectively, by the general relations (II.39), viz.,

$$
\mathbf{F} = -\nabla U_g \tag{IV.11a}
$$

and

$$
\mathbf{g} = -\nabla V_g \ . \tag{IV.11b}
$$

Here, ∇ denotes the gradient vector described following Eq. (II.39b). Given either the vector or the scalar field, one can calculate the other.

B Kepler's Laws of Planetary Motion

Newton's Law of Universal Gravitation supplies the physical principle governing our orbital-mechanics problem. The laws—that is, *descriptions*—of the motions of bodies in space were discovered decades before Newton by Johannes Kepler.[13] Kepler was the research assistant of the Danish astronomer Tycho Brahe[14]—the last great naked-eye astronomer. The raging scientific controversy of that day was the debate over the heliocentric versus geocentric models of the solar system, as described in Ch. I Sec. D previously. Brahe proposed a compromise model in which the planets revolved around the Sun, the entire arrangement moving around the Earth. This, perhaps unsurprisingly, satisfied no one. Brahe's more enduring accomplishment was the compilation of a vast collection of planetary observations[15]—angular positions on the sky as functions of time. Kepler inherited these data upon Brahe's death, and in succeeding years, in breathtaking feats of trigonometric skill, deduced from the data the proper orbits of the planets. Kepler was attempting to determine the three-dimensional trajectories of the planets with only two-dimensional measurements (reliable measurements of distances in the solar system would not be available until the eighteenth century and later) made from a moving platform (the Earth), the trajectory of which Kepler also had to determine. His conclusions are expressed as Kepler's Three Laws of Planetary Motion.

Kepler's First Law

Kepler's First Law—which, apparently, he discovered second—describes the shapes of planetary orbits. Kepler had to break free of the long-standing belief that heavenly bodies must travel on geometrically "perfect" paths, i.e., circles. At one point, he succeeded in constructing a circular orbit for Mars, which matched all the observed positions of the planet with errors no greater than *eight minutes of arc* ($0°.133$) at any point. But Kepler knew that Brahe's data were precise to *four* minutes of arc. His model wasn't good *enough* and had to be discarded. This was a turning point in the emergence of modern science— an hypothesis was discarded for failing to agree with observation to within the known precision of the data. Eventually, by noting the coincidence of certain angles in the planetary data, Kepler was able to deduce a relation for the distance r of a planet from the Sun as a function of angle θ from its position of closest approach or *perihelion*:

$$r(\theta) = \frac{a(1 - \epsilon^2)}{1 + \epsilon \cos\theta} , \qquad (\text{IV.12})$$

where a and ϵ are constants, with $0 \leq \epsilon < 1$. This is the formula, in polar coordinates, of an ellipse with the origin (the Sun) at one focus. But at the

---

[13] Johannes Kepler, 1571–1630

[14] Tycho Brahe, 1546–1601

[15] The King of Denmark gave Brahe a private island on which to set his observatory, one of the first government research grants in history.

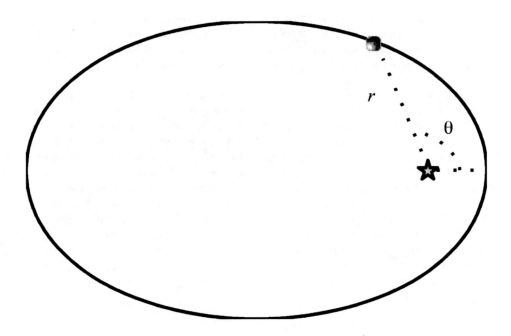

Figure IV.6: A Keplerian elliptical orbit, with the parent body at one focus. The orbits of the planets of the solar system are not this *eccentric*, i.e., elongated.

time, René Descartes had not finished inventing analytic geometry, the branch of mathematics that associates such formulae with geometric figures. Kepler was unable to interpret this result. So he decided to discard it in favor of a new hypothesis: an ellipse, with the Sun at one focus.

Kepler's First Law, then, states:

> The orbit of a satellite is an *ellipse,* with the parent body at one *focus.*

It is phrased this way because it applies, e.g., to the moons of Jupiter or the natural and artificial satellites of the Earth, as well as to the planets of the solar system. A Keplerian orbit is illustrated in Fig. IV.6. The parent body is located at one focus of the ellipse, not at its center. (There is nothing in particular at the other focus.) If the major or long axis of the ellipse has length $2a$ and the minor or short axis $2b$, then the two foci are at distance $c = (a^2 - b^2)^{1/2}$ from the center.[16] The *eccentricity* of the ellipse is $\epsilon = c/a$. A circle is an

---

[16]The sum of the distances from any point on the ellipse to the two foci is equal to $2a$.

ellipse with $\epsilon = 0$; the larger the value of ϵ, with $\epsilon < 1$, the more elongated the ellipse. The eccentricity of the ellipse in Fig. IV.6 is $\epsilon = 0.745$. The orbits of the actual planets are much less eccentric, with values from $\epsilon = 0.007$ for Venus to $\epsilon = 0.206$ for Mercury. The Earth's orbit has eccentricity $\epsilon_\oplus = 0.017$; a graph of the Earth's orbit on the scale of the figure would be difficult to distinguish from a circle.

The First Law can be extended to bodies moving on unbound trajectories. These are portions of hyperbolae, with the parent body at one focus; the trajectory of an object moving barely fast enough to be unbound, i.e., at *escape speed*, is a parabola, again with the parent at the focus of the curve.[17] The ellipse (including the circle), the parabola, and the hyperbola were known centuries before Kepler's time as *conic sections*: They are the curves formed by the intersection with a cone of a plane more horizontal than the generators of the cone, parallel to the generators, and more vertical than the generators, respectively.[18] That this simple family of geometric curves should be just those necessary for describing planetary orbits is an unearned gift of nature to humanity.

The First Law can be derived from Newton's Second Law, with the Law of Universal Gravitation as force law. Both the central and inverse-square features of the force come into play in obtaining Kepler's results.

Kepler's Second Law

Kepler's Second Law deals with the speeds of the planets in their orbits. In his derivation of this law, Kepler made an incorrect physical assumption about planetary speeds, applied an invalid mathematical procedure, miscopied some of Brahe's data, then made an elementary school arithmetic mistake. The errors canceled each other, and Kepler obtained the correct result.[19] Kepler's Second Law states:

> The radius vector from parent to satellite sweeps out area at a constant rate, i.e., sweeps out equal areas in equal times.

The satellite is moving fastest near the parent, slowest far away. The basis of this law is already familiar. The geometry of the sweeping vector is illustrated in Fig. IV.7. The area $\delta\mathcal{A}$ swept out in time interval δt is given by

$$\delta\mathcal{A} = \tfrac{1}{2}r(v\,\delta t)\sin\phi$$
$$= \tfrac{1}{2}|\mathbf{r} \times \mathbf{v}|\,\delta t\,, \qquad\qquad \text{(IV.13a)}$$

---

[17]Gravity does not "snap like a string" and vanish when an object attains escape speed. This is a common misconception.

[18]A lamp with an old-fashioned shade open at the top is handy for demonstrating these curves. If the lamp is tilted at various angles, the pattern of light cast on a wall—the intersection of the wall with the cone of light—displays the various conic sections.

[19]The reader should not be misled by this. One must be very, very good to be that lucky—to be so conversant with every aspect of a problem that one can make such mistakes and yet stay on the right track.

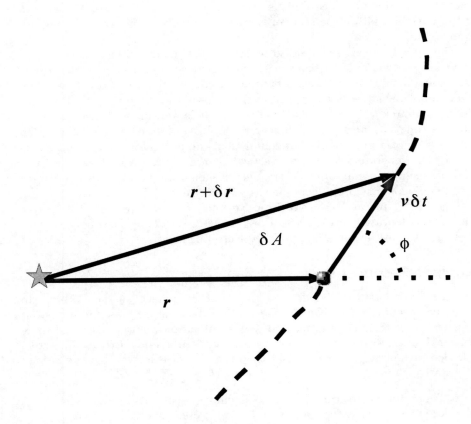

Figure IV.7: Radius vector **r** from parent to satellite sweeps out area $\delta\mathcal{A}$ in time δt. The satellite has velocity **v** along its orbit.

since the first line is the magnitude of the cross product. The instantaneous *rate* at which area is swept out, then, is

$$
\begin{aligned}
\frac{d\mathcal{A}}{dt} &= \tfrac{1}{2}|\mathbf{r} \times \mathbf{v} \\
&= \frac{1}{2m}\,|\mathbf{r} \times \mathbf{p}| \\
&= \frac{|\boldsymbol{\ell}|}{2m}\,,
\end{aligned}
\tag{IV.13b}
$$

with m the mass, $\mathbf{p}$ the momentum, and $\boldsymbol{\ell}$ the angular momentum of the satellite, as per Eq. (III.78a). The sweep rate is constant because *the angular momentum of the satellite is conserved*, because *the central gravitational force exerts no torque on the satellite.* Any central force, then, would give rise to motions obeying Kepler's Second Law.

The Second Law too can be extended, because under the influence of a central force angular momentum is conserved *as a vector.* The constant *direction* of $\boldsymbol{\ell}$ implies that vectors $\mathbf{r}$ and $\mathbf{v}$ lie in a fixed plane, through the parent body and perpendicular to $\boldsymbol{\ell}$. That is, the orbit must be planar.

Kepler's Third Law

Kepler found his Third Law after some further years pursuing the conviction that as the influence of the Sun on the planets must diminish with distance, the rates at which the planets cover their orbits must vary with the sizes of those orbits. Ultimately Kepler's Third Law took the form:

> For all orbits about a given parent body, $T^2 \propto a^3$, i.e., the square of the orbital period T is proportional to the cube of the semimajor axis a of the (elliptical) orbit.

More precisely, the proportionality is

$$
T^2 = \frac{4\pi^2}{GM_*}\,a^3\,,
\tag{IV.14}
$$

where G is Newton's Constant and M_* the mass of the parent body.[20] Kepler knew neither of these quantities, although he would have observed that the complete proportionality constant for the solar system, in yr^2/AU^3—the AU (Astronomical Unit) being the semimajor axis of the Earth's orbit—must be unity. Only after an independent measurement of G, such as that of Cavendish, can the mass of the Sun or other parent body be deduced from this relation. In fact, this is the only means we have to determine the masses of astronomical objects.

---

[20]Even more precisely, a detailed treatment of the motion of two bodies under the influence of a central force takes account of the motion of both bodies. Such an analysis shows that the mass M_* here should be the *total* mass of the two bodies. Hence, the proportionality constant is slightly different for each planet in the solar system, say. But since even Jupiter has mass only 0.955×10^{-3} times that of the Sun, the differences are very small.

Law (IV.14) is readily derived for *circular* orbits by setting the centripetal acceleration (III.35) of the satellite equal to the gravitational acceleration produced by the parent body. A more detailed analysis is required to establish this result for elliptical orbits of any eccentricity. Both the central and inverse-square features of the gravitational force come into play. In fact, result (IV.14) implies that a central force giving rise to such orbits must be inverse-square.

In Newton's day, the question of the consequences of an inverse-square influence had been considered for some time. Apparently, when Edmund Halley put the question to Newton, Newton immediately replied that such a force produces elliptical orbits obeying Kepler's Three Laws. A stunned Halley asked for proofs; Newton, unable to find them among his notes, asked Halley to return the next day. When Halley did, Newton gave him new proofs of these results. Unable to find his originals, Newton had reconstructed the proofs from scratch overnight. The fact that Kepler's Laws, derived from observation, follow as logical consequences from Newton's Laws of Motion and the Law of Universal Gravitation is a discovery of world-transforming impact. It demonstrated that Heaven and Earth were governed by mechanical laws that the human mind could grasp and harness. All the scientific advances that would change Newton's world into ours ultimately flowed from that realization.

C   Orbital dynamics

We now have the tools to tackle the calculation of the LOI burn. First, the speed v_0 of our spacecraft in low Earth orbit, at altitude $h = 200.00$ km above the Earth's surface, say, follows from the requirement that the centripetal acceleration (III.35) be provided by the gravitational field (IV.6) of the Earth. This implies

$$-\frac{v_0^2}{r_0} = -\frac{GM_\oplus}{r_0^2} \, , \tag{IV.15a}$$

with $r_0 = R_\oplus + h$ the radius of the orbit measured from the center of the Earth, which we treat here as a sphere of radius $R_\oplus$. (In actual practice, the Earth and its field would be treated more precisely.) This result gives orbital speed

$$v_0 = \left(\frac{GM_\oplus}{r_0}\right)^{1/2}$$

$$= \left(\frac{GM_\oplus}{R_\oplus + h}\right)^{1/2}$$

$$= \left(\frac{[6.67259 \times 10^{-11} \text{ m}^3/(\text{kg s}^2)](5.9763 \times 10^{24} \text{ kg})}{(6370.949 + 200.00) \times 10^3 \text{ m}}\right)^{1/2}$$

$$v_0 = 7.7902 \text{ km/s} \, . \tag{IV.15b}$$

The precision of this calculation is important. The product $GM_\oplus$ is actually known to some twenty significant figures from measurements of orbiting satellites. The precision to which the mass $M_\oplus$ itself is known is limited, then, by the precision of laboratory measurements of G. Since the altitude h is specified to tens of meters, the radius r_0 is likewise known to that precision, i.e., to six significant figures. The precision of result (IV.15b) could be pushed higher, but this—five significant figures—will do; our approximate treatment of the Earth merits no better. Speed v_0 is determined here to *tens of centimeters per second*.

The *total energy per unit satellite mass* is an important orbit parameter, as it is conserved along the orbit. For objects orbiting the Earth, it is given by

$$\mathcal{E} = \tfrac{1}{2}v^2 - \frac{GM_\oplus}{r} \, , \tag{IV.16}$$

using potential energy (IV.8), with $C = 0$, or gravitational potential (IV.10), with $V_\infty = 0$, and the result that the gravity outside a spherical Earth is the same as that of a point source. For noncircular orbits, speed v and radius r vary along the orbit, but $\mathcal{E}$ does not. The algebraic sign of $\mathcal{E}$ determines the character of the orbit: It is elliptical (a bound orbit), parabolic (defining *escape speed*) or hyperbolic (unbound), when $\mathcal{E}$ is negative, zero, or positive, respectively. For $\mathcal{E} < 0$, radius r cannot grow arbitrarily large. For $\mathcal{E} = 0$, it can: $r \to +\infty$ as $v \to 0^+$. And for $\mathcal{E} > 0$, radius r diverges to $+\infty$ as speed v approaches a positive limit value.

The *angular momentum per unit satellite mass*, i.e., $\mathbf{l} \equiv \boldsymbol{\ell}/m$, is another key, conserved orbit parameter. At the *apsides* of an orbit—the points of nearest and furthest approach to the parent body[21]—and only there, its magnitude is related to satellite radius (from the parent body) and speed by

$$
\begin{aligned}
l &= r_1 v_1 \\
&= r_2 v_2 \, ,
\end{aligned}
\tag{IV.17}
$$

where subscripts 1 and 2 denote the two apsides of an elliptical orbit. Hence, the energy per unit mass, angular momentum per unit mass, and apsidal radii are related by

$$
\begin{aligned}
\mathcal{E} &= \frac{l^2}{2r_1^2} - \frac{GM_\oplus}{r_1} \\
&= \frac{l^2}{2r_2^2} - \frac{GM_\oplus}{r_2}
\end{aligned}
\tag{IV.18}
$$

for an orbit of the Earth, with similar expressions for an orbit of any other parent body.

There is a third conserved quantity peculiar to the inverse-square central force. For a satellite of the Earth, it is the vector

$$
\begin{aligned}
\boldsymbol{\mathcal{M}} &= \frac{\mathbf{p} \times \boldsymbol{\ell}}{m} - \frac{GM_\oplus m}{r}\mathbf{r} \\
&= m\left(\mathbf{v} \times \mathbf{l} - \frac{GM_\oplus}{r}\mathbf{r} \right) \, ,
\end{aligned}
\tag{IV.19}
$$

---

[21] These have names: perigee and apogee for orbits of the Earth; perihelion and aphelion for orbits of the Sun; periastron and apastron for orbits of any other star; pericynthion and apocynthion for orbits of the Earth's Moon; pericenter and apocenter in general.

directed along the major axis of the orbit. This is called the Laplace[22]-Runge[23]-Lenz[24] (LRL) vector, the Runge-Lenz vector, or the Lenz vector, depending on whom one wishes to credit. Its conservation corresponds to the fact that a Keplerian orbit maintains a fixed orientation in space and is a closed curve.

The LOI rocket-engine burn is a brief application of thrust in the direction of the spacecraft's motion in its circular orbit. The thrust increases the speed, energy (because it does positive work), and angular momentum (because it applies torque about the center of the Earth) of the spacecraft to the values corresponding to the elliptical transfer orbit, as illustrated in Fig. IV.2. The radial velocity of the spacecraft is still zero at the burn point, so this becomes the perigee of the new orbit. At the apogee of the transfer orbit, a second burn could be used to boost the spacecraft into the (nearly) circular orbit of the Moon. In practice, the maneuver was used to bring the spacecraft to the orbit of the Moon when the Moon was at that point. It was necessary to incorporate the effects of the Moon's gravity into the calculations and maneuver the spacecraft into an orbit around the Moon. Such calculations require numerical analysis[25] beyond the scope of this text. We shall consider only the boost onto the transfer orbit to the Moon. Substituting the first of expressions (IV.17) into Eq. (IV.18) yields

$$\tfrac{1}{2}v_1^2 - \frac{GM_\oplus}{r_1} = \tfrac{1}{2}v_1^2\,\frac{r_1^2}{r_2^2} - \frac{GM_\oplus}{r_2}\;. \tag{IV.20a}$$

This gives the speed v_1 on the transfer orbit at its perigee, viz.,

$$
\begin{aligned}
v_1 &= \left(\frac{2GM_\oplus}{r_1}\,\frac{r_2}{r_1 + r_2}\right)^{1/2} \\
&= (7.7902\ \text{km/s})\left(\frac{2(3.84400\times 10^5\ \text{km})}{(3.84400\times 10^5 + 6570.95)\ \text{km}}\right)^{1/2} \\
&= 10.924\ \text{km/s}\;.
\end{aligned}
\tag{IV.20b}
$$

Here r_1 is the radius r_0 of the original circular orbit, as in Eq. (IV.15b), and r_2 is the radius of the Moon's orbit. (The mean value is used here.) The change in speed Δv (called "Delta Vee" in documents of the time) required of the LOI burn is

$$
\begin{aligned}
\Delta v &= v_1 - v_0 \\
&= (10.924 - 7.7902)\ \text{km/s} \\
&= 3.134\ \text{km/s}\;,
\end{aligned}
\tag{IV.20c}
$$

where the subtraction has cost us a significant figure. The need for precision is clear: Taking the limit $r_2 \to +\infty$ in Eq. (IV.20a) or (IV.20b) yields *escape*

---

[22] Pierre-Simon de Laplace, 1749–1827
[23] Carl David Tolmé Runge, 1856–1927
[24] Wilhelm Lenz, 1888–1957
[25] This was done with 1960s-era computing technology!

speed

$$v_{\text{esc}} = \left(\frac{2GM_\oplus}{r_1} \right)^{1/2}$$
$$= 2^{1/2}(7.7902 \text{ km/s})$$
$$= 11.017 \text{ km/s} , \qquad \text{(IV.20d)}$$

using the result of calculation (IV.15b). This corresponds to a parabolic trajectory that does not return to Earth. To three significant figures, speeds v_1 and v_{esc} are indistinguishable, but the difference between them is life and death.

We can also calculate the appropriate time or "launch window" for our LOI burn. It would be embarrassing to send a spacecraft to the orbit of the Moon, only to find that the Moon was elsewhere in its orbit when the spacecraft arrived. The transit time from low Earth orbit to the orbit of the Moon (neglecting, for simplicity, the gravitational influence of the Moon) is half the period of the transfer orbit, as given by Kepler's Third Law (IV.14), viz.,

$$\Delta t = \tfrac{1}{2}T$$
$$= \pi \left(\frac{(r_1 + r_2)^3}{8GM_\oplus} \right)^{1/2}$$
$$= \pi \left(\frac{[(6570.95 + 3.84400 \times 10^5) \times 10^3 \text{ m}]^3}{8[6.67259 \times 10^{-11} \text{ m}^3/(\text{kg s}^2)](5.9763 \times 10^{24} \text{ kg})} \right)^{1/2}$$
$$= 4.2999 \times 10^5 \text{ s}$$
$$= 4.9767 \text{ days} , \qquad \text{(IV.21)}$$

since the semimajor axis of the transfer orbit is $a = \tfrac{1}{2}(r_1 + r_2)$ and one day is 86400. s. The period of the Moon is $T_{\text{Moon}} = 27.321661$ days—the precision is real. If the Moon's orbit were circular—an approximation far less accurate than the precision of that datum—then the LOI burn would have to be made when the Moon was at angle

$$\Delta\theta = 360° \frac{\Delta t}{T_{\text{Moon}}}$$
$$= 360° \frac{4.9767 \text{ days}}{27.321661 \text{ days}}$$
$$= 65°.575 \qquad \text{(IV.22)}$$

in its orbit away from the proposed rendezvous point, again disregarding the gravitational effects of the Moon itself on the approaching spacecraft. The factor $360°$ in this calculation is *exact*, i.e., of arbitrary precision; the result is good to five significant figures as shown.

The maneuver we have examined here—a brief tangential acceleration to shift from a circular orbit to the pericenter of an elliptical orbit to reach a more distant circular orbit—is called a *Hohmann*[26] *Transfer*. It was first described by

---

[26] Walter Hohmann, 1880–1945

Hohmann in 1925 but not actually performed until the first Soviet Luna mission in 1959. It is the least energy-consuming way to move from one circular orbit to a higher circular orbit around a given parent body. Now it is a standard technique for interplanetary as well as lunar spaceflight. The trajectories of interplanetary spacecraft, however, can be very elaborate, often involving close encounters with several planetary bodies. The calculation of such orbits requires more sophistication than we can attain here.

D   Einsteinian gravitation

Newton's Law of Universal Gravitation remains useful for many applications. But our current understanding of gravitation is a considerable enhancement of Newton's. It is based on Albert Einstein's 1915 *General* Theory of Relativity. A logical starting point for this theory is a question left unanswered by the Galilean relativity described in Ch. I Sec. D and Einstein's Special Theory, introduced in Ch. I Sec. H: If all frames of reference related to an inertial reference frame by Galilean, or by Lorentz, transformations are physically equivalent inertial frames, how does one identify that first inertial frame? A frame tied to the Earth will not do—the Earth rotates and revolves around the Sun, both accelerated motions. The traditional answer is a reference frame at rest, or moving uniformly, with respect to "the distant stars" or "the fixed stars." But we have known since the early twentieth century that the stars are arrayed in great wheeling galaxies, and since 1929 that the galaxies themselves are flying away from one another—which, then, are these "fixed stars"? Put another way: Consider two frames of reference—sets of observers—moving *non*uniformly with respect to one another. In an otherwise empty universe, how would one decide, for dynamical purposes, which frame (if either) was inertial and which accelerated? Einstein's resolution of this long-standing conundrum was to "cut the Gordian[27] knot": He declined to make the distinction. The laws of physics, Einstein proposed, should properly be formulated so as to apply in *any* frame of reference.

Of course, there is a price to be paid for such a "democratization" of the notion of reference frame. The corresponding mechanics must include inertial forces, as in Eq. (I.81b). The effects of such forces are manifest in some frames. Hence, Einstein realized, mechanics must also incorporate gravitation, *because gravitation is also an inertial force*. This insight is known as the *Principle of Equivalence of Gravitation and Inertia* (often just *Principle of Equivalence*). It is usually considered the foundation of Einstein's General Theory.

How can gravitation—a "real" force—be an inertial force? An inertial force has two unfailing signatures: First, it is always proportional to the inertial mass of the body upon which it acts, since it is an $m\mathbf{a}$ term in Newton's Second Law moved over to the $\mathbf{F}$ side. Gravitation is proportional to the *passive gravitational* mass of its object—*which just happens to be equal to inertial mass* to better than parts in 10^{13}. There is no staggering coincidence to explain from Einstein's

---

[27]Gordius, King of Gordium, ninth or eighth century BCE?

viewpoint: The two measures of mass are equal because they are, in fact, the same. The second signature of an inertial force is that it can be eliminated simply by switching to an inertial reference frame. Can gravity be eliminated by changing reference frames? Yes, it can. We have already seen, in results (I.42), that gravity vanishes from the description of projectile motion as seen from another projectile. Perhaps the most famous illustration of this is Einstein's Elevator, sketched in Fig. IV.8. With the elevator in free fall, the observer will register zero weight if he steps on a scale. The apples he has placed about him will not drop to the floor. It is as if he were in free space, far from any mass.[28] From Einstein's perspective, it is such freely falling reference frames that are truly inertial, in the sense that truly free bodies move in straight lines at constant speed.

But now we are well and truly caught: If the new mechanics must include gravitation, it cannot be encompassed within the "flat" spacetime geometry Hermann Minkowski introduced to describe the kinematics and dynamics of the Special Theory of Relativity. Gravitation produces effects that reveal that the geometry of spacetime must be curved. For example, if two imaginary observers were to station themselves, say, ten kilometers apart on the Earth's equator and march due north—heedless of the mountains, jungles, oceans, et cetera, in the way—they would find themselves converging, accelerating toward one another, even though neither deviated from their meridians of longitude, the straightest possible tracks they could walk on the surface of the Earth. Eventually, they would collide at the North Pole. They could invoke an attractive force proportional to their inertial masses, or they could acknowledge that the surface of the Earth is curved. They do not need to travel into three-dimensional space to see this—ant observers could do it. Now the apples placed by the observer in the elevator in Fig. IV.8 do not remain motionless as the elevator falls. The apple below the observer's feet falls—accelerates downward, according to observers at rest in the building—slightly faster than the observer does, as it is closer to the center of the Earth. The observer will see it slowly accelerate downward. The apple above his head is slightly farther from the center of the Earth; he will see it slowly accelerate upward. All are converging radially toward the center of the Earth, so the observer will see the apples at his sides accelerate slowly toward him. The pattern of apples is squeezed horizontally, stretched vertically, though the apples are in free fall. (This is the same effect, entirely within the compass of Newtonian mechanics, that creates the tides in the Earth's oceans.) This is the same phenomenon—called *geodesic deviation*—as the covergence of our northbound observers' meridians. The latter signals the curvature of the Earth; the former signals the curvature of spacetime.

The understanding of gravitation conveyed by Einstein's General Theory, then, is as the interaction between matter and the geometry of spacetime. That geometry determines the trajectories on which bodies, particles, even light rays move; Einsteinian orbital mechanics derives from this. The geometry of space-

---

[28]Since Einstein's time, it has become possible to perform this experiment in orbiting spacecraft, which are in free fall around the Earth. Any given observer can perform the actual elevator experiment only once.

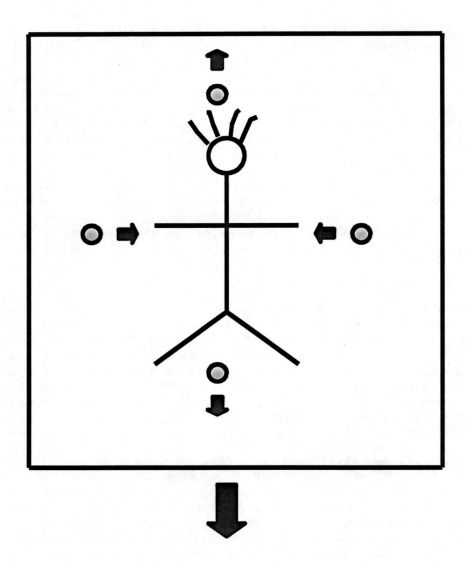

Figure IV.8: Einstein's Elevator. With the elevator in free fall, the observer feels no effects of gravity. His apples do not fall to the floor. The gradual (tidal) shifts in their positions reveal the curvature of spacetime.

time, in turn, is determined by its content—by the energy, momentum, mass, and force densities that are the sources of gravitation. The connection is expressed via the Einstein Field Equations, an equality of tensors linking certain curvature measures with the source densities. These are ten partial differential equations for ten functions of the four dimensions of spacetime. Written out completely, with no simplifying symmetries in play, the Field Equations contain some 44,000 terms. Until quite recently, simply identifying a single new solution of the equations was a publishable result. But for all its potential complexity, Einstein's theory has revealed to us features of nature far beyond the reach of Newton's description.

General relativity agrees with Newtonian gravitation in the limits in which Newton's laws give correct answers—where speeds are small compared to the speed of light and gravitational fields are weak enough not to accelerate objects to greater speeds than that. But Einstein's description provides corrections to Newton's. It took decades after Einstein announced his theory to make measurements of sufficient sensitivity to confirm his predictions for bodies within the solar system. Now technology has advanced to the point that Einstein's corrections *must* be included for predictions to match measurements. The Global Positioning System (GPS), which determines positions via timing measurements of signals from orbiting satellites, would not work to the precision it does if GPS units did not include Einsteinian effects in their programming.

General relativity also works in realms where Newtonian physics simply does not. It allows us to describe black holes—collapsed stellar cores so compact their spacetime curvature can trap light—accurately. (Black holes were actually introduced in the context of Newtonian physics by John Michell[29] in 1784, but the details were all wrong.) We now know such objects are actually found in astronomical x-ray sources and in the cores of galaxies. General relativity also includes a feature Newton's Law of Universal Gravitation does not: time dependence. It describes the propagation of *changes* in gravitational fields through space as waves. The effects of these waves on binary star systems have been observed and are in accord with the predictions of Einstein's theory. The direct detection of the waves is an ongoing scientific effort. Finally, general relativity affords us a description of the structure and dynamics, origin and evolution, of the universe as a whole—an understanding that eluded Newton. Theoretical and observational exploration of cosmology, the study of the universe as a whole, is currently one of the hottest areas in science. For the details of these fruits of the study of gravitational physics, the reader must explore more advanced texts.

References

Newton's Law of Universal Gravitation and orbital mechanics are foundational topics covered in detail in physics texts at every level. Orbital mechanics, also

---

[29] John Michell, 1724–1793

called *astrodynamics,* is the subject of specialized texts such as those of Herrick [Her71, Her72] and Bate, Mueller, Saylor, and White [BMSW13].

The Hohmann transfer maneuver is described in the intermediate-level texts of Barger and Olsson [BO95], Taylor [Tay05], and Thornton and Marion [TM04]. That fifth edition of Thornton and Marion devotes an entire section to the Hohmann transfer, although the second edition (published by Marion in 1965) does not mention it—interplanetary space travel was not commonplace back then.

The General Theory of Relativity is only briefly introduced in introductory, intermediate, and even advanced physics texts. Specialized texts at the undergraduate level, such as those of Hartle [Har03], Rindler [Rin06], and Schutz [Sch09], are now available. Among advanced texts, Misner, Thorne, and Wheeler (MTW) [MTW73] is renowned; Adler, Bazin, and Schiffer [ABS75], Landau and Lifshitz [LL80], Wald [Wal84], and Weinberg [Wei72] are also authoritative.

The full story of Tycho Brahe, Johannes Kepler, and the discovery of the Laws of Planetary Motion probably could not be sold in Hollywood as a movie script—it's too melodramatic. (In 2010, Brahe's body was exhumed from his tomb in Prague Cathedral to use modern methods to test whether Brahe was poisoned by Kepler for his data collection.[30] Apparently, insufficient poison was found to substantiate a murder charge.) One account is provided by Arthur Koestler in *The Sleepwalkers* [Koe90]. Excerpts from this and many other sources can be found in the remarkable anthology edited by Young [You71].

Problems

1. A satellite of mass m orbits a parent body of mass M_* in a circular orbit of radius r_0. Calculate the ratio of the satellite's kinetic energy to its potential energy. Take the zero of potential energy to be at infinity.

2. Consider two bodies of masses m_1 and m_2, with positions $\mathbf{r}_1(t)$ and $\mathbf{r}_2(t)$, respectively. The bodies move under the influence of their mutual (Newtonian) gravitational attraction; no other force is in play.

   (a) Write the (differential) equations of motion for $\mathbf{r}_1$ and $\mathbf{r}_2$ in terms of the given masses and positions and Newton's Gravitational Constant.

   (b) Using results (a), write differential equations of motion for the center-of-mass position $\mathbf{r}_{cm}$ and the relative position $\mathbf{r} \equiv \mathbf{r}_2 - \mathbf{r}_1$ of the two bodies. Solve the first of these. Characterize the solutions of the second.

3. **Space Chase: Plan A.** You are traveling in a small spacecraft in a circular orbit about the Earth. You wish to effect a rendezvous—a challenging maneuver—with another spacecraft in the same orbit, but 180° around the

---

[30]The modern image of scientists as unkempt but mild-mannered nebbishes is rubbish. For much of his life, Brahe wore a brass nose—he let people think it was gold—because the original had been lost in a sword duel.

orbit from you. The quickest way to do this is to apply brief braking thrust opposite to your motion, putting you at the apogee of a elliptical orbit with half the period of your original orbit. When you have completed this orbit, the other spacecraft will have completed half an orbit to meet you. (You will then have to match speed with it, a separate maneuver.) The perigee of your elliptical orbit must not be lower than 100.00 km above the Earth's surface, or you will crash. Calculate the minimum radius and altitude your original circular orbit must have for this maneuver to work. Treat the Earth as a sphere of radius equal to its mean radius $R_\oplus$.

4. **Space Chase: Plan B.** If your original circular orbit is too low to use the rendezvous maneuver of Plan A, an alternative is to apply brief forward thrust, putting you at the perigee of an elliptical orbit with period three-halves that of the original orbit. Calculate the radius of the apogee of your new orbit. If your original orbit had altitude between 100.00 km and the limit calculated in the preceding problem, in what range of values will this apogee radius lie?

5. The geometric features of a Keplerian orbit are determined by its dynamical features, and vice versa. Relation (IV.18) can be solved for the apsidal radii of the orbit in terms of its energy per unit mass and angular momentum per unit mass. (Note that these are properties of the orbit, not of any particular satellite.) Using these solutions, find the semimajor axis a and eccentricity ϵ of the orbit in terms of $\mathcal{E}$ and l.

6. Prove that for an object moving in a Keplerian orbit around the Earth, the LRL vector (IV.19) has zero time derivative, i.e., it is conserved. This is a vigorous exercise in dynamics and calculus.

7. Consider the LOI maneuver analyzed in Sec. C of the text. Suppose the spacecraft has final mass 100.0 metric tons after the burn. (The Apollo spacecraft had mass 43. metric tons, not including the third stage of the Saturn V launch vehicle used for its LOI burn.) If the rocket engine used is an advanced design with nozzle speed $U = 4.00$ km/s, what fuel mass is required for the LOI burn?

8. **Space Chase: Fuel Economy.** Consider the rendezvous maneuvers of problems 3 and 4 preceding. Assuming that the original orbit radius r_0 is large enough that both maneuvers are feasible, calculate and compare the fuel mass required for each maneuver in terms of given spacecraft (final) mass M_f, radius r_0, and rocket engine exhaust speed U.

9. Consider a Hohmann Transfer maneuver such as described in this chapter, but to take a spacecraft from the Earth's orbit around the Sun to that of Mars. Calculate the Delta Vee required and the fuel mass needed for final spacecraft mass 250. metric tons (an Apollo-sized spacecraft would not be adequate for a trip to Mars), with engine exhaust speed $U = 4.00$ km/s. Calculate also the transit time from the orbit of the Earth to that of Mars.

Disregard motion relative to the Earth itself, i.e., take initial speed v_0 to be that of the Earth's (approximately circular) orbit around the Sun. You need not consider here the maneuvering required upon reaching the vicinity of Mars, though that would have to be considered for an actual voyage.

Chapter V

The Harmonic Oscillator

- *Solve the Simple Harmonic Oscillator Initial Value Problem.*
- *Understand the Damped Harmonic Oscillator.*
- *Analyze the Driven Oscillator and Resonance.*
- *Examine Coupled Oscillators and Normal Modes.*

There is no cute title or clever project for this chapter, because the harmonic oscillator itself is of such surpassing importance. The harmonic oscillator, such as the mass-and-spring system shown in Fig. V.1, is to physics what the hydrogen atom is to chemistry: the elementary system through which a great many secrets are revealed.

A   Universality of the harmonic oscillator

One reason the harmonic oscillator—a system governed by a Hooke-Law force, such as that of Eq. (II.33a)—is of such importance is its universality: *Almost every physical system with a stable equilibrium configuration will behave as an harmonic oscillator if disturbed slightly from that configuration.* The phrase "almost every" in that statement is code language familiar from mathematics. It means every such system except for "a set of measure (i.e., probability) zero." Consider, for example, a body moving in one dimension, subject to force $F(x)$, with a stable equilibrium point (see Ch. II Sec. E) at position x_{eq}. A remarkable and far-reaching discovery of Maclaurin[1] and Taylor[2] reveals that any well-behaved function can be approximated by a sequence of polynomials, the coefficients of which are the derivatives of the function:

$$F(x) = F(x_{eq}) + \frac{dF}{dx}(x_{eq})\,(x - x_{eq}) + \tfrac{1}{2}\frac{d^2F}{dt^2}(x_{eq})\,(x - x_{eq})^2 + \cdots \; ; \quad \text{(V.1a)}$$

the approximation can be made as accurate as desired by including enough terms. Such a sequence is called a Taylor series[3] (a Maclaurin series in the case $x_{eq} = 0$). But here, x_{eq} is an equilibrium point, at which the force is zero: $F(x_{eq}) = 0$. And it is a *stable* equilibrium point: If the body is disturbed slightly from equilibrium, the force should accelerate it back toward x_{eq}. Hence, the coefficient of the first-order term must be negative: $dF/dx(x_{eq}) = -k$, with k a positive constant, as this term will dominate over the others if $x - x_{eq}$ is small enough. The force law takes the form

$$F(x) = -k(x - x_{eq}) + \dots \, , \quad \text{(V.1b)}$$

and there must be some interval about x_{eq} in which the first-order (Hooke-Law) term dominates over all the others. The only exceptions to this behavior would arise in cases in which the first and second derivatives of F, i.e., the coefficients of the first- and second-order terms, were *exactly* zero, the "set of measure zero" to which the "almost every" in the opening statement of this section refers.

Hence, for displacement $\Delta x \equiv x - x_{eq}$ sufficiently small—where "sufficiently" is determined by the coefficients in expansion (V.1a) and the accuracy desired—

---

[1]Colin Maclaurin, 1698–1746

[2]Brook Taylor, 1685–1731

[3]This is often the least popular, but most important, part of the second-term calculus course. Without it, mathematical physics would have ground to a halt in the nineteenth century once all the exactly soluble problems were done.

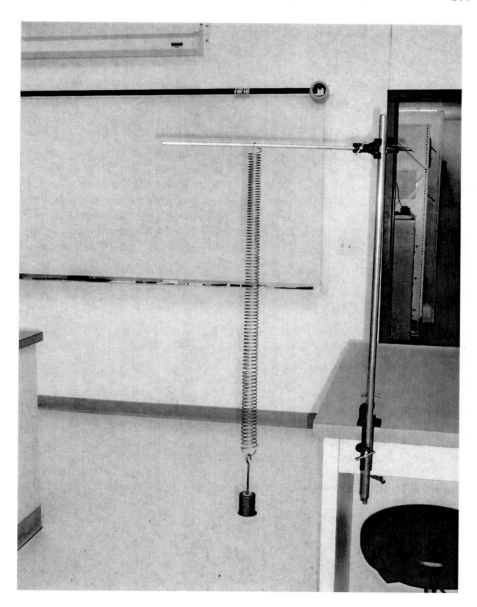

Figure V.1: Does the sight of this apparatus send the reader into transports of delight? It should, for if one understood everything the harmonic oscillator has to reveal, that would be about three-quarters of physics. (Photograph by author.)

Newton's Second Law implies for this system the equation of motion

$$\frac{d^2\Delta x}{dt^2} + \omega_0^2\,\Delta x = 0\ ,\qquad\qquad\text{(V.2a)}$$

with

$$\omega_0 \equiv \left(\frac{k}{m}\right)^{1/2}\ ,\qquad\qquad\text{(V.2b)}$$

where m is the mass of the body. This—the differential equation (V.2a)—is the *simple harmonic oscillator equation*. It is the second-most important of all differential equations.[4]

The classic example of such a system is the mass and spring shown in Fig. V.1. In this case, the spring obeys the Hooke Law for quite large displacements from equilibrium. If the mass of the spring itself can be treated as negligible compared to that of the suspended mass m, then Newton's Second Law implies that the vertical position y of the mass satisfies

$$m\,\frac{d^2y}{dt^2} = -ky - mg\ ,\qquad\qquad\text{(V.3a)}$$

where y is taken positive upward, with origin $y = 0$ at the unstretched and uncompressed position of the spring. This implies equation of motion

$$\frac{d^2\Delta y}{dt^2} + \omega_0^2\,\Delta y = 0\ ,\qquad\qquad\text{(V.3b)}$$

with

$$\Delta y \equiv y - \frac{-mg}{k}\ ,\qquad\qquad\text{(V.3c)}$$

and ω_0 as in Eq. (V.2b). This is the same as Eq. (V.2a); the only effect of gravity here is to shift the equilibrium point to $y_{\text{eq}} = -mg/k$. The actual apparatus shown in Fig. V.1 has spring constant $k = 10.0$ N/m and mass $m = 0.550$ kg, hence,

$$\begin{aligned}
\omega_0 &= \left(\frac{k}{m}\right)^{1/2}\\[4pt]
&= \left(\frac{10.0\ \text{N/m}}{0.550\ \text{kg}}\right)^{1/2}\\[4pt]
&= 4.26\ \text{s}^{-1}\ .
\end{aligned}\qquad\qquad\text{(V.3d)}$$

The significance of this will be made clear shortly.

Another famous example is the simple pendulum sketched in Fig. V.2. If the weight of the shaft (or string) is negligible, then the weight of the bob alone exerts restoring torque $\tau = -mg\ell\sin\theta$ about the suspension point. The rotational

---

[4]First place goes to the equation $dx/dt = \lambda x$, with λ constant, for exponential growth or decay. It is first because of the great variety of contexts in which it appears.

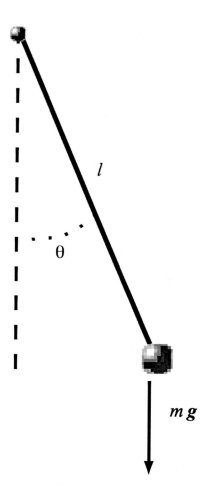

Figure V.2: A simple pendulum of length l. The weight of the bob exerts a restoring torque about the suspension point.

form of Newton's Second Law then implies that the angular displacement θ satisfies

$$m\ell^2 \frac{d^2\theta}{dt^2} + mg\ell \sin\theta = 0 \; , \tag{V.4a}$$

or

$$\frac{d^2\theta}{dt^2} + \omega_0^{(\mathrm{p})2} \sin\theta = 0 \; , \tag{V.4b}$$

with

$$\omega_0^{(\mathrm{p})} \equiv \left(\frac{g}{\ell}\right)^{1/2} \; . \tag{V.4c}$$

This is *not* the simple harmonic oscillator equation. However, for small angular displacements $|\theta| \ll 1$ (in radians), the sine function satisfies $\sin\theta \cong \theta$. (This is the first term in the Maclaurin expansion for $\sin\theta$.) Hence, the pendulum is described by the *approximate* equation of motion

$$\frac{d^2\theta}{dt^2} + \omega_0^{(\mathrm{p})2} \theta \cong 0 \; , \tag{V.4d}$$

i.e., it behaves approximately as a simple harmonic oscillator for small angular displacements.

B   The Initial-Value Problem

The harmonic oscillator illuminates many features of an important class of problems known as *initial-value problems*. These call for determining the future behavior of a system given information about its present state. Initial-value problems constitute a major component of the mathematics of differential equations.

The simple harmonic oscillator equation of motion (V.2a) is an *ordinary differential equation* (ODE), because it involves one independent variable—the time t. It is a *second-order* ODE, because the highest derivative in it is the acceleration, the second derivative of Δx. And it is a *linear* second-order ODE, because the dependent variable Δx appears only to the first power in each term. It has an infinity of solutions, because any constant multiple of a solution, or linear combination of different solutions with constant coefficients, is another solution. But if we supplement the equation of motion with the two *initial conditions*

$$\Delta x(0) = x_0 \tag{V.5a}$$

and

$$\frac{d\Delta x}{dt}(0) = v_0 \; , \tag{V.5b}$$

for any given *initial values* x_0 and v_0, then the problem has exactly one—one and only one—solution. The number of initial conditions is equal to the order of the ODE. This is not accidental.

Mathematicians have elegant theorems to establish that an initial-value problem posed this way has a solution—"existence theorems"—and that there is only one—"uniqueness theorems." We shall leave these to texts on differential equations and consider a simple illustration. Suppose we knew nothing about differentiation or integration except what a derivative means. We could write a computer algorithm to solve the initial-value problem by generating a sequence of values x_i and v_i for the function $\Delta x(t)$ and its time derivative, so:

$$x_1 = x_0 + v_0 \Delta t$$
$$v_1 = v_0 + a_0 \Delta t$$

$$x_2 = x_1 + v_1 \Delta t$$
$$v_2 = v_1 + a_1 \Delta t$$

$$\vdots$$

$$(V.6)$$

where the accelerations a_i at each stage are obtained from the equation of motion and the position and velocity at that stage. Computer scientists and numerical analysts would gag at this: It's crude and inefficient; much more effective algorithms are available. But it would work and could be made as accurate as desired by taking the time step Δt small enough. As long as we have initial values x_0 and v_0 to start the program, it will generate a unique solution.

This mathematics has a profound implication: *The initial position and velocity of any system with a second-order equation of motion determine its future behavior for all time.* (They also determine its past behavior. The formalism does not distinguish between the two. What, if anything, does is a separate issue.) The logical consequences of this are often associated with the French mathematician Pierre-Simon de Laplace and are called *Laplacian determinism*: Do the present positions and velocities of all the particles in the universe, plus Newton's Laws, determine the future of everything for all time? If so, then what is free will? Do human beings actually make any choices? If we do not, then why do we regret some actions and condemn others if all are simply the workings of mechanical laws? It does not matter whether this author or the reader can do the calculations or not. The solutions exist if the initial values exist.

Two possible resolutions are usually offered for this dilemma. One relies on *chaos,* or more properly *deterministic dynamics*: Systems with nonlinear equations of motion can exhibit *exponential sensitivity* to their initial data. That is, trajectories of the system that are initially close together diverge exponentially in time. Predicting the behavior of the system over time, then, requires impossible precision in the initial data; a Laplacian calculation of the future of everything is out of reach. (The weather is the classic example of a physical system that behaves this way.) The other resolution is more radical: It invokes quantum mechanics, which asserts that only half of the required initial data actually exist—an implementation of the Heisenberg Uncertainty Relations. Quantum mechanics does not admit the calculation of trajectories of bodies, rather of *wave functions* from which probabilities of future outcomes are extracted. As a matter of *opinion,* it is not clear to this author whether or not either of these resolutions suffices to "save" the notions of free will and responsibility. Classical chaotic systems are, after all, still *determined*: Even if we cannot calculate their future behavior to sufficient precision, the solutions of their equations of motion still exist. The quantum resolution still allows for the calculation of future probabilities. Are we then not free to choose outcomes, the future probabilities of which—calculated now—are zero? What is clear is that the ideas of physics have a far wider sweep than is commonly appreciated.

C   Solving the *linear* SHO IVP

That's the Simple Harmonic Oscillator Initial-Value Problem—the alphabet soup flows like wine. It is not assumed in this text that the reader has completed an introductory Differential Equations course. Such techniques as are needed are developed here.

Solution of the IVP

The linearity of equation of motion (V.2a) greatly simplifies the task. If we can find two functions $x^{(1)}(t)$ and $x^{(2)}(t)$ that satisfy that differential equation and are *independent* (i.e., not constant multiples of each other), then a solution of the form $\Delta x(t) = ax^{(1)}(t) + bx^{(2)}(t)$, with a and b constant, can be made to satisfy initial conditions (V.5). That is, those conditions would constitute two (linear) equations for the two unknown coefficients a and b, a guaranteed win. The result must be *the* correct solution of the initial-value problem, since the existence and uniqueness theorems assure us that there is one and only one.

For all the wonderful array of techniques one might learn in a Differential Equations course, there are fundamentally only two ways to solve a differential equation: integrate or guess. In Ch. I Sec. E we integrated; here we shall guess. Specifically, we shall try a function of the form $x(t) = e^{\lambda t}$, where λ is a constant. We do not guess blindly: The behavior of the exponential function enables us to transform the *differential* equation (V.2a) into an *algebraic* equation we can solve. We obtain

$$\frac{d^2x}{dt^2} + \omega_0^2\, x = (\lambda^2 + \omega_0^2)\, e^{\lambda t}$$
$$= 0 \ . \tag{V.7}$$

Failure? Hardly: $\lambda = \pm i\omega_0$, where $i = \sqrt{-1}$ is the "imaginary unit," will give us two independent solutions of the differential equation.

Exponentials of imaginary or complex arguments are interpreted via a remarkable identity known as the *Euler*[5] *relation*:

$$e^{i\theta} = \cos\theta + i\sin\theta \ . \tag{V.8}$$

This can be verified by noting that both sides solve the first-order initial-value problem $df/d\theta = if$, with $f(0) = 1$; as such a problem has a unique solution,

---

[5]Leonhard Euler, 1707–1783

the two sides must be equal. Or readers in the know can write the Maclaurin series for $e^{i\theta}$, separate out the real and imaginary terms, and recognize the Maclaurin series for $\cos\theta$ and $\sin\theta$, respectively. The Euler relation implies[6] that any complex number $z = x + iy$ can be written in the polar form

$$z = \rho\, e^{i\theta} \,, \tag{V.9a}$$

with

$$\rho = (x^2 + y^2)^{1/2} \tag{V.9b}$$

and

$$\theta = \tan^{-1}(y/x) \,. \tag{V.9c}$$

This form is particularly useful for multiplying and taking powers of complex numbers.

It is convenient to choose for our two independent solutions the real combinations

$$
\begin{aligned}
x^{(1)}(t) &= \tfrac{1}{2}\left(e^{+i\omega_0 t} + e^{-i\omega_0 t}\right) \\
&= \cos(\omega_0 t)
\end{aligned}
\tag{V.10a}
$$

and

$$
\begin{aligned}
x^{(2)}(t) &= \tfrac{1}{2i}\left(e^{+i\omega_0 t} - e^{-i\omega_0 t}\right) \\
&= \cos(\omega_0 t) \,.
\end{aligned}
\tag{V.10b}
$$

As linear combinations of solutions with constant coefficients, these too are solutions of the differential equation.

The *general solution* of the simple harmonic oscillator equation of motion is

$$\Delta x(t) = a\,\cos(\omega_0 t) + b\,\sin(\omega_0 t) \,, \tag{V.11}$$

with a and b arbitrary constants. Initial conditions (V.5) then take the forms $x_0 = a$ and $v_0 = \omega_0 b$, so the solution of the full initial-value problem is

$$\Delta x(t) = x_0\,\cos(\omega_0 t) + \frac{v_0}{\omega_0}\,\sin(\omega_0 t) \,. \tag{V.12}$$

This gives the trajectory of the oscillator for all time given any initial values x_0 and v_0.

Using elementary trigonometric identities, we can transform this solution into the alternative form

$$\Delta x(t) = A\,\cos(\omega_0 t - \delta) \,, \tag{V.13a}$$

with

$$A = \left(x_0^2 + \frac{v_0^2}{\omega_0^2}\right)^{1/2} \tag{V.13b}$$

---

[6] It also implies the delightful equation $e^{i\pi} + 1 = 0$, which contains the Five Most Important Numbers and no others.

and

$$\delta = \tan^{-1}\left(\frac{v_0}{\omega_0 x_0}\right) , \qquad\qquad \text{(V.13c)}$$

where the appropriate quadrant for δ is determined by the algebraic signs of x_0 and v_0. The features of the oscillator's motion are apparent from this form. This is periodic motion, with period $T_0 = 2\pi/\omega_0$ and frequency $\nu_0 = \omega_0/(2\pi)$. (The parameter ω_0 is the *angular frequency* of the oscillator, measured in s^{-1}, while the *frequency* ν is measured in Hz. This is analogous to the usage for uniform circular motion shown in Ch. III Sec. D.) These are dynamical features independent of the initial conditions. For example, the oscillator of Fig. V.1 has period

$$\begin{aligned} T_0 &= \frac{2\pi}{\omega_0} \\ &= \frac{2\pi}{4.26 \text{ s}^{-1}} \\ &= 1.47 \text{ s} \end{aligned} \qquad\qquad \text{(V.14a)}$$

and corresponding frequency

$$\begin{aligned} \nu_0 &= \frac{\omega_0}{2\pi} \\ &= \frac{4.26 \text{ -1}}{2\pi} \\ &= 0.679 \text{ Hz} , \end{aligned} \qquad\qquad \text{(V.14b)}$$

using result (V.3d). In constrast, the amplitude A (center-to-peak; the peak-to-peak or peak-to-trough excursion is $2A$) of the motion and the *phase shift δ* depend on the initial data. This last describes the displacement of the peak of the motion from the start time $t = 0$, i.e., a peak of the oscillations is reached at time $t_{\text{peak}} = \delta/\omega_0$.

The independence of the period and frequency of the motion from its amplitude and initial conditions is a feature peculiar to the simple harmonic oscillator, i.e., to the Hooke-Law force. Galileo is reputed to have observed this by watching lamps suspended from a cathedral ceiling and timing their motion with his own pulse. Huygens[7] is credited with first using the pendulum to regulate a clock in 1656. It is a misconception, however, that the amplitude-independence of the pendulum period is used to regulate clocks, and this misconception reveals an important aspect of approximations in physics. The simple pendulum of Fig. IV.2 and Eqs. (V.4) is approximately a simple harmonic oscillator for small angles. The discrepancies or *anharmonicities* are of relative magnitude θ^2 for θ in radians. For a pendulum swinging through a degree or so, its amplitude in θ is of the order 10^{-2} radians, its anharmonicity of the order 10^{-4}. Errors of that magnitude would constitute excellent accuracy in a teaching lab, say, *where one might be measuring a few periods* of the motion. But a clock pendulum swings thousands of times a day, every day. Such an error is of the order

---

[7]Christian Huygens, 1629–1695

ten seconds a day, a minute a week—unacceptable for a precision timepiece. Actual pendulum clocks operate *at fixed amplitude,* and the actual period of the pendulum—anharmonicities and all—is incorporated into the gearing of the clock movement.[8] This example shows that the accuracy or validity of any approximation in science depends sensitively on the situations and measurements to which it is applied.

The simple-pendulum equation of motion (V.4b) can be solved *exactly* using Jacobian[9] elliptic functions rather than circular functions. The menagerie of functions—even considering just those with names—is large. Unfortunately, opportunities to meet some of its wilder members are becoming rare at the undergraduate level.

Energy and the SHO

The simple harmonic oscillator is a conservative system, with potential energy given by Eq. (II.33b). The total energy is

$$
\begin{aligned}
E &= \tfrac{1}{2}m\left(\frac{d\Delta x}{dt}\right)^2 + \tfrac{1}{2}k(\Delta x)^2 \\
&= \tfrac{1}{2}m\omega_0^2 A^2 \sin^2(\omega_0 t - \delta) + \tfrac{1}{2}kA^2 \cos^2(\omega_0 t - \delta) \\
&= \tfrac{1}{2}kA^2 \,,
\end{aligned}
\tag{V.15}
$$

where results (V.13a) and (V.2b) are used. The energy is transferred back and forth between the kinetic- and potential-energy terms of this expression, the total remaining constant.

The conservation of energy E is equivalent to the equation of motion for the system, i.e., to Newton's Second Law. The constancy of E implies the equation of motion, thus:

$$
\begin{aligned}
0 &= \frac{dE}{dt} \\
&= \frac{d}{dt}\left[\tfrac{1}{2}m\left(\frac{d\Delta x}{dt}\right)^2 + \tfrac{1}{2}k(\Delta x)^2\right] \\
&= m\frac{d\Delta x}{dt}\left(\frac{d^2\Delta x}{dt^2} + \omega_0^2\,\Delta x\right) \,,
\end{aligned}
\tag{V.16a}
$$

which implies Eq. (V.2a), since $d\Delta x/dt$ is not identically zero. Conversely, multiplying Eq. (V.2a) by $m\,(d\Delta x/dt)$ and integrating both sides yields

$$
\int_0^t 0\,dt' = \int_0^t \left(m\frac{d\Delta x}{dt'}\frac{d^2\Delta x}{dt'^2} + k\,\Delta x\,\frac{d\Delta x}{dt'}\right)dt'
$$

$$
0 = \left[\tfrac{1}{2}m\left(\frac{d\Delta x}{dt'}\right)^2 + \tfrac{1}{2}k(\Delta x)^2\right]\Bigg|_0^t \,,
\tag{V.16b}
$$

---

[8]That is, for *real* pendulum clocks, not electronic clocks with ornamental pendula.
[9]Karl Gustav Jakob Jacobi, 1804–1851

i.e.,

$$\tfrac{1}{2}m\left(\frac{d\Delta x}{dt}\right)^2 + \tfrac{1}{2}k(\Delta x)^2 = \tfrac{1}{2}mv_0^2 + \tfrac{1}{2}kx_0^2$$

$$= E \,, \tag{V.16c}$$

a constant. By virtue of the latter derivation, the energy E is called a *first integral of the motion*. A dynamical system that possesses as many first integrals[10] as dependent variables or *degrees of freedom* is called *integrable*. Hence, any one-dimensional conservative system is integrable. The name is apt: Here, if we solve Eq. (V.15) or (V.16c) for $d\Delta x/dt$, the result is a first-order differential equation that can be *separated* in the manner of the equations of motion for vertical flight with drag in Ch. I Sec. E. We could then obtain result (V.13a) by integration rather than by guessing.

D   The damped harmonic oscillator

The apparatus of Fig. V.1 is only approximately a simple harmonic oscillator. Its motion is not exactly periodic: The amplitude of its oscillations gradually decreases, and it eventually comes to rest. In addition to its Hooke-Law and gravitational forces, the system is subject to aerodynamic drag and dissipative forces within the spring. Because of the low speeds involved, it is convenient to describe these with a linear drag force

$$f_D = -B\,\frac{d\Delta x}{dt}$$

$$\equiv -2m\beta\,\frac{d\Delta x}{dt} \,, \tag{V.17}$$

using the notation of Eqs. (V.1) and (V.2). Here, B and β are positive constants with suitable units. Included in Newton's Second Law with the other forces, this yields equation of motion

$$\frac{d^2\Delta x}{dt^2} + 2\beta\,\frac{d\Delta x}{dt} + \omega_0^2\,\Delta x = 0 \tag{V.18}$$

for the *damped harmonic oscillator*. Constant β is the *damping parameter*.

This equation defines an initial-value problem as before, with a unique solution for given initial position and velocity values. The general solution to this ODE can be found by making the same guess $x(t) = e^{\lambda t}$ as before. Once again, this reduces the equation of motion to an algebraic equation:

$$\lambda^2 + 2\beta\lambda + \omega_0^2 = 0 \,, \tag{V.19a}$$

with solutions

$$\lambda_\pm = -\beta \pm (\beta^2 - \omega_0^2)^{1/2} \,. \tag{V.19b}$$

Three cases, with solutions of different characters, arise:

---

[10]For example, for the Keplerian system of Ch. IV, the energy, angular momentum, and LRL vector are all first integrals of the motion.

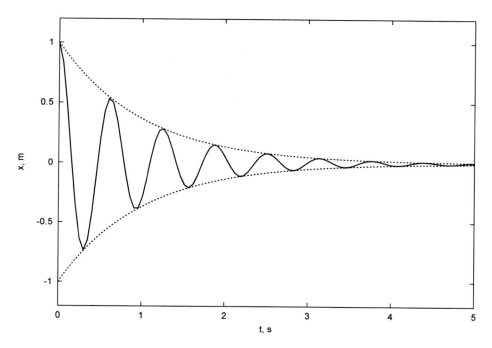

Figure V.3: Motion of a damped harmonic oscillator, underdamped case. This oscillator has damped "angular frequency" $\omega_\beta = 10.0$ s^{-1} and damping parameter $\beta = 1.00$ s^{-1}. The exponential envelope of the oscillations is shown (dotted curves).

- Case I (Underdamped): $\beta < \omega_0$. The exponents λ take the form

$$\lambda_\pm = -\beta \pm i\omega_\beta , \qquad (V.20a)$$

with real

$$\omega_\beta \equiv (\omega_0^2 - \beta^2)^{1/2} . \qquad (V.20b)$$

Two independent, real solutions can be constructed in the manner of Eqs. (V.10). The corresponding general solution is

$$\begin{aligned} \Delta x(t) &= a\,e^{-\beta t}\,\cos(\omega_\beta t) + b\,e^{-\beta t}\,\sin(\omega_\beta t) \\ &= A\,e^{-\beta t}\,\cos(\omega_\beta t - \delta) . \end{aligned} \qquad (V.21)$$

This solution has two constants a and b (or A and δ) with which to match two initial values x_0 and v_0, although the relations between these constants and the initial values will be more complicated than those of Eqs. (V.12) and (V.13). It describes *damped oscillations,* as illustrated in Fig. V.3, with reduced "angular frequency" $\omega_\beta < \omega_0$. (Terms such as angular frequency, frequency, and period should be used cautiously here, as this motion is not periodic.) The amplitude of the oscillations decays by a factor $1/e = 0.3679\ldots$ every time interval $t_{1/e} = 1/\beta$. For example, the oscillations of the actual mass and spring shown in Fig. V.1 diminish to approximately one-third of their original amplitude after some one hundred oscillations. Hence, it has damping parameter

$$
\begin{aligned}
\beta &= -\frac{\ln(1/3)}{t_{1/3}} \\
&\cong -\frac{\ln(0.33)}{100(2\pi/\omega_0)} \\
&\cong -\frac{\ln(0.33)}{200\pi/(4.26\ \text{s}^{-1})} \\
&\cong 7.4 \times 10^{-3}\ \text{s}^{-1}\ ,
\end{aligned}
\tag{V.22}
$$

to two significant figures, using result (V.3d). Clearly, this system is underdamped, and the discrepancy between ω_β and ω_0—disregarded here—is beneath the precision of the data.

- Case II (Overdamped): $\beta > \omega_0$. In this case the exponents take the real values

$$
\lambda_\pm = -\beta \pm \Omega_\beta\ ,
\tag{V.23a}
$$

with

$$
\Omega_\beta \equiv (\beta^2 - \omega_0^2)^{1/2}\ .
\tag{V.23b}
$$

The general solution takes the form

$$
\Delta x(t) = e^{-\beta t}(a\, e^{+\Omega_\beta t} + b\, e^{-\Omega_\beta t})\ ,
\tag{V.24}
$$

again with two arbitrary constants to match the two initial-data values. This describes nonoscillatory motion; with $\Omega_\beta < \beta$, both terms represent decaying exponential functions. The system either relaxes back to equilibrium from its initial displacement or crosses through equilibrium once before relaxing. It can pass through $\Delta x = 0$ at most once. Overdamped "oscillators" have their uses: Door-closing mechanisms with springs attached to pistons or dashpots moving through oil or compressed air are designed to be overdamped so that the door closes suitably slowly.

- Case III (Critically damped): $\beta = \omega_0$. This is the boundary case between underdamping and overdamping. The two exponents $\lambda_\pm$ are equal, so the guess $x(t) = e^{\lambda t}$ gives only one solution, insufficient to solve the initial-value problem. A second solution can be obtained in the manner of Eq. (V.10b), dividing by the difference $\lambda_+ - \lambda_-$, then taking the limit as the λ values converge, i.e., *differentiating* the exponential with respect to λ. The resulting general solution is

$$
\Delta x(t) = e^{-\beta t}[a + b(\beta t)]\ ,
\tag{V.25}
$$

where the factor β is included in the second term so that constants a and b—again, sufficient to match two initial values—have the same dimensions. This too describes nonoscillatory motion, which can cross the equilibrium point $\Delta x = 0$ at most once. In general, a critically damped "oscillator" returns to the vicinity of the equilibrium point faster than a comparable underdamped or overdamped system. This is why the spring-and-shock-absorber assemblies (e.g., MacPherson[11] struts) in automobile suspensions are designed for critical damping; the struts are repaired or replaced if the components deviate from critical-damping specifications.

The energy E of Eqs. (V.15) and (V.16c) is not a first integral of equation of motion (V.18). It is not constant for the damped harmonic oscillator—it decays in time as $e^{-2\beta t}$—since the damping force does negative work on the system.

E   Driven oscillator and resonance

Most harmonic oscillators are found coupled to other physical systems. The motion of such an oscillator is subject to external driving force $F(t)$, say, in addition to its Hooke-Law and damping forces. It is described by the equation of motion

$$\frac{d^2\Delta x}{dt^2} + 2\beta\,\frac{d\Delta x}{dt} + \omega_0^2\,\Delta x = \frac{F(t)}{m}\ ,\qquad\text{(V.26)}$$

which follows, as before, from Newton's Second Law. This is an *inhomogeneous,* linear, second-order ODE because of the driving-force term. By contrast, equation of motion (V.18) is *homogeneous,* because all its terms are first-order in Δx and its derivatives. However, the linearity of these equations offers a simplification: The *difference* of any two solutions of Eq. (V.26) is a solution of the homogeneous version (V.18). Hence, the general solution of this new equation of motion is

$$\begin{aligned}\Delta x(t) &= x_p(t) + x_h(t)\\ &= x_p(t) + a x^{(1)}(t) + b x^{(2)}(t)\ .\end{aligned}\qquad\text{(V.27)}$$

Here, subscript p stands for "particular"; term $x_p(t)$ is the *particular solution.* Subscript h stands for "homogeneous"; term $x_h(t)$ is the *homogeneous solution,* i.e., a solution of the homogeneous version of the equation of motion. Functions $x^{(1)}(t)$ and $x^{(2)}(t)$ are the two independent solutions from which the general solution of the homogeneous equation are constructed, as in Secs. C and D preceding. The homogeneous solution contains the two arbitrary constants a and b, so that solution (V.27) can be made to fit any initial-data values—thus solving the initial-value problem for this system. It does not matter which solution of the full equation of motion we find for $x_p(t)$; the difference between any two of them can be absorbed into $x_h(t)$.

As a demonstration, we can find a particular solution for a *sinusoidal* driving force $F(t) = \mathcal{F}\cos(\omega t)$, where $\mathcal{F}$ is a constant force amplitude. (We can

---

[11] Earle Steele MacPherson, 1891–1960

also anticipate the result of Fourier[12] that *any* reasonable driving force can be given as a superposition of sinusoidal functions.) The angular frequency ω of the driving force here need bear no relation to ω_0 or ω_β; it is an independent parameter. The exponential-guessing procedure of the preceding sections might be useful if the driving force were an exponential function. But by the Euler relation, it is the *real part* of an exponential function! Hence, $x_p(t) = \Re z_p(t)$ will provide the desired solution for any solution $z_p(t)$ of the equation

$$\frac{d^2z}{dt^2} + 2\beta\frac{dz}{dt} + \omega_0^2 z = \frac{\mathcal{F}}{m}\,e^{i\omega t}\,, \qquad (V.28)$$

as can be seen by equating the real parts of both sides.[13] Guessing a solution of form $z_p(t) = Z\,e^{i\omega t}$ reduces this to an algebraic equation for the constant complex amplitude Z. It has solution

$$\begin{aligned} Z &= \frac{\mathcal{F}/m}{\omega_0^2 - \omega^2 + 2i\beta\omega} \\ &= \frac{\mathcal{F}/m}{[(\omega_0^2 - \omega^2)^2 + 4\beta^2\omega^2]^{1/2}}\,e^{-i\Delta}\,, \end{aligned} \qquad (V.29a)$$

with

$$\Delta = \mathrm{Cos}^{-1}\frac{\omega_0^2 - \omega^2}{[(\omega_0^2 - \omega^2)^2 + 4\beta^2\omega^2]^{1/2}}\,, \qquad (V.29b)$$

where the second expression for Z is obtained by casting the complex denominator into polar form (V.9). This yields complex solution

$$z_p(t) = \frac{\mathcal{F}/m}{[(\omega_0^2 - \omega^2)^2 + 4\beta^2\omega^2]^{1/2}}\,e^{i(\omega t - \Delta)} \qquad (V.30a)$$

of Eq. (V.28) and real solution

$$x_p(t) = \frac{\mathcal{F}/m}{[(\omega_0^2 - \omega^2)^2 + 4\beta^2\omega^2]^{1/2}}\,\cos(\omega t - \Delta) \qquad (V.30b)$$

of Eq. (V.26).

Hence, the general solution for the cosine-driven, (under)damped oscillator is

$$\begin{aligned} \Delta x(t) = &\frac{\mathcal{F}/m}{[(\omega_0^2 - \omega^2)^2 + 4\beta^2\omega^2]^{1/2}}\,\cos(\omega t - \Delta) \\ &+ A\,e^{-\beta t}\cos(\omega_\beta t - \delta)\,, \end{aligned} \qquad (V.31)$$

with Δ given by Eq. (V.29b) and ω_β by Eq. (V.20b). The first term is the *steady-state response* of the oscillator to the driving force. The second term is the *transient response*. It contains the two constants A and δ, with which the

---

[12] Jean Baptiste Joseph Fourier, 1768–1830

[13] A particular solution for a sine-function driving force is given by the *imaginary* part of the complex solution for free.

solution can be made to match the initial-data values (though not via the same relations as Eqs. (V.13b) and (V.13c)), and it decays away in time as $e^{-\beta t}$. The first, steady-state term bears no relation to the initial data; observed after a long time ($t \gg 1/\beta$), the system "forgets" its initial conditions. This is characteristic of systems with dissipative forces. The mathematical separation of the solution into particular and homogeneous parts (Eq. (V.27)) embodies this physical behavior.

Solutions (V.30b) and (V.31) are problematic for an *undamped* ($\beta = 0$) oscillator driven at its natural frequency ($\omega = \omega_0$). For that case, a particular solution can be obtained via the limit

$$
\begin{aligned}
x_p(t) &= \lim_{\omega \to \omega_0} \left(\frac{(\mathcal{F}/m)\cos(\omega t)}{\omega_0^2 - \omega^2} - \frac{(\mathcal{F}/m)\cos(\omega_0 t)}{\omega_0^2 - \omega^2} \right) \\
&= -\frac{\mathcal{F}/m}{2\omega_0} \left. \frac{d}{d\omega}[\cos\omega(t)] \right|_{\omega = \omega_0} \\
&= \frac{\mathcal{F}}{2m\omega_0} t \sin(\omega_0 t) .
\end{aligned}
\tag{V.32}
$$

The function in parentheses in the first line is a particular solution of the undamped equation of motion, as the second term is a solution of the homogeneous equation. An undamped oscillator driven at $\omega = \omega_0$ does not attain a steady state: It oscillates with ever-increasing amplitude. Also, for such an oscillator, the transient response never decays away.

This driven oscillator displays the phenomenon of *resonance: the transfer of energy from one system to another, enhanced by a matching of frequencies.* The amplitude of the steady-state term in solution (V.31), i.e., of particular solution (V.30b), viz.,

$$
\mathcal{A}(\omega) = \frac{\mathcal{F}/m}{[(\omega_0^2 - \omega^2)^2 + 4\beta^2\omega^2]^{1/2}} ,
\tag{V.33}
$$

is graphed in Fig. V.4. This is called a *resonance curve* or *resonance peak,* or specifically a *Lorentzian peak* (named for the same Lorentz as the Lorentz transformations of Ch. I Sec. H). The low-frequency value of the amplitude is $\mathcal{A}(0) = \mathcal{F}/(m\omega_0^2) = \mathcal{F}/k$; this is just a static displacement of the oscillator, like that due to gravity in Eqs. (V.3). But the amplitude near the peak is the much larger value

$$
\begin{aligned}
\mathcal{A}(\omega_0) &= \frac{\mathcal{F}}{2m\beta\omega_0} \\
&= \frac{\omega_0}{2\beta} \mathcal{A}(0) \\
&= Q\frac{\mathcal{F}}{k} ,
\end{aligned}
\tag{V.34}
$$

which defines the *quality factor Q* of the oscillator. This is an example of resonant amplification, as when a playground swing is driven to large amplitudes when pushed or pumped at its natural frequency. The larger the value of Q,

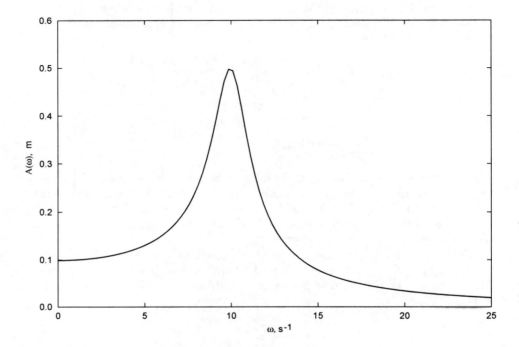

Figure V.4: Resonance curve: Steady-state amplitude for a sinusoidally driven oscillator. This oscillator has the same parameters as that of Fig. V.3: $\omega_\beta = 10.0$ s^{-1} and $\beta = 1.00$ s^{-1}.

the higher and narrower the resonance peak: The full width, in ω, of the peak at $1/\sqrt{2}$ of its maximum height (at which the total energy of the oscillations is half the maximum value) is approximately 2β for a lightly damped ($\beta \ll \omega_0$) oscillator.

The oscillator represented in Figs. V.3 and V.4 has quality factor

$$
\begin{aligned}
Q &= \frac{\omega_0}{2\beta} \\
&= \frac{(\omega_\beta^2 + \beta^2)^{1/2}}{2\beta} \\
&= \frac{[(10.0 \text{ s}^{-1})^2 + (1.00 \text{ s}^{-1})^2]^{1/2}}{2(1.00 \text{ s}^{-1})} \\
&= 5.02 \,,
\end{aligned}
\tag{V.35a}
$$

for purposes of illustration. But the actual oscillator of Fig. V.1 has factor

$$
\begin{aligned}
Q &= \frac{\omega_0}{2\beta} \\
&= \frac{4.26 \text{ s}^{-1}}{2(7.4 \times 10^{-3} \text{ s}^{-1})} \\
&= 290 \,,
\end{aligned}
\tag{V.35b}
$$

using results (V.3d) and (V.22). Much larger values are possible—some crystals can vibrate with quality factors of the order 10^7.

The phase shift Δ of the steady-state oscillations, given by Eq. V.29b, also exhibits behavior characteristic of resonance. It is graphed as a function of ω in Fig. V.5 for the same oscillator as in Figs. V.3 and V.4. For larger values of Q, the change near $\omega = \omega_0$ is steeper: The limiting case for an undamped ($\beta = 0$) oscillator is a discontinuous step from $\Delta = 0$ below ω_0 to $\Delta = \pi$ above. The shift from oscillations in phase with the driving force below resonance to oscillations opposite to the driving force above it is a most sensitive indicator of the resonant frequency.

Resonance is arguably the most important phenomenon or "metaphenomenon"—class of phenomena—in physics.[14] Chemical reactions, nuclear reactions, and elementary-particle reactions are resonances; short-lived particles are identified by resonance peaks in scattering measurements. Human vision and hearing are resonant phenomena. Musical instruments resonate, as do electrical circuits. Even the planets and their moons exhibit resonances in their orbits.

F Coupled oscillators and normal modes

A physical systems of many variables near a stable equilibrium configuration behaves as a collection of harmonic oscillators coupled to each other. The simplest such system is illustrated in Fig. V.6: two masses m, each connected to

---

[14]Only wave-interference phenomena are of comparable significance.

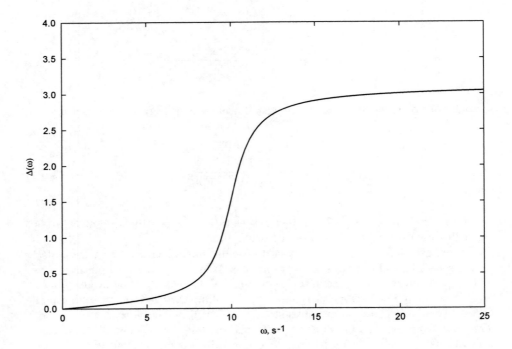

Figure V.5: Steady-state phase shift, in radians, for a sinusoidally driven oscillator. The oscillator has the same parameters as in Figs. V.3 and V.4: $\omega_\beta = 10.0$ s^{-1} and $\beta = 1.00$ s^{-1}.

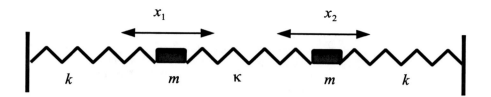

Figure V.6: Two coupled oscillators: Two identical masses connected to three springs in a symmetric arrangement. The masses move horizontally without friction.

a fixed anchor by a spring with spring constant k, connected to each other by a spring with spring constant κ, otherwise free to move frictionlessly in one dimension. The masses of the springs themselves are negligible. Newton's Second Law supplies equation of motion for each mass

$$m \frac{d^2 x_1}{dt^2} = -k x_1 - \kappa(x_1 - x_2) \tag{V.36a}$$

and

$$m \frac{d^2 x_2}{dt^2} = -k x_2 - \kappa(x_2 - x_1) , \tag{V.36b}$$

or

$$\frac{d^2 x_1}{dt^2} + \frac{k}{m} x_1 + \frac{\kappa}{m}(x_1 - x_2) = 0 \tag{V.36c}$$

and

$$\frac{d^2 x_2}{dt^2} + \frac{k}{m} 2_1 + \frac{\kappa}{m}(x_2 - x_1) = 0 , \tag{V.36d}$$

where the displacements x_1 and x_2 are measured from their respective equilibrium positions, as indicated in the figure.

These are *coupled*, linear, second-order, ordinary differential equations. The motion of each mass depends on the trajectory of the other. However, it is possible to decouple the equations by combining them appropriately: Averaging Eqs. (V.36c) and (V.36d) yields

$$\frac{d^2}{dt^2}\left(\frac{x_1 + x_2}{2}\right) + \frac{k}{m}\left(\frac{x_1 + x_2}{2}\right) = 0 \tag{V.37a}$$

for the center-of-mass or *common-mode* (CM) motion of the system—the coupling terms cancel. Subtracting Eq. (V.36c) from Eq. (V.36d) yields

$$\frac{d^2}{dt^2}(x_2 - x_1) + \frac{k + 2\kappa}{m}(x_2 - x_1) = 0 \tag{V.37b}$$

for the relative motion of the two masses.[15] These are *independent* simple-harmonic-oscillator equations of motion for their respective quantities.

The solutions of these equations are of the same form as, e.g., Eqs. (V.13), specifically

$$\frac{x_1 + x_2}{2} = A_+ \cos(\omega_+ t - \delta_+) , \tag{V.38a}$$

with

$$\omega_+ \equiv \left(\frac{k}{m}\right)^{1/2} , \tag{V.38b}$$

and

$$x_2 - x_1 = A_- \cos(\omega_- t - \delta_-) , \tag{V.38c}$$

---

[15]In a more systematic treatment the combinations would be defined slightly differently, more symmetrically. But these are simple and easily interpreted and will suffice here.

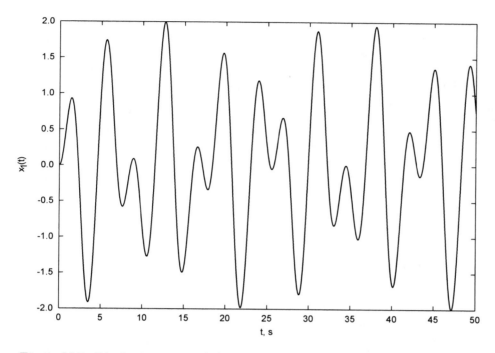

Figure V.7: Displacement x_1 of the oscillator in Fig. V.6, with $\kappa = k$, $A_- = 2A_+$, and $\delta_\pm = 0$. Spring constants and masses are chosen to give $\omega_+ = 1$ s^{-1} and $\omega_- = \sqrt{3}$ s^{-1}. The apparent repetition is deceptive, as the motion is nonperiodic.

with

$$\omega_- \equiv \left(\frac{k + 2\kappa}{m} \right)^{1/2} . \tag{V.38d}$$

These imply solutions

$$x_1(t) = A_+ \cos(\omega_+ t - \delta_+) - \tfrac{1}{2} A_- \cos(\omega_- t - \delta_-) \tag{V.38e}$$

and

$$x_2(t) = A_+ \cos(\omega_+ t - \delta_+) + \tfrac{1}{2} A_- \cos(\omega_- t - \delta_-) \tag{V.38f}$$

for the individual displacements. These solutions contain *four* arbitrary constants $A_\pm$ and $\delta_\pm$, so they can match two initial-data values for x_1 and two for x_2. The motions described by solutions (V.38e) and (V.38f) are complicated in general and nonperiodic if the ratio of the two angular frequencies ω_- / ω_+ is an irrational number, i.e., if the two are *incommensurable*. For example, a solution for x_1 is plotted[16] in Fig. V.7 for the case of three identical springs,

---

[16]The plotted trajectory cannot actually be nonperiodic, because no finite digital representation of a number can be irrational.

i.e., with $\kappa = k$ and $\omega_- / \omega_+ = \sqrt{3}$. Nonetheless, the combinations or configurations described by solutions (V.38a) and (V.38c) oscillate in independent simple-harmonic motion.[17]

A configuration of the motion of any system in which all parts of the system oscillate with the same frequency is called a *normal mode* of the system. A system described by N dynamical variables or degrees of freedom has N normal modes.[18] And any motion of any system executing small oscillations about some equilibrium can be decomposed into—i.e., can be represented by—a superposition of normal-mode motions. Human-built structures, molecules, force and matter fields—all can be described in this way. Since the normal modes behave as independent harmonic oscillators, the harmonic oscillator is central to any such description.

G   The quantum harmonic oscillator

The harmonic oscillator retains its importance beyond the realm of classical mechanics, where quantum mechanics holds sway. A simple harmonic oscillator is a bound system, like an atom. Consequently, it has discrete energy levels, as the electrons in atoms do. However, unlike the energy levels of atomic electrons, those of the simple harmonic oscillator are evenly spaced, given by

$$E_n^{(\text{SHO})} = \left(n + \tfrac{1}{2}\right) \hbar \omega_0 \qquad \text{for} \qquad n = 0, 1, 2, \dots , \qquad (\text{V.39})$$

where $\hbar = 1.05457266 \times 10^{-34}$ J s is *Planck's*[19] *reduced constant* or *Planck's reduced universal quantum of action*. The energy excitations between adjacent levels are all identical. When the harmonic oscillator in question is a normal mode of a quantum wave field, those excitations are elementary particles, as they are understood in the context of quantum field theory. The harmonic oscillator thus lies at the foundation of our best current understanding of matter and of existence itself.

References

Simple harmonic motion, damped harmonic motion, driven oscillations, and resonance are fundamental topics covered in physics textbooks at every level, although the degree of detail varies. For example, Bauer and Westfall [BW14], Fishbane, Gasiorowicz, and Thornton [FGT96], and Resnick, Halliday, and Krane [RHK02] treat all these topics. Intermediate-level texts such as Barger and Olsson [BO95], Taylor [Tay05], and Thornton and Marion [TM04] offer more detailed treatment of both linear and nonlinear oscillations, as do advanced-level

---

[17]The *symmetric* combination of Eq. (V.38a) has the lower angular frequency ω_+. This is always the case.

[18]The identification of normal modes and normal-mode frequencies takes the form of a matrix-diagonalization problem.

[19]Max Karl Ernst Ludwig Planck, 1858–1947

physics texts and specialized engieering texts on small vibrations. Nonlinear vibrations and chaos, not treated in this text, are areas of active research with an extensive literature of their own.

Problems

1. The potential energy graphed in Fig. II.5 is given by

$$U(x) = \frac{1}{x^3} - \frac{11}{6x^2} + \frac{1}{x} \; ,$$

where x is in nanometers (nm) and U in *electron-volts* (eV), with 1.000 eV $= 1.602 \times 10^{-19}$ J. Treat the numerical coefficients in this expression as exact. An ion of mass $m = 1.67 \times 10^{-26}$ kg moves with this potential energy. It is *slightly* perturbed away from the stable equilibrium point b shown in that figure. Calculate the *frequency* ν with which it oscillates.

2. To understand the effects of prolonged space flight on human health, it is important to measure the body masses of astronauts in weightless (i.e., free-fall) conditions. This is done with precision timers and a chair attached to springs. The chair oscillates with measured period T_0 when empty and with period T_1 with an astronaut strapped into it. If the chair has mass m_0, what is the mass of the astronaut in terms of these quantities?

3. Prove that a simple harmonic oscillator is *not* a chaotic system. That is, consider two trajectories of a given oscillator with different initial conditions. Show that the difference between the trajectories remains bounded, with a fixed bound for all time.

4. Consider a simple harmonic oscillator consisting of a mass M connected to a spring of *non*-negligible mass m and spring constant k, the other end of which is fixed. The mass is otherwise free to move in one dimension without friction.

   (a) Assuming that the spring stretches and compresses uniformly, calculate the total energy of the mass/spring combination in terms of the position Δx and velocity $d\Delta x/dt$ of the mass M.

   (b) Using the fact that this energy is a first integral of the motion, find the equation of motion for this oscillator.

   (c) Determine the angular frequency ω_0 for this oscillator in terms of M, m, and k. That is, determine the correction to ω_0 associated with the mass of the spring.

5. Consider a body of mass m, moving frictionlessly in one dimension with potential energy

$$U(x) = \frac{\gamma}{2n} x^{2n} \; ,$$

where n is a positive integer and γ a positive constant of suitable dimensions. The body's motion is periodic but not simple-harmonic except in the $n = 1$ case.

(a) Construct an integral expression for the *period* of the body's motion in terms of its amplitude A and the parameters m, γ, and n. The integral should be a definite integral. It is not necessary to evaluate the integral.

(b) Determine the dependence of the period of the motion on its amplitude.

6. Show explicitly that the rate of change of the total (kinetic plus potential) energy of a damped harmonic oscillator, such as is described by Eq. (V.18), is equal to the rate at which the damping force does work on the system.

7. Calculate the fractional energy loss $-\delta E/E$ per oscillation for an underdamped oscillator in terms of its angular frequency and damping parameter.

8. Consider a damped harmonic oscillator with mass m, spring constant k, and damping parameter β (i.e., damping force $f = -2m\beta\,dy/dt$) suspended *vertically*, with gravitational acceleration g. Treat this as a *driven* oscillator, with the weight of the mass a constant driving force. Write the general solution for the motion of this oscillator in terms of a particular (steady-state) solution and an homogeneous (transient) solution. Is the long-term behavior of this solution correct?

9. The resonant amplitude $\mathcal{A}(\omega)$ of the sinusoidally driven oscillator, given in Eqs. (V.31) and (V.33) and graphed in Fig. V.4, does not quite attain its maximum value at $\omega = \omega_0$. Calculate the value of ω at which the resonance curve actually attains its peak value. How much does this differ from ω_0 for the oscillator of Figs. V.3, V.4, and V.5 and the actual oscillator of Fig. V.1?

10. After a long time, the transient response of a damped, driven oscillator decays away, leaving only the steady-state motion. The driving force must do work on the oscillator sufficient to replace the energy lost to damping. For the sinusoidally driven oscillator described by soluton (V.31), calculate the work done per cycle by the driving force in the steady-state limit.

11. Consider an array of masses and springs as in Fig. V.7, but with n identical masses m connected to $n + 1$ identical springs of spring constant k and negligible mass, fixed at the outer ends as in the figure. The displacement x_i, with $i = 1, 2, \ldots, n$, of each mass is measured from its equilibrium position. Damping is absent. Write the array of equations of motion for all the masses.

Chapter VI

Catch a Wave (A Knotty Problem)

- *A long string, with linear mass density μ and tension τ, has a knot in it of mass M. A wave traveling on the string is partially reflected and partially transmitted at the knot. Find the reflection and transmission coefficients, i.e., the ratios of reflected and transmitted power to incident power, as functions of the wave number, angular frequency, μ, M, and/or τ.*

- *This is the classic wave-scattering problem reduced to one dimension of propagation. This problem arises everywhere, from radar mapping to atmospheric optics to electron propagation in solids to seismology to elementary particle physics.*

Figure VI.1: Standing waves on a stretched string. Wave phenomena appear in nature on every scale from the elementary particle to the cosmos. (Photograph by author.)

The problem proposed for this chapter seems quite trivial at first glance— a knot on a string? But this is the simplest, one-dimensional version of the classic problem of wave scattering—in this case, visible waves such as those in Fig. VI.1—at a point center. The language and techniques we shall examine here apply throughout physics at every scale. Much of what we know about the world around us is gleaned from wave-scattering processes; human vision is itself a wave-scattering phenomenon.

Waves are fundamental to our understanding of nature in almost every context. The notion has superseded the point particle as the basic unit of our description of matter. At the heart of quantum mechanics, of which classical mechanics is a limit case, is the understanding that every particle is to be described by means of waves. And classical waves are found in great variety. Our first task, then, is to establish a language with which to classify and describe waves.

A Waves defined

The diversity of wave phenomena makes a general definition a challenge. This one should cover most cases: *A wave is a disturbance, from some equilibrium, which propagates in space and time.* Necessarily, then, a wave must be described by *fields,* in the sense used for the gravitational field in Ch. IV Sec. A preceding.

It is useful to group wave phenomena into several broad categories:

- Mechanical waves—waves in which matter *oscillates* as the wave pattern *propagates.* Da Vinci[1] is credited with the observation that wave patterns on a wheat field propagate great distances, while the wheat itself remains rooted to the ground. Wave trains may cross the Pacific Ocean from Japan to California, but the water that washes up on the beach came from just offshore. Mechanical waves can be divided into two subcategories:

  - *Transverse* waves are waves in which the disturbance or oscillation is perpendicular to the direction of propagation. Waves on a stretched string, such as those shown in Fig. VI.1, are transverse. Gravity waves—waves in which the restoring force of the oscillations is provided by gravity—on water or air are likewise transverse.

  - *Longitudinal* waves consist of a disturbance or oscillation parallel to the direction of propagation. Compression waves on a stretched spring are longitudinal, as are sound waves in gaseous, liquid, and solid media.

- Electromagnetic waves—waves consisting of coupled electric and magnetic force fields propagating in vacuum or material media. These are the subject of Ch. XI. Electromagnetic waves are transverse in character.

- Gravitational[2] waves—ripples in the geometry of spacetime produced by accelerating masses, as described by Einstein's General Theory of Relativity, mentioned in Ch. IV Sec. D preceding. These waves are also transverse.

- Quantum or matter waves—waves that embody the quantum-mechanical nature of matter particles, often interpreted as complex *probability amplitudes.* The distinction between transverse and longitudinal waves is not drawn for these.

Mechanical waves, the first category, are our chief concern in this chapter. But the descriptions we shall use can be applied to the others as well.

---

[1] Leonardo da Vinci, 1452–1519

[2] These are sometimes called "gravity waves," but that is a misnomer. The distinction drawn here is the correct usage.

B   Wave kinematics

Certain features characterize all waves: *amplitude,* the magnitude of the displacement or disturbance; the *shape* of the wave pattern; and the *velocity* of wave propagation. Transverse waves possess an additional property, *polarization,* the choice of direction of the wave disturbance in the two-dimensional space of directions perpendicular to the direction of propagation.

In general, a wave is described by a *wave function,* a notion not limited to quantum mechanics. The wave function is a solution of the *wave equation* for the physical system considered. For example, a stretched string extending in the x direction could propagate displacements in the y direction. A pattern of such displacements might be described by the wave function

$$y(x,t) = f(x - vt) + g(x + vt) . \tag{VI.1}$$

This description displays a *right-moving wave* of a shape given by the function $f(\phi_+)$, with *phase*[3] $\phi_+ = x - vt$, and a *left-moving wave* of shape $g(\phi_-)$, with phase $\phi_- = x + vt$, in linear superposition. The first term is right-moving because a point of given phase ϕ_+ moves in the $+x$ direction with velocity v; a point of given phase ϕ_- for the second term moves in the opposite direction. Velocity v, the *phase velocity* of this wave, is dynamically determined—it follows from the wave equation of which $y(x,t)$ is a solution. The linear superposition (sum) of the two terms does not follow from some general principle. Rather, it is a feature of waves governed by a linear wave equation, for which the sum of two solutions is also a solution. We shall examine the wave equation for this particular system in Sec. C.

A first special case of the general wave is the *periodic* wave. This is a wave pattern that repeats exactly after a fixed interval, both in space at fixed time and in time at fixed position. For this case only, the language of periodic phenomena can be used: The *wavelength* or *spatial period* λ is the distance, in the direction of propagation, over which the wave repeats. The *temporal period* or just *period* $T = \lambda/v$ is the time interval over which the wave would be seen to repeat by an observer at a fixed location. The *frequency* ν is the number of cycles through which the wave repeats, at fixed location, per unit time interval. As with frequencies describing uniform circular motion or simple harmonic motion, it is measured in Hz, algebraically s^{-1}. Frequency is related to period, wavelength, and velocity via relations

$$\nu = \frac{1}{T} = \frac{v}{\lambda}. \tag{VI.2a}$$

That is, the relation

$$v = \lambda\nu \tag{VI.2b}$$

connects velocity, wavelength, and frequency for periodic waves.

---

[3]The term *phase,* beloved of science-fiction writers, here simply means the argument of a wave function. That is, it is the quantity specifying the location on a wave pattern.

A second, most important special case—a subcase of periodic waves—is the sinusoidal or *monochromatic* wave. For example, wave function

$$f(x - vt) = A \sin\left[\frac{2\pi}{\lambda}(x - vt) + \phi_0\right]$$
$$= A \sin(kx - \omega t + \phi_0) \qquad \text{(VI.3a)}$$

describes a sinusoidal wave with *amplitude A, wave number*[4]

$$k = \frac{2\pi}{\lambda}, \qquad \text{(VI.3b)}$$

angular frequency

$$\omega = \frac{2\pi}{T}$$
$$= 2\pi\nu$$
$$= kv, \qquad \text{(VI.3c)}$$

measured in s^{-1} (not Hz), and phase shift ϕ_0, a constant. As the sine and cosine functions differ only by a phase shift of $\pi/4$, a cosine function can equally well be used to describe such a wave. The term *monochromatic*—literally, "single color"—refers to the fact that in optics, color is associated with frequency or wavelength (as is pitch for sound waves). That is, the term monochromatic means *single-frequency*. All other shapes are multiple-frequency: They can be constructed from superpositions of sinusoidal waves of various frequencies. Hence, sinusoidal or monochromatic waves are the "building blocks" of all the others.[5] This is an astounding mathematical result of great significance.

Combinations of sinusoidal waves can display many important phenomena. For example, consider the superposition

$$f(x,t) = A \sin(k_1 x - \omega_1 t) + A \sin(k_2 x - \omega_2 t) \qquad \text{(VI.4a)}$$

of two sinusoidal waves of equal amplitude, with slightly different wave numbers k_1 and k_2 and angular frequencies ω_1 and ω_2. Defining the difference $\Delta k \equiv k_2 - k_1$ and the average $\bar{k} \equiv \frac{1}{2}(k_1 + k_2)$, and similarly for the angular frequencies, then writing the individual quantities as the average plus or minus half the difference and using angle sum and difference identities, we can transform this into the form

$$f(x,t) = 2A \cos[\tfrac{1}{2}(\Delta k\, x - \Delta\omega\, t)] \sin(\bar{k}x - \bar{\omega}t). \qquad \text{(VI.4b)}$$

This describes an underlying wave, the sine factor, modulated by an envelope, the cosine factor. Such a combination is graphed in Fig. VI.2; the underlying wave is bunched into "wave packets." These are called *beats*.[6] The *beat*

---

[4]This is sometimes called *angular spatial frequency*.

[5]More formally, the sinusoidal waves are basis vectors for an infinite-dimensional vector space of wave functions.

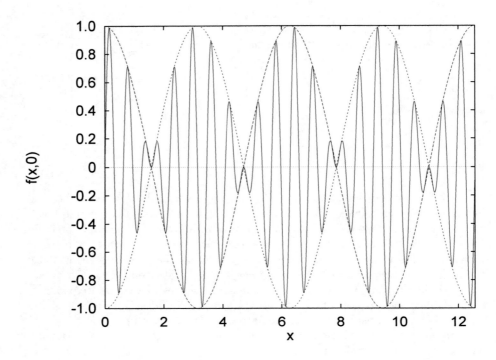

Figure VI.2: "Snapshot" at time $t = 0$ of a beat pattern formed by the super-position of two sinusoidal waves of slightly different wavelength and frequency. The solid curve is the underlying wave; the dotted curves are the wave-packet "envelope."

frequency, the rate at which wave packets pass, is given by

$$\nu_B = \left(\frac{\pi}{\frac{1}{2}\Delta\omega} \right)^{-1}$$

$$= \Delta\nu \,, \tag{VI.5}$$

because the envelope reaches successive peaks when the phase of the cosine factor changes by π radians. Sound waves combined in this way produce an audible "wobble" at this frequency. A traditional piano tuner uses this effect: The tuner strikes a tuning fork and the piano key for the same note and adjusts the tension of the piano string until the wobble is too slow to hear, i.e., until the piano's frequency approaches that of the fork.[7]

In general, the underlying wave of pattern (VI.4) and the envelope of the wave packets need not travel at the same speed. The underlying wave travels at *phase velocity*

$$v_p = \frac{\omega}{k} \,. \tag{VI.6a}$$

The envelope, i.e., a wave packet, travels at *group velocity*

$$v_g = \frac{d\omega}{dk} \,, \tag{VI.6b}$$

where both of these expressions follow from Eq. (VI.4b) in the $\omega_2 \to \omega_1$ limit. The distinction between phase and group velocities is a feature of all waves. Both velocities are determined by the dynamical wave equation for the system considered, as will be seen in Sec. C following.

The representation of any wave form as a superposition of sinusoidal waves is called *Fourier analysis,* a specific example of *harmonic analysis.* The super-position is called a *Fourier series* or *Fourier integral,* depending on whether a discrete sum over frequencies (for a periodic waveform) or a continuous sum (for a nonperiodic form) is needed. The assertion that *any* (not-too-pathological) function can be represented in this way is so outlandish that its discoverers did not believe it. The result follows from discoveries by Leonhard Euler and Daniel Bernoulli in the eighteenth century, but those worthies did not accept the implications of their work. Jean Baptiste Joseph Fourier did, but his 1807 paper proposing the idea was rejected as insufficiently rigorous. Not until 1831 did Dirichlet[8] provide an acceptable proof, and the mathematics underlying Fourier analysis remains an area of active research. No proof will be attempted here, but an example is persuasive: The series

$$f(\phi) = \frac{4}{\pi} \sum_{n=0}^{\infty} \frac{1}{2n+1} \sin[(2n+1)\phi] \tag{VI.7}$$

---

[6]Beats are an *interference pattern,* distributed in time as well as in space. Wave interference is examined in more detail in Ch. XII.

[7]A modern piano tuner does this with electronic aids.

[8]Peter Gustav Lejeune Dirichlet, 1805–1859

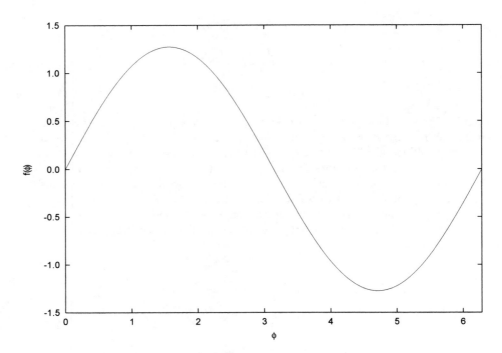

Figure VI.3: Fourier series (VI.7) for a square wave, first (fundamental) term.

represents a unit-amplitude square wave as a function of phase ϕ. The techniques by which one determines this are the province of a specialized math course. But graphs of the sums of the first one, three, ten, and one hundred terms (*harmonics*) in this series, shown in Figs. VI.3, VI.4, VI.5, and VI.6, illustrate the result. The spike on each side of the points where the square wave makes a discontinuous jump, an 18% over/undershoot known as the *Gibbs*[9] *overshoot,* never goes away, but clearly the approximation to a square wave becomes very good as the number of terms increases. This is why "square-wave response" is a performance specification for high-end audio equipment—it measures the ability of the amplifier to reproduce high frequencies as well as low frequencies.

Dispersion is a general feature of waves that plays an important role when monochromatic waves are combined to form signals. If the phase velocity v_p is independent of k or ω, the waves (or the wave medium) are said to be *nondispersive.* In such case, the terms in a series like Eq. (VI.7) will propagate in lockstep, and the shape of the waveform is preserved. This occurs if and only if angular frequency ω is a linear function of wave number k, i.e., the *dispersion relation* $\omega = \omega(k)$ is linear. This is also equivalent to the equality of phase and group velocities: $v_p = v_g$, as defined by Eqs. (VI.6). The contrary conditions are also equivalent: v_p depends on k and ω if and only if the dispersion relation

---

[9]Josiah Willard Gibbs, 1839–1903

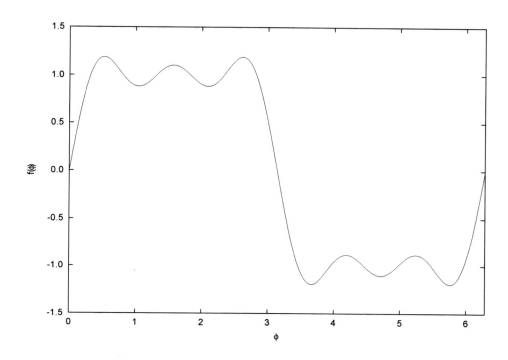

Figure VI.4: Fourier series (VI.7) for a square wave, first three terms (harmonics).

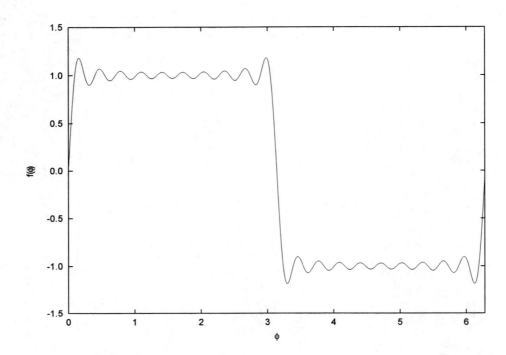

Figure VI.5: Fourier series (VI.7) for a square wave, first ten harmonics.

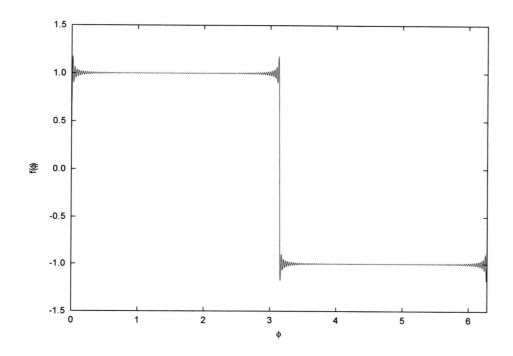

Figure VI.6: Fourier series (VI.7) for a square wave, first one hundred harmonics.

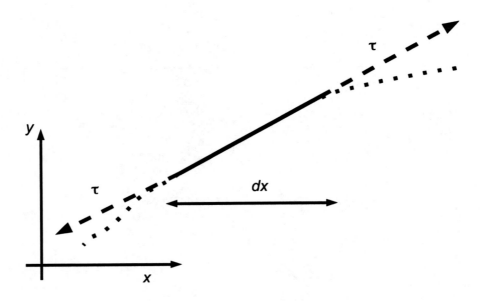

Figure VI.7: Forces on an infinitesimal segment of a stretched string with transverse displacement $y(x, t)$.

$\omega = \omega(k)$ is nonlinear, if and only if the phase and group velocities are unequal $(v_p \neq v_g)$, in which case the wave medium is *dispersive,* and the shapes of wave forms change—disperse—as they propagate. The most famous example of dispersion is that of light waves in materials such as water and glass. This gives rise to the rainbow and the breaking of white light into colors by a prism. For any waves, the governing wave equation determines the dispersion relation and its implications.

C   Wave dynamics

Wave equation for a stretched string

The wave equation for transverse waves on a stretched string follows from Newton's Second Law, applied to each infinitesimal segment of the string. The geometry is depicted in Fig. VI.7, but it is assumed that the displacement y

of the string is small compared to a wavelength ($|y| \ll \lambda$), so that the angle of inclination of the string is small. Hence, to first order in this angle, its sine and tangent are equal to the angle (in radians), and its cosine is unity. If the string carries constant tension τ, the segment shown experiences upward force $\tau\, \partial y/\partial x$ on the right and downward force $-\tau\, \partial y/\partial x$ on the left. (Here $\partial y/\partial x$ is the slope of the string, the tangent of the inclination angle—equal to its sine, with $|\partial y/\partial x| \ll 1$. Partial derivatives, such as those described in Ch. II Sec. E, must be used here, since y is a function both of distance x along the string and time t.) The net vertical force on the segment is the sum of these, i.e., $\tau\,(\partial^2 y/\partial x^2)\, dx$. In the same approximation, the net horizontal force is zero. If the string has uniform linear mass density μ, the mass of the segment is $\mu\, dx$. Newton's Second Law implies

$$\tau\, \frac{\partial^2 y}{\partial x^2}\, dx = \mu\, dx\, \frac{\partial^2 y}{\partial t^2} \tag{VI.8a}$$

for the segment, where the (vertical) acceleration has also been written in terms of partial derivatives. This yields

$$\frac{\partial^2 y}{\partial x^2} - \frac{\mu}{\tau}\, \frac{\partial^2 y}{\partial t^2} = 0\ , \tag{VI.8b}$$

the desired wave equation for transverse waves on the string.

In some texts, this is called "*the* wave equation," which gives the mistaken impression that there is only one. More accurately, it is *a* wave equation, one of many. Specifically, it is a second-order *partial differential equation* (PDE). Equations of this type can be written in the general form

$$\sum_{i,j} a_{ij}\, \frac{\partial^2 u}{\partial x_i\, \partial x_j} + \cdots = 0\ , \tag{VI.9}$$

where x_i, x_j are the various independent variables. The character of the equation is set by the matrix (a_{ij}) of coefficients of the second-derivative terms. The equation is termed *elliptic, parabolic,* or *hyperbolic,* when the determinant $|a_{ij}|$ is positive, zero, or negative, respectively. This character determines the nature of the solutions of the equation and the type of problems to which it applies. Elliptic equations describe equilibrium configurations, such as potentials in Newtonian gravitation (Ch. IV) or electrostatics (Ch. VII). Hyperbolic equations describe waves. Parabolic equations describe diffusion processes, such as heat flow.[10] Wave equation (VI.8b) is hyperbolic, as its two second-derivative terms have opposite signs. It is a *linear* equation—both terms are first-order in y—so superposition is valid: Any linear combination of solutions with constant coefficients is another solution.

The general solution of wave equation (VI.8b) can be written

$$y(x,t) = f(x - vt) + g(x + vt)\ , \tag{VI.10a}$$

---

[10]The Schrödinger wave equation of quantum mechanics breaks this rule. In form it is parabolic, but because it has an imaginary coefficient, its solutions propagate as waves.

a superposition of a right-moving wave and a left-moving wave of any (and independent) shapes. This is a solution of wave equation (VI.8b) for any twice-differentiable functions f and g, provided only that the velocity v is

$$v = \left(\frac{\tau}{\mu}\right)^{1/2} . \tag{VI.10b}$$

For example, for the apparatus shown in Fig. VI.1, the string might be under tension $\tau = 9.8$ N (obtained by suspending a 1.0 kg mass from one end) and possess mass density $\mu = 2.0$ g/m. The wave velocity is

$$\begin{aligned}
v &= \left(\frac{\tau}{\mu}\right)^{1/2} \\
&= \left(\frac{9.8 \text{ N}}{2.0 \times 10^{-3} \text{ kg/m}}\right)^{1/2} \\
&= 70. \text{ m/s} .
\end{aligned} \tag{VI.10c}$$

In terms of the angular frequency and wave number of sinusoidal waves, condition (VI.10b) takes the form of dispersion relation

$$\omega(k) = \left(\frac{\tau}{\mu}\right)^{1/2} k . \tag{VI.10d}$$

As this is linear, the string is a *nondispersive* medium; velocity v, independent of k and ω, is both phase and group velocities. As stated in the preceding section, all these features follow from the wave equation.

Energy transport

Waves transport energy, a phenomenon observed in many contexts. For example, a transverse wave on the stretched string has kinetic energy per unit length

$$\frac{dK}{dx} = \tfrac{1}{2}\mu \left(\frac{\partial y}{\partial t}\right)^2 , \tag{VI.11a}$$

since $\partial y/\partial x$ is the transverse velocity of each segment of string. And if a segment of string is stretched transversely by the transverse component of the tension, it stores potential energy like a stretched spring (Eq. (II.33b)). Hence, the wave has potential energy per unit length

$$\frac{dU}{dx} = \tfrac{1}{2}\tau \left(\frac{\partial y}{\partial x}\right)^2 , \tag{VI.11b}$$

since $\tau \, \partial y/\partial x$ represents the spring force ($k \, dy$) and $dy = (\partial y/\partial x)\, dx$ its displacement. Propagation of the wave transports these energy densities at the wave speed. The corresponding energy-transport rate or power for a wave traveling in one direction is

$$P = \left(\frac{dK}{dt} + \frac{dU}{dx} \right) v$$

$$= \tfrac{1}{2} \left[\mu \left(\frac{\partial y}{\partial t} \right)^2 + \tau \left(\frac{\partial y}{\partial x} \right)^2 \right] v$$

$$= \mu v \left(\frac{\partial y}{\partial t} \right)^2 . \tag{VI.11c}$$

The two terms are equal because the displacement y is a function of $x - vt$, say, implying $\partial y/\partial t = -\partial y/\partial x$, with v given by Eq. (VI.10b). This is the instantaneous power or energy-transport rate at a given location and a given instant. It is often useful to measure the energy-transport rate averaged over one or more cycles of a periodic wave. For example, for a sinusoidal wave with wave function $y(x,t) = A \sin(kx - \omega t)$, this average power is

$$\langle P \rangle = \mu v \left\langle \left(\frac{\partial y}{\partial t} \right)^2 \right\rangle$$

$$= \mu v \omega^2 A^2 \langle \cos^2(kx - \omega t) \rangle$$

$$= \tfrac{1}{2} \mu v \omega^2 A^2 . \tag{VI.11d}$$

The average value $\langle \cos^2(kx - \omega t) \rangle = \tfrac{1}{2}$—a useful quantity to remember—can be obtained by integrating over $2\pi/\omega$ and dividing by that time interval, or by noting that the average over one or more periods must equal $\langle \sin^2(kx - \omega t) \rangle$, while the two averages must sum to unity. Instantaneous power P and average power $\langle P \rangle$ both vary as the square of the wave amplitude, A^2, a feature common to many types of waves.

Boundary conditions

The behavior of waves can be modified by *boundary conditions* imposed on the medium in which they propagate. Because the equation of motion for waves is a partial differential equation, conditions on the solutions at the boundaries of the domain of the spatial variables, which hold for all time—as well as initial conditions imposed at a particular time—are needed for a well-posed dynamical problem, one with a unique solution. For example, a stretched string of finite length might have a fixed end at, say, $x = L$, corresponding to the condition $y(L, t) = 0$ for all t. Or it might have a free end—an end tied to a massless ring free to slide frictionlessly along a post. Since the vertical force on the ring must be zero (else its acceleration would be infinite), this corresponds to the boundary condition $\partial y/\partial x(L, t) = 0$ for all t. (These have names: The fixed end is an example of a *Dirichlet* boundary condition on the wave function itself; the free end imposes a *Neumann*[11] boundary condition on its derivative.) These

---

[11]Carl Gottfried Neumann, 1832–1925

conditions give rise to *reflections* at the boundary: At a fixed end, the force exerted on the string gives rise to a reflected wave opposite in displacement to the incoming wave and propagating in the opposite direction; the superposition of the two waves enforces the boundary condition. At a free end, the reflected wave displaces in the same direction as the incoming while propagating in the opposite direction; the superposition of these maintains zero slope at the end point.

The superposition of waves traveling in opposite directions, such as a wave and its reflection at an end point, can create *standing waves.* These are patterns of oscillation that do not propagate, such as those shown in Fig. VI.1. For example, two sinusoidal waves of equal amplitude and wave number, propagating in opposite directions, yield the pattern

$$y(x,t) = A \sin(kx - \omega t) + A \sin(kx + \omega t)$$
$$= 2A \sin(kx) \cos(\omega t) , \qquad (VI.12)$$

as follows from the angle sum/difference identities for the sine functions. Each segment of the string oscillates with angular frequency ω, but the pattern remains fixed in space. Points at which the oscillation amplitude is zero are *nodes,* here at $x = n\pi/k$ for all integers n. Points at which that amplitude is maximal are *antinodes,* here midway between the nodes at $x = (n + \frac{1}{2})\pi/k$.

Sinusoidal standing waves on a string are normal modes of the string, as described in Ch. V Sec. F previously: All parts of the string execute simple harmonic motion with a common frequency. For example, for a string with fixed ends at $x = 0$ and $x = L$, these take the form

$$y_n(x,t) = 2A \sin\left(\frac{n\pi}{L} x\right) \cos(\omega_n t) \qquad (VI.13)$$

for $n = 1, 2, 3, \ldots$. The choice of wave numbers $k_n = n\pi/L$ enforces the boundary conditions $y(0,t) = 0$ and $y(L,t)$. The corresponding angular frequencies are $\omega_n = n\pi v/L$, with v from Eq. (VI.10b). That is, the boundary conditions at the ends of the string force the normal-mode wave numbers, angular frequencies, and frequencies to take discrete values.[12] The normal modes of the string are also called *harmonics,* like the terms in the Fourier series (VI.7). The $n = 1$ mode is the *fundamental* mode; its wave function has one *loop* and no interior nodes. The higher harmonics, with $n > 1$, have n loops and $n - 1$ interior nodes. The index n runs to infinity only if the string is approximated as continuous. For a real string made of atoms, modes with wavelengths shorter than the interatomic spacing are not meaningful.

Any oscillation of this string can be described as a superposition of its normal-mode motions or harmonics, as indicated in Ch. V Sec. F, a fact demonstrated by Daniel Bernoulli in 1753. Combined with the 1748 result of Euler that the string could be made to vibrate in any shape, this implies that an arbitrary function can be represented by sinusoids—the proposal, as mentioned in Sec. B preceding, advanced by Fourier half a century later.

---

[12]This feature is the origin of *energy levels* in quantum mechanics.

The harmonics of musical instruments determine their individual sounds. For example, the seventh harmonic of a note strikes the ear as discordant. For this reason the hammers of a piano are designed to strike the strings one-seventh of their lengths from one end, suppressing that harmonic by preventing a node there. A B♭ clarinet essentially creates a vibrating air column with one fixed and one free end. The length of the instrument is $\frac{1}{4}, \frac{3}{4}, \frac{5}{4}, \ldots$ the wavelength of the successive harmonics; the corresponding frequencies form an odd-integer progression. As suggested by Fourier series (VI.7), the clarinet produces tones of approximately square-wave form. In contrast, the harmonics of tympani (kettledrums) are described by wave functions of two spatial dimensions and time, constructed not of sinusoids but of Bessel[13] functions (other members of the menagerie of functions). The corresponding normal-mode frequencies do not form an integer progression at all.

D Scattering at a knot

We are now poised to tackle the basic wave-scattering problem introduced at the start of this chapter. We consider a very long string, extending far to the left and right of a knot of mass M at position $x = 0$. Boundary conditions at the distant ends of the string can be disregarded. A traveling wave impinges on the knot from the left, moving in the $+x$ direction. It is partially reflected and partially transmitted at the knot. How much of each?

The waves on the string consist of three components: The *incident wave*, on the left side of the string, has wave function

$$y_{\mathcal{I}}(x, t) = \alpha \, \sin(kx - \omega t) \qquad \text{for } x < 0. \tag{VI.14a}$$

The *reflected wave,* moving in the $-x$ direction on the left side of the string, has wave function

$$y_{\mathcal{R}}(x, t) = \mathcal{A} \, \sin(kx + \omega t + \delta_{\mathcal{R}}) \qquad \text{for } x < 0. \tag{VI.14b}$$

The displacement on the left side of the string is the superposition of these two waves. The *transmitted wave,* moving in the $+x$ direction on the right side of the string, has wave function

$$y_{\mathcal{T}}(x, t) = \mathcal{B} \, \sin(kx - \omega t + \delta_{\mathcal{T}}) , \qquad \text{for } x > 0. \tag{VI.14c}$$

There is no left-moving, incident wave on the right side of the string. These forms allow for phase differences $\delta_{\mathcal{R}}$ and $\delta_{\mathcal{T}}$ between the reflected and transmitted waves, respectively, and the incident wave.

Boundary conditions at the knot determine the dynamics of this scattering process. For simplicity, we take the knot to be of negligible *size.* Then the string must be *continuous* at the knot, i.e., the condition

$$\lim_{x \to 0^-} y(x, t) = \lim_{x \to 0^+} y(x, t) \tag{VI.15a}$$

---

[13] Friedrich Wilhelm Bessel, 1784–1846

implies

$$-\alpha \sin(\omega t) + \mathcal{A} \sin(\omega t + \delta_{\mathcal{R}}) = -\mathcal{B} \sin(\omega t - \delta_{\mathcal{T}}) \qquad \text{(VI.15b)}$$

for all time t. The slope of the string is *not* continuous. The discontinuity in the slope at $x = 0$ supplies a net vertical force, accounting for the acceleration of the knot mass. Newton's Second Law for the knot takes the form

$$\tau \left(\lim_{x \to 0^+} \frac{\partial y}{\partial x} - \lim_{x \to 0^-} \frac{\partial y}{\partial x} \right) = M \left. \frac{\partial^2 y}{\partial t}^2 \right|_{x=0} , \qquad \text{(VI.16a)}$$

which implies condition

$$k\tau[\mathcal{B} \cos(\omega t - \delta_{\mathcal{T}}) - \alpha \cos(\omega t) - \mathcal{A} \cos(\omega t + \delta_{\mathcal{R}})] = M\omega^2 \mathcal{B} \sin(\omega t - \delta_{\mathcal{T}}) , \qquad \text{(VI.16b)}$$

again for all time t. This looks like a loss: two conditions for the four unknowns $\mathcal{A}$, $\mathcal{B}$, $\delta_{\mathcal{R}}$ and $\delta_{\mathcal{T}}$. (Parameters α, k, and ω are inputs.) Not so, because the two conditions must hold at the knot at all times. Using angle-sum formulae for the sine and cosine functions in the conditions yields

$$(\mathcal{A} \sin \delta_{\mathcal{R}} - \mathcal{B} \sin \delta_{\mathcal{T}}) \cos(\omega t) + (\mathcal{A} \cos \delta_{\mathcal{R}} + \mathcal{B} \cos \delta_{\mathcal{T}} - \alpha) \sin(\omega t) = 0 \qquad \text{(VI.17a)}$$

from Eq. (VI.15b) and

$$[k\tau(\mathcal{B} \cos \delta_{\mathcal{T}} - \alpha - \mathcal{A} \cos \delta_{\mathcal{R}}) + M\omega^2 \mathcal{B} \sin \delta_{\mathcal{T}}] \cos(\omega t)$$
$$+ [k\tau(\mathcal{B} \sin \delta_{\mathcal{T}} + \mathcal{A} \sin \delta_{\mathcal{R}}) - M\omega^2 \mathcal{B} \cos \delta_{\mathcal{T}}] \sin(\omega t) = 0 \qquad \text{(VI.17b)}$$

from Eq. (VI.16b). But $\cos(\omega t)$ and $\sin(\omega t)$ are *independent* functions—not constant multiples of each other. Any linear combination of the two with constant coefficients can only be zero for all t if those coefficients are individually zero. That implies the four conditions

$$\mathcal{A} \sin \delta_{\mathcal{R}} - \mathcal{B} \sin \delta_{\mathcal{T}} = 0 ,$$
$$\mathcal{A} \cos \delta_{\mathcal{R}} + \mathcal{B} \cos \delta_{\mathcal{T}} = \alpha ,$$
$$\mathcal{B} \cos \delta_{\mathcal{T}} - \mathcal{A} \cos \delta_{\mathcal{R}} = \alpha - \frac{M\omega^2}{k\tau} \mathcal{B} \sin \delta_{\mathcal{T}} ,$$
$$\text{and} \quad \mathcal{B} \sin \delta_{\mathcal{T}} + \mathcal{A} \sin \delta_{\mathcal{R}} = \frac{M\omega^2}{k\tau} \mathcal{B} \cos \delta_{\mathcal{T}} \qquad \text{(VI.18)}$$

for the four unknowns. These can be combined to yield solutions

$$\delta_T = \text{Tan}^{-1}\left(\frac{M\omega^2}{2k\tau}\right)$$

$$\delta_R = \text{Tan}^{-1}\left(\frac{2k\tau}{M\omega^2}\right)$$

$$\mathcal{A} = \frac{\alpha \dfrac{M\omega^2}{2k\tau}}{\left(1 + \dfrac{M^2\omega^4}{4k^2\tau^2}\right)^{1/2}}$$

$$\mathcal{B} = \frac{\alpha}{\left(1 + \dfrac{M^2\omega^4}{4k^2\tau^2}\right)^{1/2}} \tag{VI.19}$$

for the phase shifts and amplitudes.

The reflection and transmission coefficients are the ratios of the average power of the reflected and transmitted waves, respectively, to that of the incident wave. Result (VI.11d) implies that these are given by

$$\begin{aligned}
\mathcal{R} &= \frac{\langle P_R \rangle}{\langle P_I \rangle} \\[2mm]
&= \frac{\mathcal{A}^2}{\alpha^2} \\[2mm]
&= \frac{\dfrac{M^2\omega^4}{4k^2\tau^2}}{1 + \dfrac{M^2\omega^4}{4k^2\tau^2}} \\[4mm]
&= \frac{M^2\omega^2}{4\tau\mu + M^2\omega^2}
\end{aligned} \tag{VI.20a}$$

for reflection and

$$
\begin{aligned}
\mathcal{T} &= \frac{\langle P_\mathcal{T} \rangle}{\langle P_\mathcal{I} \rangle} \\
&= \frac{\mathcal{B}^2}{\alpha^2} \\
&= \frac{1}{1 + \dfrac{M^2 \omega^4}{4 k^2 \tau^2}} \\
&= \frac{4 \tau \mu}{4 \tau \mu + M^2 \omega^2}
\end{aligned}
\tag{VI.20b}
$$

for transmission. Here, $\omega/k = v$ and $v = (\tau/\mu)^{1/2}$ from Eq. (VI.10b) were used to simplify the expressions.

These results satisfy some simple checks. Energy is conserved, as the reflection and transmission coefficients satisfy $\mathcal{R} + \mathcal{T} = 1$ for all parameter values. In the limits $M \to 0$ (no knot) or $\omega \to 0$ (quasistatic displacement of the string), there is no reflection and perfect transmission. This is displayed by the limit values $\mathcal{R} \to 0$, $\mathcal{T} \to 1$, and $\delta_\mathcal{T} \to 0$. In the opposite limits $M \to \infty$ (immovably massive knot) or $\omega \to \infty$ (oscillations too fast for the knot to respond), the knot acts as a fixed end. This yields $\mathcal{R} \to 1$, $\delta_\mathcal{R} \to 0$, and $\mathcal{T} \to 0$—perfect reflection and no transmission.

Another success! As with previous chapter projects, a fair amount of detailed work was needed to reach fairly simple results. But this toy calculation exemplifies a breathtaking variety of wave-scattering problems. The language of reflection, transmission, and phase shifts is used in analyses from radar mapping of distant planets, to atmospheric optical phenomena, to human vision, to sonar, to x-ray crystallography, to Computerized Axial Tomography (CAT scans), to ultrasonic imaging, to nuclear and particle scattering experiments where quantum waves are in play. Just how much of what we know about the world around us comes down to wave scattering?

E   Sound waves

Before leaving the subject, we can apply these principles of wave kinematics and dynamics to the description of sound waves. These are longitudinal compression/rarefaction waves in solids, liquids, and gases. The wave disturbance is the displacement from equilibrium of the medium in the direction of propagation, e.g.,

$$
\Delta x(x, t) = A \sin(kx - \omega t) \ . \tag{VI.21a}
$$

The displacement amplitudes of sound waves in the range of human hearing are remarkably small: At the *threshold of hearing,* at frequency $\nu \approx 1000$ Hz in air, that amplitude is of order $A \approx 10^{-11}$ m—an order of magnitude smaller than atomic diameters.[14] Sound waves at the *threshold of pain* have amplitudes of order $A \approx 10^{-5}$ m, some six orders of magnitude larger.

Sound waves can equivalently be described as pressure disturbances from equilibrium. The variation in displacement Δx in the direction of propagation corresponds to the compression or rarefaction of segments of the medium. This is described by pressure wave function

$$\Delta p(x,t) = -B\,\frac{\partial \Delta x}{\partial x}$$
$$= -BkA\,\cos(kx - \omega t) \qquad \text{(VI.21b)}$$

for the previous displacement wave function. Here, B is the *bulk modulus* of the medium, the ratio of pressure increment to fractional change in volume, as per

$$dP = -B\,\frac{dV}{V}\ . \qquad \text{(VI.21c)}$$

The negative sign corresponds to the fact that an increase in pressure yields a decrease in volume, so that modulus B is a positive quantity. Forms (VI.21a) and (VI.21b) show that the displacement wave has nodes where the pressure wave has antinodes, and vice versa. The *pressure amplitude* is related to the displacement amplitude via $\Delta p_{\max} = BkA$. This too is very small for audible waves. Waves in air at the threshold of hearing, at $\nu \approx 1000$ Hz, have pressure amplitude roughly $\Delta p_{\max} \approx 30\ \mu$Pa. Those at the threshold of pain have $\Delta p_{\max} \approx 30$ Pa, again six orders of magnitude larger. And the pascal[15] (1 Pa $\equiv$ 1 N/m^2) is a small unit of pressure: Standard atmospheric pressure is 101.325 kPa. The sensitivity of the human ear, both to displacement and to pressure, is truly extraordinary.

A wave equation for sound, similar in form to Eq. (VI.8b), can be derived by applying Newton's Second Law to infinitesimal segments of the medium, accelerated by the differential of the pressure displacement. Like Eq. (VI.8b), this equation implies the speed of the sound waves. The speed of sound is given by

$$v_s = \left(\frac{B}{\rho}\right)^{1/2} \qquad \text{(VI.22)}$$

in general, with B the bulk modulus and ρ the volume mass density of the medium. For example, this gives sound speed in water $v_s^{(\mathrm{H_2O})} = 1480$ m/s, and in steel $v_s^{(\mathrm{Steel})} = 4530$ m/s. The bulk moduli of gases are sensitive to the details of the compression/rarefaction process. For sound waves in the audible frequency range,[16] this process is so fast that heat transfer in the gas is negligible, i.e., the oscillations are *adiabatic*. The bulk modulus associated with such processes is given by

$$B = \gamma p\ , \qquad \text{(VI.23a)}$$

---

[14]Since the air is a gas of molecules, this must represent an average of not-quite-random molecular displacements.

[15]Blaise Pascal, 1623–1662

[16]This is roughly from $\nu = 20$ Hz to 20 kHz for healthy young people.

where p is the equilibrium gas pressure and γ is the *ratio of specific heats.* Readers may recall from an introductory chemistry or thermodynamics course that this is $\gamma = (n+2)/n$, where n is the number of dynamical degrees of freedom of a gas molecule. A monatomic molecule can only move in three dimensions (n=3); a diatomic molecule can move in three dimensions and rotate about two directions perpendicular to the molecule (n=5); a triatomic or more complex molecule can move in three dimensions and rotate about any of three directions (n=6). Hence, monatomic gases have ratio $\gamma = \frac{5}{3}$, diatomic gases $\gamma = \frac{7}{5}$, all others $\gamma = \frac{4}{3}$. The speed of sound in the gas is

$$v_s = \left(\frac{\gamma p}{\rho} \right)^{1/2} . \qquad \text{(VI.23b)}$$

Air is a mixture mostly of diatomic gases, N_2 and O_2. The resulting speed of sound in air is $v_s^{(\text{air})} = 331.45$ m/s at Standard Temperature and Pressure—0.00° C and 101.325 kPa.

Energy transport by sound waves is measured by both *intensity* and *loudness.* Intensity is average energy flux, i.e., average energy transport per unit time (power) per unit area. In terms of wave functions VI.21, this is given by

$$
\begin{aligned}
I &= \left\langle \left[\tfrac{1}{2}(-\Delta p) \frac{\partial \Delta x}{\partial x} + \tfrac{1}{2}\rho \left(\frac{\partial \Delta x}{\partial t} \right)^2 \right] v_s \right\rangle \\
&= \tfrac{1}{2} \left\langle B \left(\frac{\partial \Delta x}{\partial x} \right)^2 + \rho \left(\frac{\partial \Delta x}{\partial t} \right)^2 \right\rangle v_s \\
&= \tfrac{1}{2} B k \omega A^2 \\
&= \tfrac{1}{2} \frac{v_s}{B} \left(\Delta p_{\text{max}} \right)^2 ,
\end{aligned}
\qquad \text{(VI.24)}
$$

the last two expressions appropriate to sinusoidal waves as illustrated in Eqs. (VI.21). The first term in the first two lines is a Hooke's Law-type potential-energy density, the second a kinetic-energy density, similar to the terms appearing in Eq. (VI.11c). Loudness or *decibel*[17] *level* is a measure of sensory response. It is defined as

$$\beta \equiv (10.0 \text{ dB}) \log_{10} \left(\frac{I}{I_0} \right) , \qquad \text{(VI.25a)}$$

with

$$
\begin{aligned}
I_0 &\equiv 1 \times 10^{-10} \ \mu\text{W/cm}^2 \\
&\equiv 1 \times 10^{-12} \ \text{W/m}^2 ,
\end{aligned}
\qquad \text{(VI.25b)}
$$

an exact value, approximately the intensity at the threshold of human hearing. A logarithmic measure of sensory response is not unusual: The modern magnitude scale for gauging the brightness of astronomical objects is also logarithmic in their light intensity. Presumably, human senses have logarithmic responses

---

[17]Alexander Graham Bell, 1847–1922

because this offers much greater dynamic (functional) range than would be possible with linear senses.

Sound waves exhibit the *Doppler*[18] *effect*, a change in frequency due to the motion of source or observer. For example, if a source of sound waves is moving, relative to the medium, toward a stationary observer at velocity v, then successive wave crests will be emitted at spatial intervals less than the wavelength produced by a stationary source by the distance covered by the moving source in a period of the waves. This implies that the frequency perceived by the observer will be increased by the factor

$$\nu_{\text{obs}} = \nu_{\text{stat}} \, \frac{v_s}{v_s - v} \, . \tag{VI.26a}$$

This result applies for a source receding from a stationary observer as well; for $v < 0$ the frequency is shifted lower. This is readily observed, e.g., by listening to the horn of an approaching and receding car or locomotive. The apparent divergence in the limit $v \to v_s$ corresponds to the formation of a shock wave or "sonic boom" by a transsonic or supersonic object. Similarly but not identically, an observer moving with respect to the medium toward a stationary source with velocity v (or away for $v < 0$) will encounter the oncoming wave crests at shorter (longer) intervals than will an observer at rest. The observed frequency is increased (decreased) by factor

$$\nu_{\text{obs}} = \nu_{\text{stat}} \, \frac{v_s + v}{v_s} \, . \tag{VI.26b}$$

In either case, the relative frequency shift $\Delta\nu/\nu_{\text{stat}}$ is approximately v/v_s for low speeds $|v| \ll v_s$.

There is a similar Doppler effect for light waves, but in this case there is no medium. The moving-source and moving-observer situations are dynamically equivalent, related by a Lorentz transformation like those described in Ch. I Sec. H. (Another contributing effect, as described by those transformations, is that source and observer measure time intervals differently.) For source and observer approaching in one spatial dimension at relative velocity v (receding for $v < 0$), the observed frequency of the waves is altered by factor

$$\nu_{\text{obs}} = \nu_{\text{stat}} \left(\frac{c + v}{c - v} \right)^{1/2} , \tag{VI.27}$$

where c is the speed of light. This effect allows astronomers to measure, for example, the velocities of the stars toward or away from the Earth.[19] Here, too, for low speeds $|v| \ll c$, the relative frequency shift $\Delta\nu/\nu_{\text{stat}}$ is approximately v/c. Radar (microwave) and infrared-laser speed detectors rely on this: The signal reflected from a moving vehicle is combined with the source signal to form a

---

[18]Christian Johann Doppler, 1803–1853

[19]Many textbooks describe the shift in frequency of light from distant galaxies, which signals the expansion of the universe, as a Doppler shift. This is a misconception—that shift is an entirely different phenomenon.

beat pattern. The beat frequency, counted electronically, is proportional to the vehicle's speed. Typical beat frequencies are in the kHz range, so the speed can be measured very quickly. Understanding this mechanism, the reader can deduce the author's secret method for avoiding speeding tickets.[20]

References

Mechanical waves and sound are fundamental topics. Most physics texts at both introductory and intermediate levels devote one or more entire chapters to them.

At one time, efforts were made to promote wave mechanics to a more central role in the introductory physics curriculum. For example, the Physical Science Study Committee (PSSC) at MIT produced materials for high school physics classes and labs, with enhanced emphasis on wave physics, used extensively in the 1960s and 1970s.

The problem of the scattering of waves on a stretched string at a point mass featured in this chapter appears as an exercise (Problem 13-20) in the intermediate-level text by Thornton and Marion [TM04]. The results to be shown in that exercise are equivalent to those presented here.

Problems

1. Pictured in Fig. VI.1 is a string with two essentially fixed ends, oscillating in its $n = 3$ standing-wave mode with displacement amplitude 1.0 cm. As calculated in Eq. (VI.10c), the wave speed on the string is 70. m/s. The frequency of the oscillations, determined by reading the frequency of the driving oscillator at resonance with this mode, is $\nu = 46$. Hz. These are the actual values for the apparatus in the photograph.

   (a) Write the standing-wave form (VI.13) of the wave function for this pattern with numerical values for all coefficients. For simplicity, take the $t = 0$ phase of the oscillations to be zero.

   (b) Write the same wave function as a superposition of two traveling sinusoidal waves moving in opposite directions. Again, detemine numerical values for all coefficients in the resulting expression.

2. Cosmic string, if it exists, bears tension equal to its linear mass density times the vacuum speed of light squared: $\tau = \mu c^2$. Calculate the speed of transverse waves on such string, assuming it obeys wave equation (VI.8b). Suppose a 1.00 m length of cosmic string with fixed ends (how???) is set vibrating in its fundamental mode with displacement amplitude 1.00 mm. What is the maximum speed of any portion of the string? Is this a Newtonian or an Einsteinian system?

---

[20]Don't speed.

3. In one spatial dimension, the Schrödinger equation for the quantum wave function $\psi(x,t)$ describing a free *Newtonian* particle of mass m is

$$-\frac{\hbar^2}{2m}\frac{\partial^2\psi}{\partial x^2} = i\hbar\frac{\partial\psi}{\partial t} \ ,$$

where $\hbar$ is Planck's reduced constant. This is a true wave equation despite its appearance. Find the dispersion relation for monochromatic waves satisfying this equation; here, monochromatic waves are represented by the complex form

$$\psi(x,t) = \alpha\,e^{i(kx-\omega t)}$$

rather than by real sinusoidal functions. Calculate the phase and group velocities of the waves. Are these waves dispersive or nondispersive? If the phase and group velocities are different, which corresponds to the classical velocity of the particle if its classical momentum is $p = \hbar k$?

4. In one spatial dimension, the Klein[21]-Gordon[22] equation for the quantum wave function $\varphi(x,t)$ describing a free *Einsteinian*particle of mass m is

$$\frac{\partial^2\varphi}{\partial x^2} - \frac{1}{c^2}\frac{\partial^2\varphi}{\partial t^2} - \frac{m^2c^2}{\hbar^2}\varphi = 0 \ ,$$

where c is the vacuum speed of light and $\hbar$ is Planck's reduced constant. Find the dispersion relation for monochromatic waves satisfying this equation, and calculate the phase and group velocities of the waves. Are these waves dispersive or nondispersive? If the phase and group velocities are different, which corresponds to the classical velocity of the particle?

5. The normal modes of the array of n coupled oscillators described in Problem 11 of Ch. V are essentially discrete longitudinal standing waves of form

$$x_j(t) = a_r\sin\left(\frac{\pi r j}{n+1}\right)\cos(\omega_r t - \delta_r) \ ,$$

with $j = 1, 2, \ldots, n$ for the n masses and $r = 1, 2, \ldots, n$ for the n normal modes. Parameters a_r and δ_r are the (constant) amplitude and phase shift, respectively, for each normal mode. These displacements satisfy the array of equations of motion sought in Ch. V Problem 11. Determine the normal-mode angular frequencies ω_r for this system.

6. The initial-value problem for wave equation (VI.8b) is similar to that for the harmonic oscillator, except that since the equation of motion is a partial differential equation, the initial data must be functions of x rather than single values. Consider a stretched string governed by that equation,

---

[21]Oskar Klein, 1894–1977
[22]Walter Gordon, 1893–1939

with fixed ends at $x = 0$ and $x = L$. At time $t = 0$ the displacement of the string is given by

$$y(x, 0) = \sum_{n=1}^{\infty} a_n \sin\left(\frac{n\pi x}{L}\right) ,$$

and the transverse velocity distribution on the string is given by

$$\frac{\partial y}{\partial t}(x, 0) = \sum_{n=1}^{\infty} b_n \sin\left(\frac{n\pi x}{L}\right) ,$$

with specified constant coefficients $\{a_n\}$ and $\{b_n\}$. Find the displacement $y(x, t)$ of the string for all time.

7. A long string with linear mass density μ_1 bearing tension τ, extending to the left, is joined smoothly at $x = 0$ to a second string with linear density μ_2 and the same tension τ, extending to the right. A sinusoidal wave incident from the left will be partially reflected and partially transmitted at the junction. Calculate the reflection and transmission coefficients and the phase shifts of the reflected and transmitted waves. Note that both the displacement and slope of the string must be continuous at $x = 0$; there is neither break nor kink in the string. Note also that waves on the two portions of the string must have the same time dependence, but not necessarily the same position dependence. This is the one-dimensional, mechanical equivalent of reflection and refraction at the boundary of two media.

8. Consider the two-component string of the previous problem. Is it possible to attach a bead of suitable mass M at the junction (like the knot in the problem in this chapter) so as to arrange that there be *no reflected wave* at the junction? What parameter values are required? This is a one-dimensional, mechanical equivalent of antireflective coatings in optics.

9. An *ideal gas* satisfies the Ideal Gas Law

$$pV = nRT ,$$

where p is the equilibrium gas pressure, V its volume, n the number of moles of the gas, R the Universal Gas Constant, and T the absolute temperature of the gas. (Real gases satisfy this law approximately.)

(a) Determine the speed of sound in an ideal gas of molar mass $\mathcal{M}$ as a function of its absolute temperature.

(b) Air is a diatomic gas mixture with effective molar mass $\mathcal{M}_{\mathrm{air}} = 28.9$ g/mol The speed of sound in air at temperatures near room temperature (293. K) can be approximated as

$$v_s^{(\mathrm{air})}(\delta T) = v_{s0} + \eta\,\delta T + \cdots ,$$

with $\delta T \equiv T - (293.\ \mathrm{K})$ and $|\delta T| \ll 293.$ K, where the ellipsis refers to terms of order δT^2 and higher. Evaluate v_{s0} and η.

Chapter VII

The Classical Electron

- *Develop and evaluate a model of the electron as a solid sphere of electric charge.*

- *Determine the "classical radius of the electron."*

With this chapter, we move from the study of mechanics to the exploration of electricity and magnetism. Known in ancient times, these forces were incorporated into the modern science of physics in the eighteenth and nineteenth centuries. Their study can be divided into four topics: electrostatics, the behavior of electric charges and forces not in motion; electric currents and circuits, electric charge in motion; magnetism[1]; and electromagnetic induction, which connects the two. These are the subjects of this chapter and Chs. VIII, IX, and X, respectively. The development of these topics culminates in the grand synthesis of James Clerk Maxwell, featured in Ch. XI. Maxwell's ideas subsume the entire science of optics—the subject of Ch. XII—into electromagnetic physics. A warning: These topics are generally found to be harder than introductory mechanics. The concepts are less tangible and less easily visualized. And they place greater demands on our facility with the tools of calculus and vector mathematics, many of which were invented particularly for these fields.

J. J. Thomson[2] is credited with discovering the electron as a separate constituent of matter in 1897. He measured its charge-to-mass ratio with an apparatus like that in Fig. VII.1 and found it to be some 1800 times greater than that of any known ion. This suggested that atoms, despite their name (which means "uncuttable"), have substructure. Some consider Thomson's discovery to be the actual beginning of the quantum-physics revolution, predating the introduction of the quantum hypothesis by Planck in 1900. We regard the electron as a fundamental particle of electricity. Although real electrons are governed by quantum mechanics, in this chapter we shall see how far we can push a simple classical model of the electron—a uniform-density sphere or ball of electric charge, say—to understand some of the observed features of the actual particle. To do so, we shall need to explore the notions of *electrostatics,* i.e., electricity not in motion.

A   Electrostatic "amber" force

The words *electricity* and *electron* derive from the Greek word ελεκτρον, amber—fossilized conifer sap. This follows from the observation, known to Aristotle and his contemporaries, that nonmetallic materials such as amber,[3] rubbed with cloth or fur, could be made to exert forces on each other through space without touching. Hence, the amber or electric (electrostatic) force.

Even simple demonstrations of this reveal important features of the force:

- *Like repels like,* i.e., identical objects treated identically repel each other. Hence, this is not some form of gravity—that produces attraction between identical objects.[4]

---

[1]The term *magnetostatics* is sometimes used, but this is a bit of a misnomer, as magnetism would be absent in a truly static situation.

[2]Joseph John Thomson, 1856–1940

[3]Modern demonstrations use, e.g., hard rubber rubbed with fur or glass rubbed with silk. Amber is a semiprecious stone, and physics departments aren't *that* well-funded.

[4]To impress people at parties, one observes that this is because the photon is a spin-one

Figure VII.1: In 1897, J. J. Thomson used an apparatus like this to measure the charge-to-mass ratio of the electron, discover that atoms had moving parts, and fire the opening salvo of the quantum revolution. (Photograph by author.)

- All the materials that display the electric force fall into two and only two categories. For example, hard rubber rubbed with fur is in one category, glass rubbed with silk in the other. All objects in each category repel one another and attract all objects in the other category. The categories are designated *positively* and *negatively charged*. This can be attributed to two different "flavors" of electric character or charge (two-fluid model) or to surplus and deficit of a single "flavor" (one-fluid model). Benjamin Franklin[5] is credited with recognizing that these two models are logically equivalent—any phenomenon described by one can equally be described by the other—and designating positive and negative charges. By Franklin's convention, rubber rubbed with fur acquires negative charge (the fur acquires positive charge), while glass rubbed with silk acquires positive charge (the silk becomes negative).[6]

- The electric force is a *central* force, as described in Ch. II Sec. E and Ch. IV Sec. A.

- The force increases in magnitude with decreasing distance between the objects, and vice versa.

Beyond such basic observations, however, remarkably little seems to have been done with the electric force for two millenia after Aristotle's time.[7] Only in the eighteenth century did scientists, well aware of the successes of Newton's mechanics and his Law of Universal Gravitation, seek to incorporate electricity into the new physics.

B　The Coulomb Law

Franklin and his contemporaries knew they needed a force law for the electric force to be used in Newton's Second Law of Motion—akin to the Law of Universal Gravitation. The form of this law was obtained from experiments performed by Cavendish and Coulomb[8] in 1784. (The law is named for Coulomb because he published his results first.[9]) They used torsion balances similar to that sketched in Fig. IV.4 with electric charges placed on the spheres. The charge on a sphere could be adjusted, e.g., by placing an identical, uncharged conducting sphere in contact with a charged sphere, reducing the charge on the latter by half. The distances between charges could be varied by moving the outer spheres on levers.

---

particle, the graviton spin-two.

[5] Benjamin Franklin, 1706–1790

[6] As will be seen later, Franklin made the wrong choice, though this would not be demonstrated for nearly a century after his time.

[7] The Battery of Baghdad, a medieval pottery vessel containing residues consistent with its use as an electrochemical wet cell, indicates that medieval Arab scientists may have experimented with electricity. If so, such research was either not pursued very far or not widely disseminated.

[8] Charles Augustin de Coulomb, 1736–1806

[9] Some aspects of science were the same then as now. Still, Cavendish did all right.

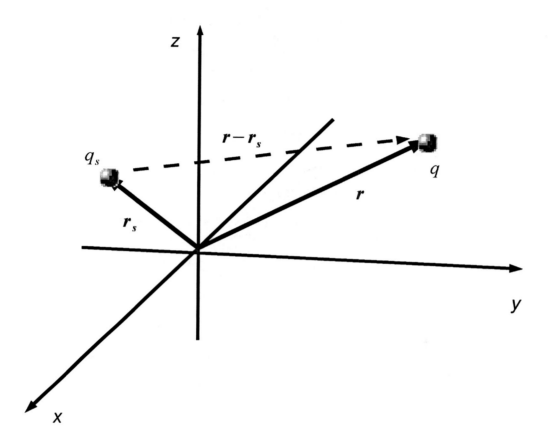

Figure VII.2: Electric charges and position vectors for the Coulomb Law.

The result of their measurements can be expressed in terms of the vectors of
Fig. VII.2: The force **F** on charge q, produced by source charge q_s, is

$$\mathbf{F} = +\frac{1}{4\pi\epsilon_0}\frac{qq_s(\mathbf{r}-\mathbf{r}_s)}{|\mathbf{r}-\mathbf{r}_s|^3}\ , \qquad\qquad (\text{VII.1})$$

with a force equal in magntiude and opposite in direction exerted on q_s by
charge q. Like the Law of Universal Gravitation (IV.2), this describes an inverse-
square, central force, and for the same reason: The electrostatic force is an
influence spreading outward from a point source in three-dimensional space.
But since charges q and q_s can be positive or negative independently, this force
law can describe either attraction or repulsion.

The *Coulomb constant* leading this expression is sometimes labeled k in
introductory treatments; the more elaborate form here is used to simplify other

expressions, which we shall see later. Its value depends on the units for electric charge. The SI unit of charge is the *coulomb* (C), the value of which we shall examine here, although its precise SI definition must wait until Ch. IX. Other subsets of the metric system use different units. For example, in *Gaussian* units, charge is measured in *electrostatic units* (esu), force in dynes ($g\,cm/s^2$), and distance in centimeters, setting the value of the Coulomb constant to unity. Fortunately, there are no electrical units from the so-called British system in use. To avoid confusion, we shall adhere to SI units throughout this text. The corresponding value of the Coulomb constant is

$$k = \frac{1}{4\pi\epsilon_0}$$

$$= 8.9875517873681764 \times 10^9 \ \frac{N\,m^2}{C^2} \ . \qquad\qquad \text{(VII.2a)}$$

The constant $\epsilon_0 = 1/(4\pi k)$ has a name; it is the *permittivity of free space* or *vacuum permittivity*. Its value in SI units is

$$\epsilon_0 = 8.854187817\ldots \times 10^{-12} \ \frac{C^2}{N\,m^2} \ . \qquad\qquad \text{(VII.2b)}$$

All those digits are not the outcome of some extraordinarily precise experiment. Value (VII.2a) is *exact,* emerging from the definitions of the SI units. It's $c^2 \times 10^{-7}$, and the vacuum speed of light c is a defined, exact integer value in SI. Hence, vacuum permittivity ϵ_0 is a nonterminating, nonrepeating decimal value because of the π factor, but it can be calculated to as many digits as desired.

The coupled questions of the strength of the electrostatic force and the size of the coulomb unit can be illuminated by some examples. If one were to hold a one-coulomb charge in each hand, one meter apart, then by force law (VII.1), the magnitude of the force between them would be approximately nine billion newtons. This is roughly equal to the weight of nine hundred thousand metric tons of mass, e.g., nine full-sized CVA or CVN aircraft carriers. Readers who have seen such a vessel may have noticed that they are large. This author's ability to lift very heavy objects is redoubtable [see Eq. (II.9) and the accompanying discussion], but nine aircraft carriers is a bit much. Perhaps the coulomb is an outsize unit of charge. But coulombs are not rare—a pocketful of brass keys contains some *one hundred thousand* coulombs of movable charge. How, then, could anyone carry an ordinary key ring? Because the one hundred thousand coulombs of movable negative charge in the valence electrons of the metal atoms of the keys are balanced to extraordinary precision by positive charges in the nuclei of those atoms. The naked Coulomb force of Eq. (VII.1) is rarely encountered directly by human beings. Most of the forces of our everyday experience—other than the Earth's gravity—are *residual* electric forces left over after the pure Coulomb forces have almost, but not completely, canceled out.

Owing to the famous oil-drop experiments of R. A. Millikan,[10] we now know—as Franklin and his contemporaries did not—that electric charge is not

---

[10] Robert Andrews Millikan, 1868–1953

a continuous fluid but comes in discrete packets, like matter itself. The *elementary charge* e, i.e., the charge of a proton or the negative of the charge on an electron,[11] has measured value

$$e = (1.6021773 \pm 0.0000005) \times 10^{-19} \text{ C}$$

or $\qquad 1.000000 \text{ C} = (6.241506 \pm 0.000002) \times 10^{18} \, e \; .$ $\qquad$ (VII.3)

Here, care must be taken to avoid confusion with the number $e = 2.718281828\ldots$, the base of the natural logarithm. The number of elementary charges in a coulomb seems large, but it is much smaller than the Avogadro[12] number, familiar from chemistry. The Avogadro number, i.e., a mole of elementary charges, is a unit of charge called the faraday[13], with value

$$1.000000 \text{ faraday} = 9.648531 \times 10^4 \text{ C} \; . \qquad \text{(VII.4)}$$

This is the amount of charge exchanged in one mole of a univalent chemical reaction and is the basis of the estimate for the keys described here. Nature has no difficulty assembling large numbers of coulombs, but as also noted, *separated* charges of this magnitude are not encountered in ordinary experience. The total net charge on the Earth is of the order of several faradays. (See Problem 1 at the end of this chapter.) Lightning strikes typically transfer between one and thirty colombs of charge. Classroom demonstrations with rubber and glass, such as those described previously, might involve tens of nanocoulombs.

To understand the strength of the electrostatic force, we should consider systems found in nature rather than human-defined quantities such as the coulomb and the faraday. Nature's favorite atom, the ordinary hydrogen atom (some 90% of the atoms in the universe are of this type), consists of an electron bound to a proton. The two particles exert both electric and gravitational forces on each other. If we treat the atom as a classical planetary system—not a very accurate model, but adequate for the present purpose—we can calculate the ratio of the magnitudes of the two forces from force laws (IV.2) and (VII.1). The force ratio is the dimensionless number

$$\frac{F_e}{F_g} = \frac{\dfrac{e^2}{4\pi\epsilon_0}}{G m_p m_e}$$

$$= \frac{\left(8.99 \times 10^9 \; \dfrac{\text{N m}^2}{\text{C}^2}\right)(1.602 \times 10^{-19} \text{ C})^2}{\left(6.673 \times^{-11} \; \dfrac{\text{m}^3}{\text{kg s}^2}\right)(1.67 \times 10^{-27} \text{ kg})(9.11 \times 10^{-31} \text{ kg})}$$

$$= 2.27 \times 10^{39} \; , \qquad \text{(VII.5)}$$

---

[11] The *quarks* that make up protons, neutrons, and other particles have charges $\pm\frac{1}{3}$ or $\pm\frac{2}{3}$ of this value, but they have never been observed on their own. All observed charges, to date, are multiples of e.

[12] Amedeo Avogadro, 1776–1856

[13] Michael Faraday, 1791–1867

where suitable values have been inserted for the force constants and the charges
and masses of the proton and electron. Because both forces are inverse-square
forces, that factor cancels, and the result is independent of any estimate of
the radius of the atom. In this system, with naturally determined masses
and charges, the electrostatic force is *thirty-nine orders of magntiude larger*
than the gravitational force. No known measurement is precise enough to detect the gravitational contribution. Apparently gravitation means nothing on
atomic scales. Yet on planetary scales and beyond, gravity is all-controlling—
electrostatic forces are negligible. (See problems 2 and 8 at the end of this
chapter.) Why? Because differences in the nature of the two forces make it possible to assemble very large masses on planetary scales but not large charges:
Since like charges repel, large accumulations of charge tend to blow themselves
apart, while large masses are bound together by gravitational attraction. Also,
the availability of charges of opposite signs means large charges tend to neutral-
ize themselves by attracting the opposite charge, but all mass is positive—there
is no mass neutralization.

As this calculation shows, the fundamental forces of nature vary widely in
strength. The fact that atomic nuclei contain positively charged protons held
together against their own electrostatic repulsion indicates that the *strong nuclear force* must be present and stronger than the electrostatic force—typically,
around one hundred times as strong. (This force is now understood to be a
residue of an even stronger force, the *color force* between quarks, of which protons and neutrons are made. The color force is described by the theory termed
quantum chromodynamics.[14]) The *weak nuclear forces* responsible for some
types of radioactive decay are usually some ten orders of magnitude weaker than
this, with gravitation more than forty orders of magnitude behind the strong
force. Understanding the hierarchy of strengths of the fundamental interactions
remains one of the great challenges of contemporary physics.

Force law (VII.1) applies between two point charges. But the Coulomb
force, like Newtonian gravitation, is linear in its sources: The force on charge q
produced by a collection of sources $\{q_s\}$, or a charge distribution treated as
continuous, with charge density $\rho(\mathbf{r})$, is given by the superpositions

$$\mathbf{F} = \frac{1}{4\pi\epsilon_0} \sum_{\text{sources}} \frac{qq_s(\mathbf{r} - \mathbf{r_s})}{|\mathbf{r} - \mathbf{r}_s|^3}$$
$$= \frac{q}{4\pi\epsilon_0} \int_{\text{source}} \frac{\mathbf{r} - \mathbf{r}'}{|\mathbf{r} - \mathbf{r}'|^3} \, \rho(\mathbf{r}') \, d\mathcal{V}' \, , \qquad (\text{VII.6})$$

as appropriate. Most introductory texts contain numerous exercises involving calculations of this type. But as they are essentially algebra or calculus
exercises—and can be evaluated exactly for only a few simple geometries—they
are not emphasized here. Two examples are examined in the following section.

The near identity of form between force laws (VII.1) and (VII.6) and
laws (IV.2) and (IV.5) has inspired scientists for centuries to seek a funda-

---

[14]The terms "color" and "chromo-" here are code language, having nothing to do with
visual color.

mental connection between the electric and gravitational forces. These include Riemann and Einstein; the latter spent much of his later career searching— unsuccessfully—for a Unified Field Theory. But such a search is ill-founded for two deep reasons:

- First, as we now understand, the electric and gravitational forces are only part of the picture. In fact, the modern Standard Model treats the electric, magnetic, and weak nuclear forces together as aspects of a single *electroweak* interaction, as described in the 1968 *electroweak theory* of Weinberg,[15] Glashow,[16] and Salam.[17] Models in which this and the chromodynamic, i.e., strong nuclear, force are treated as facets of a unified interaction are called *Grand Unified Theories* or GUTs. Efforts to identify a successful Grand Unified model have been ongoing since the 1970s. And models that combine these forces with gravitation, all as aspects of a single interaction, are called *Super Unified Theories* or *Theories of Everything,* although the last is something of a misnomer. *String theory* is the most publicized, but not the only such model, and to date, no such model has emerged as a successful description of nature.

- Second, the similarity of the electric and gravitational force laws is somewhat illusory when viewed from a broader persepective. The electric force and its magnetic counterpart, the subject of Ch. IX to follow, consist of six components in three-dimensional space. But the gravitational force is properly described by a *forty*-component object in four-dimensional spacetime. Not so alike after all.

An ultimate unification of these forces may eventually emerge, or maybe not: It is not at present certain that such a description is inevitable or necessary.

The more immediate significance of the similarity of the electrostatic and gravitational force laws is that they must yield the same motions, viz., Keplerian orbits. Bohr[18] and Sommerfeld[19] used these in their detailed, early-twentieth-century models of the atom.[20] Unbound orbits are used in the analysis of scattering of charged particles via the Coulomb force, a process known as Rutherford[21] scattering.

C   Electric fields

The Coulomb force law engenders the same question as its gravitational counterpart: How do the source and receiver "know" each other's positions to create

---

[15] Steven Weinberg, 1933–

[16] Sheldon Glashow, 1932–

[17] Mohammad Abdus Salam, 1926–1996

[18] Niels Henrik David Bohr, 1885–1962

[19] Arnold Johannes Wilhelm Sommerfeld, 1868–1951

[20] A beautiful but doomed effort, as the atom must be understood through quantum mechanics.

[21] Ernest Rutherford, 1871–1937

the right force? And it has the same resolution: Every charge exists surrounded by its *electric field,* which extends throughout space, and every charge responds to the electric field at its immediate location.

Definition

Like the gravitational field $\mathbf{g}$ in Eq. (IV.6), the electric field is defined as the limit

$$\mathbf{E}(\mathbf{r}) \equiv \lim_{q \to 0} \frac{\mathbf{F}(\mathbf{r})}{q} \; , \tag{VII.7a}$$

which casts the force law in the form

$$\mathbf{F} = q\mathbf{E} \tag{VII.7b}$$

in terms of the field $\mathbf{E}$. Form (VII.6) of the Coulomb law then implies

$$\begin{aligned}
\mathbf{E}(\mathbf{r}) &= \frac{1}{4\pi\epsilon_0} \sum_{\text{sources}} \frac{q_s(\mathbf{r} - \mathbf{r_s})}{|\mathbf{r} - \mathbf{r}_s|^3} \\
&= \frac{1}{4\pi\epsilon_0} \int_{\text{source}} \frac{\mathbf{r} - \mathbf{r}'}{|\mathbf{r} - \mathbf{r}'|^3} \, \rho(\mathbf{r}') \, d\mathcal{V}'
\end{aligned} \tag{VII.7c}$$

for the electric field of a collection of charges or a charge distribution. The SI units of the electric field are newtons per coulomb (N/C); no name is attached to this combination.

Calculation of electric fields

The electric field of a *dipole* charge distribution is an important example because it occurs so often in nature. The simplest version is illustrated in Fig. VII.3: equal-magnitude, opposite-sign point charges $\pm q$ (hence "*di*pole," two poles or point charges) separated by a fixed distance a. The electric field produced by this distribution is

$$\begin{aligned}
\mathbf{E}(\mathbf{r}) &= \frac{q}{4\pi\epsilon_0} \left(\frac{\mathbf{r} - \frac{a}{2}\hat{z}}{|\mathbf{r} - \frac{a}{2}\hat{z}|^3} - \frac{\mathbf{r} + \frac{a}{2}\hat{z}}{|\mathbf{r} + \frac{a}{2}\hat{z}|^3} \right) \\
&= \frac{q}{4\pi\epsilon_0} \left(\frac{\mathbf{r} - \frac{a}{2}\hat{z}}{\left(r^2 - ar\cos\theta + \frac{a^2}{4}\right)^{3/2}} - \frac{\mathbf{r} + \frac{a}{2}\hat{z}}{\left(r^2 + ar\cos\theta + \frac{a^2}{4}\right)^{3/2}} \right) \\
&= \frac{q}{4\pi\epsilon_0 r^3} \left\{ (\mathbf{r} - \tfrac{a}{2}\hat{z}) \left[1 + \tfrac{3}{2}\tfrac{a}{r}\cos\theta + O\left(\tfrac{a^2}{r^2}\right) \right] \right. \\
&\qquad\qquad \left. - (\mathbf{r} + \tfrac{a}{2}\hat{z}) \left[1 - \tfrac{a}{2}\tfrac{a}{r}\cos\theta + O\left(\tfrac{a^2}{r^2}\right) \right] \right\} \\
&= \frac{1}{4\pi\epsilon_0} \frac{3(\mathbf{p}\cdot\hat{r})\hat{r} - \mathbf{p}}{r^3} \left[1 + O\left(\tfrac{a}{r}\right) \right] \; , \tag{VII.8}
\end{aligned}$$

using the notation of the figure, with $\hat{r} = \mathbf{r}/r$ the unit vector radially away from the center of the dipole and *electric dipole moment* $\mathbf{p} \equiv qa\,\hat{z}$ the charge

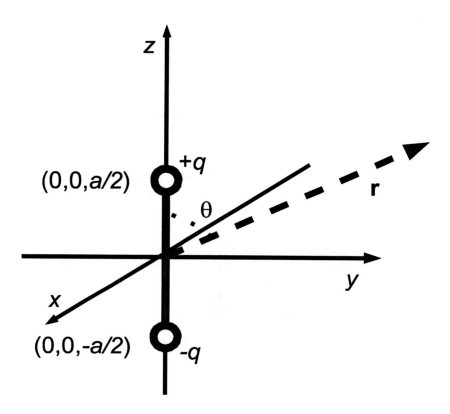

Figure VII.3: Geometry of an electric dipole. Equal-magnitude, opposite-sign point charges $\pm q$ are separated by a fixed displacement $a\,\hat{z}$. The electric field they produce is a function of position $\mathbf{r}$. Angle θ is the angle between vector $\mathbf{r}$ and the dipole $(+z)$ axis.

on one end of the dipole times the vector from the negative to the positive end. The first two lines of this expression are exact. The last two lines are obtained from Taylor-series expansions of the $-\frac{3}{2}$ powers of the denominators in the limit $r \gg a$—the useful, if loosely named, "point-dipole" approximation. The notation $O(a^2/r^2)$ in the third line refers to the collection of terms of second order or higher[22] in the small quantity a/r. These terms constitute the collected $O(a/r)$ terms of the final line, as the terms explicitly displayed there are themselves of the order a/r. The last line is the most familiar expression for the field of a dipole. The $1/r^3$ dependence is not a violation of the Coulomb law: The dipole field is the *residue* of two Coulomb fields that almost, but not completely, cancel. It falls off with distance faster than does the naked Coulomb field.

Two equal but oppositely oriented dipoles, close together, constitute a *quadrupole* charge distribution. This produces a field that falls off with distance as $1/r^4$. Two nearly canceling quadrupoles produce an *octopole* distribution, two of those a *hexadecapole,* and so on. In more advanced texts on electrostatics, it is shown that any bounded charge distribution can be treated as a superposition of monopole (i.e., point-charge) plus dipole plus quadrupole plus $\cdots$ plus 2^n-pole plus $\cdots$ contributions.[23]

A simple example of a continuous charge distribution is the ring of charge illustrated in Fig. VII.4. Calculation of the field $\mathbf{E}(\mathbf{r})$ everywhere requires elliptic integrals, appropriate to a higher-level text. But the field on the axis perpendicular to the plane of the ring, through its center, is given by the integral

$$\mathbf{E}(z\,\hat{\boldsymbol{z}}) = \frac{1}{4\pi\epsilon_0} \int_0^{2\pi} \frac{z\,\hat{\boldsymbol{z}} - (a\cos\phi\,\hat{\boldsymbol{x}} + a\sin\phi\,\hat{\boldsymbol{y}})}{(a^2 + z^2)^{3/2}} \frac{Q}{2\pi}\,d\phi$$

$$= \frac{Q}{4\pi\epsilon_0} \frac{z}{(a^2 + z^2)^{3/2}}\,\hat{\boldsymbol{z}}\,, \tag{VII.9}$$

using the notation of the figure and dividing the ring into angular increments $d\phi$. The integrals of $\cos\phi$ and $\sin\phi$ over the ring vanish; the symmetry of the situation implies that the on-axis field cannot have an x or y component. In the limit $|z| \gg a$, this field approaches that of a point charge Q, as expected. The on-axis field of a circular disk of charge could be calculated by treating the disk as nested concentric rings and integrating this result with respect to radius a. The field of a sphere on any axis, i.e., anywhere—appropriate to our proposed electron model—could then be calculated by treating the sphere as a stack of thin disks of suitable radii. However, in Sec. D, this calculation will be accomplished via a much more powerful method.

---

[22]This is mathematical code language, handy for dealing with expressions containing "higher-order" terms treated as negligible. It is read "big O of a^2/r^2," and it means that the collection of terms divided by a^2/r^2 remains bounded in the limit $a/r \to 0$. The more restrictive notation $o(a^2/r^2)$, "little o of a^2/r^2," refers to terms that, likewise divided, go to zero in that limit.

[23]The same applies when calculating the gravitational fields of mass distributions, except that in coordinates centered on the center of mass, the mass dipole contribution is zero. There are no "positive and negative masses on a stick."

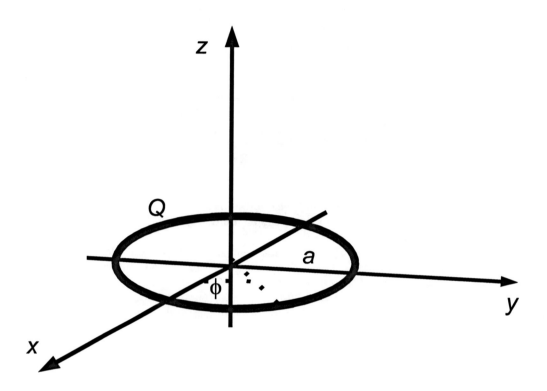

Figure VII.4: A uniform ring of electric charge of radius a and total charge Q. Angle ϕ is measured around the ring from the $+x$ axis.

Effects of electric fields

The previous examples illustrated the production of electric fields by charge distributions. The effects of electric fields follow from force law (VII.7b). For example, a dipole such as that in Fig. VII.3, immersed in a *uniform* external electric field $\mathbf{E}$, is subject to net force

$$\mathbf{F}_{\text{net}} = +q\mathbf{E}(\mathbf{r}_+) - q\mathbf{E}(\mathbf{r}_-)$$
$$= \mathbf{0} , \qquad\qquad\qquad\qquad (\text{VII.10a})$$

with $\mathbf{r}_\pm$ the positions of the positive and negative ends of the dipole, respectively, and $\mathbf{E}(\mathbf{r}_+) = \mathbf{E}(\mathbf{r}_-$ in a uniform field. Zero net force, but net *torque*

$$\boldsymbol{\tau}_{\text{net}} = \mathbf{r}_+ \times (q\mathbf{E}) + \mathbf{r}_- \times (-q\mathbf{E})$$
$$= q(\mathbf{r}_+ - \mathbf{r}_-) \times \mathbf{E}$$
$$= \mathbf{p} \times \mathbf{E} , \qquad\qquad\qquad (\text{VII.10b})$$

from definition (III.80) of torque and the definition of the electric dipole moment. The effect of this torque is to rotate the dipole moment $\mathbf{p}$ toward alignment with the electric field $\mathbf{E}$.

If the angle ϑ between the dipole moment and the external electric field changes, this torque does work on the dipole. The equivalent of a one-dimensional system with "force" dependent only on "position," this is a conservative situation. That is, there is a potential energy associated with the orientation of the dipole. This follows from the differential definition

$$dU_{\text{dip}} = -dW$$
$$= |\boldsymbol{\tau}| \, d\vartheta$$
$$= pE \sin\vartheta \, d\vartheta , \qquad\qquad (\text{VII.10c})$$

using the magnitude of the cross product for $\boldsymbol{\tau}$, and with sign determined by the fact that the work dW is positive if $d\vartheta$ is negative, i.e., if $\mathbf{p}$ rotates toward alignment with $\mathbf{E}$. This yields potential energy

$$U_{\text{dip}}(\vartheta) = -pE \cos\vartheta$$
$$= -\mathbf{p} \cdot \mathbf{E} , \qquad\qquad (\text{VII.10d})$$

with the zero of U_{dip} taken with $\mathbf{p}$ and $\mathbf{E}$ perpendicular.

These effects have a widely used application: the microwave oven. Water molecules are electrically neutral but possess electric dipole moments, the hydrogen end of the molecule being slightly positive and the oxygen end being slightly negative because of the bonds between the atoms. The molecules in liquid water bond weakly to one another via *hydrogen bonds*. But owing to torque (VII.10b), the molecules will flip back and forth in the oscillating electric fields of microwave radiation, constantly breaking and reforming these bonds in a kind of internal friction. This dissipates the energy tabulated by potential energy U_{dip}, heating the water. That's what a microwave oven does—it boils water, which is why any food cooked in a microwave oven is boiled or steamed, never baked or broiled. A water molecule has electric dipole moment of magnitude $p_{\mathrm{H_2O}} = 6.2 \times 10^{-30}$ C m. Hence, the energy associated with one flip between antialigned and aligned orientations in a typical field of, say, magnitude $E = 1.0 \times 10^7$ N/C is

$$\begin{aligned} \Delta U_{\mathrm{dip}} &= 2p_{\mathrm{H_2O}}E \\ &= 2(6.2 \times 10^{-30} \text{ C m})(1.0 \times 10^7 \text{ N/C}) \\ &= 1.2 \times 10^{-22} \text{ J} \\ &= 7.7 \times 10^{-4} \text{ eV} , \end{aligned} \qquad \text{(VII.11a)}$$

the last value in *electron volts* (eV), with conversion factor 1.000 eV = 1.602×10^{-19} J. Converted to a *temperature* change via the Boltzmann[24] constant k_B, this corresponds to

$$\begin{aligned} \Delta T &= \frac{\Delta U_{\mathrm{dip}}}{k_B} \\ &= \frac{1.2 \times 10^{-22} \text{ J}}{1.381 \times 10^{-23} \text{ J/K}} \\ &= 9.0 \text{ K} , \end{aligned} \qquad \text{(VII.11b)}$$

in kelvin[25] (K) or degrees Celsius.[26]

Field lines

A field, extended in space, is more difficult to visualize than the trajectory of a particle. A traditional method for representing vector fields graphically, more illuminating than a "forest of arrows," is adopted from fluid mechanics: the construction of *field lines,* or in the case of an electric field, say, *lines of force.* Akin to the streamlines of fluid flow, these are lines or curves drawn according to three rules:

- The tangent to an electric-field line at any point is the direction of **E** at that point.

- Field lines end only on electric charges, emerging from positive charge and entering negative charges. Elsewhere, the lines are continuous.

- The density of lines, per unit area perpendicular to the lines, is proportional to the magnitude |**E**|. More precisely, the number of lines passing through a surface of any orientation, per unit area, is proportional to the component of **E** in the direction perpendicular to that surface.

---

[24] Ludwig Boltzmann, 1844–1906
[25] William Thomson, 1st Baron Kelvin, 1824–1907
[26] Anders Celsius, 1701–1744

For example, the field lines for a positive point charge extend radially away from the charge in three dimensions, like the spines of a sea urchin. The pattern for a negative point charge is similar, the lines converging radially on the charge. (Field lines are directional.) The three rules incorporate the inverse-square nature of the Coulomb field: The lines radiating outward pass through concentric spheres with areas proportional to the square of their radii; as the lines are continuous, their density per unit area must decrease inversely as those areas, i.e., as $1/r^2$.

The field-line pattern for a dipole is illustrated in Fig. VII.5 in two-dimensional cross section. Near each charge, the lines are radial; farther away, they bend toward the other charge and connect up in this "butterfly" pattern. This pattern is worth remembering, as it appears in several different contexts.

The field-line construction predates the modern vector representation of $\mathbf{E}$, which was invented later. The modern interpretation of field lines is as a graphical device, but in older works, the lines are sometimes described almost as physical objects in their own right.

There is a field line through every point of a region where an electric field is present. Hence, only a "representative sample" is drawn. One is free to choose the "sampling rate," but only once for a given diagram: Once the proportionality constant between line density and $|\mathbf{E}|$ is set, the third rule fixes the density everywhere. Hence, in any given diagram, the number of lines coming from or to any charge (in accord with the second rule) is proportional to the magnitude of that charge, as $|\mathbf{E}|$ at any given radius is proportional to the source charge. That seems a quite trivial consequence of the third rule. In fact, it expresses an idea of colossal power.

D   The Gauss Law

Definition

This idea is based on another feature of vector fields adopted from fluid mechanics: *flux*. This measures the extent to which a vector field passes *through* a surface. If we imagine the surface subdivided into tiles of area $d\mathcal{A}$, the "throughness" of the electric field at a tile is $E_\perp\, d\mathcal{A}$. Here $E_\perp$ is the component of $\mathbf{E}$ perpendicular to the tile; the tangential components do not go through. The flux of vector field $\mathbf{E}$ through surface S, then, is the *surface integral*

$$
\begin{aligned}
\Phi_{\mathbf{E}}(S) &= \int_S E_\perp\, d\mathcal{A} \\
&= \int_S \mathbf{E} \cdot \hat{n}\, d\mathcal{A} \\
&= \int_S \mathbf{E} \cdot d\boldsymbol{\mathcal{A}}\,.
\end{aligned}
\qquad (\text{VII.12})
$$

Here $\hat{n}$ is the unit vector perpendicular to the surface at each point, so the dot product $\mathbf{E} \cdot \hat{n}$ gives $E_\perp$. A choice among the two possible directions for $\hat{n}$—the

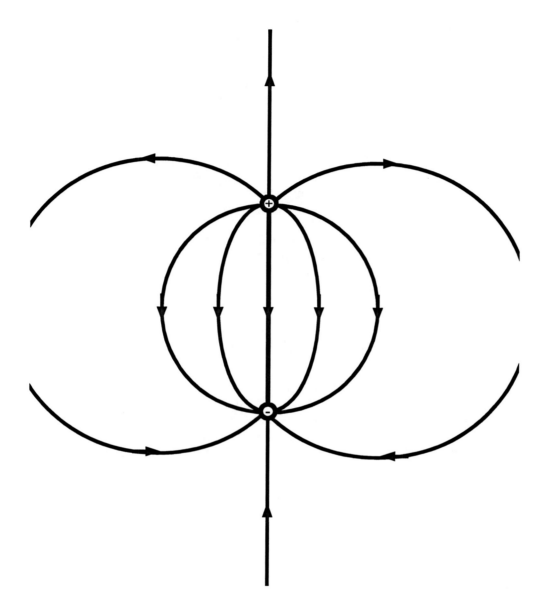

Figure VII.5: Field lines for an electric dipole. The pattern is axially symmetric about the dipole axis; it is shown here in two-dimensional cross section. The curves are not exactly accurate, but the overall shape is indicated.

"positive" perpendicular direction—must be made all over the surface to define the integral. The surface must be sufficiently well-behaved that it is possible to do this in a consistent way; such a surface is called *orientable*. (By contrast, a Möbius[27] strip—which can be envisioned as a long strip with one end turned 180°, then the ends joined—is an example of a *nonorientable* surface.) The product $\hat{n}\, dA \equiv d\boldsymbol{A}$ is the area of a tile as a vector; surface area does in fact have all the attributes of a vector quantity. In fluid mechanics, the flux of the vector $\rho\mathbf{v}$, where ρ is the density distribution of the fluid and $\mathbf{v}$ its velocity, gives the rate of mass transport through the surface. Given the interpretation of component $E_\perp$ in the third rule for field lines, here the flux $\Phi_{\mathbf{E}}(S)$ represents *the net number of field lines passing through surface S in the designated positive direction.* Except the "number of field lines" contains an arbitrarily chosen "sampling rate" factor; the flux is an unambiguously well-defined rendition of the same idea.

The observation that the number of field lines from or to a charge is proportional to the charge is realized unambiguously in the Gauss Law: *For any **closed**, orientable surface S_c, the flux of $\mathbf{E}$ through the surface and the electric charge Q_{in} enclosed within are in proportion*

$$\Phi_{\mathbf{E}}(S_c) = \frac{Q_{\mathrm{in}}}{\epsilon_0}$$

i.e.,
$$\oint_{S_c} \mathbf{E} \cdot d\boldsymbol{A} = \frac{Q_{\mathrm{in}}}{\epsilon_0} \ , \tag{VII.13}$$

where the $\oint$ integral sign identifies a surface integral over a closed surface. The surface is *closed* in the mathematical sense that it has no boundary curve and separates space into disjoint interior and exterior regions; the positive direction for $\hat{n}$ and $d\boldsymbol{A}$ is taken to be the *outward* normal direction. This form is for charges surrounded by vacuum, e.g., for matter described microscopically. The modifications needed for dealing with matter treated as continuous are discussed in Sec. G.

We can motivate this result by considering an orientable, closed surface S_c of arbitrary shape and a single point charge q. If the charge is *outside* the surface, then any of its field lines that enters S_c must also exit, as the lines are unbroken. The *net* number of lines crossing out of S_c is $\Phi_{\mathbf{E}}(S_c) = 0$, in accord with the Gauss Law. If the charge is inside the surface, we can surround it with a tiny sphere of sufficiently small radius[28] so that the field on that sphere is completely dominated by that of the given charge. If we treat surface S_c and the tiny sphere together as the boundary surface of the region between them—which contains no charge—then the previous result implies

$$\Phi_{\mathbf{E}}(S_c) - \Phi_{\mathbf{E}}(S_{\mathrm{tiny}}) = 0 \ , \tag{VII.14a}$$

the minus sign because the *outward* direction from the region between the sur-

---

[27] August Ferdinand Möbius, 1790–1868

[28] It must always be possible to do this if the point charge occupies an *interior* point of S_c.

faces, at the sphere S_{tiny}, is *into* the sphere. This in turn implies

$$\Phi_{\mathbf{E}}(S_c) = \Phi_{\mathbf{E}}(S_{\text{tiny}})$$

$$= \lim_{r \to 0} \left[4\pi r^2 \left(\frac{q}{4\pi\epsilon_0 r^2} \right) \right]$$

$$= \frac{q}{\epsilon_0} \ , \tag{VII.14b}$$

where the Coulomb law (VII.7c) is used to evaluate $\Phi_{\mathbf{E}}(S_{\text{tiny}})$. These two results for $\Phi_{\mathbf{E}}(S_c)$—for charges outside and inside S_c— plus the linearity of the $\mathbf{E}$ field in its sources imply the Gauss law (VII.13). This argument suggests that the Coulomb law implies the Gauss law, but in fact, *the Gauss law is more general.* It applies in situations where the Coulomb law does not, e.g., for moving charges. The Gauss law plus the symmetry of a static point charge implies the Coulomb law, as is shown following. The Gauss law is one of the Maxwell equations; the Coulomb law does not rise to that status.

Applications

Some introductory texts give short shrift to the Gauss law, suggesting, inaccurately, that it is of limited or only theoretical application. In fact it is one of the primary tools of electrostatics, and it applies in dynamical situations as well. Some examples will illustrate its utility.

The electric field produced by a spherical charge distribution can, as noted previously, be calculated by direct integration, but applying the Gauss law is much more efficient. The situation is illustrated in Fig. VII.6: A spherically symmetric charge distribution has radius a and total charge Q. The electric field E is sought at the indicated point, a distance r from the center of the sphere. One reason the Gauss law is such an effective tool is that the *Gaussian surface,* through which the flux is evaluated, can be freely chosen. It does not have to be the surface of anything; any closed, orientable surface will do. In this case, the surface is chosen to take advantage of the spherical symmetry of the problem, viz., sphere S_c concentric with the charge distribution, passing through the field point. The electric field on S_c must be radial; no tangential component could be consistent with the spherical symmetry.[29] Moreover, the symmetry requires that the radial component E_r be the same over the entire sphere. Hence, the electric flux through the sphere is

$$\Phi_{\mathbf{E}}(S_c) = 4\pi r^2 \, E_r(r) \ . \tag{VII.15a}$$

This isn't the Gauss law; this is only the evaluation of the flux. The Gauss law asserts that this is equal to the charge within S_c divided by ϵ_0. We consider first the case of a *thin spherical shell* of charge. In that case, the charge within

---

[29]This is part of a famous mathematical result: It is not possible to define a tangential vector field on a sphere without a singularity, i.e., "it is not possible to comb the hair on a bowling ball without leaving a bald spot."

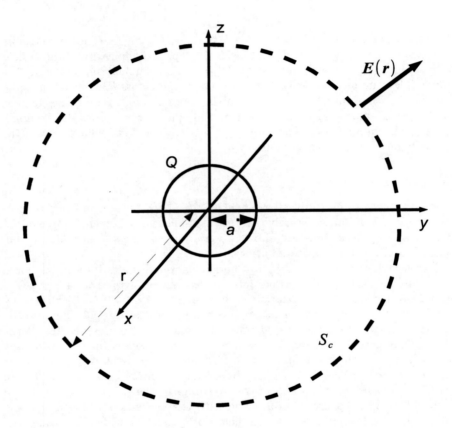

Figure VII.6: Electric field produced by spherically symmetric charge Q of radius a. Concentric sphere S_c is the chosen Gaussian surface.

is charge Q, *provided S_c is chosen with $r > a$*, i.e., *the field point is outside the charged shell*. But S_c could also have been constructed *inside* the shell, with $r < a$. In that case, the flux would still take form (VII.15a), but the charge within S_c would be zero. So the electric field produced by a spherical shell of charge, everywhere, is

$$\mathbf{E}_{\text{shell}}(\mathbf{r}) = \begin{cases} \mathbf{0} \, , & \text{for } r < a; \\ \dfrac{Q}{4\pi\epsilon_0 r^2} \, \hat{\boldsymbol{r}} \, , & \text{for } r > a. \end{cases} \qquad \text{(VII.15b)}$$

Here, $\hat{\boldsymbol{r}}$ is the unit vector radially away from the center of the spheres at any point. The electric field changes discontinuously at the shell of charge, so no value is given for $r = a$. An arbitrary spherical charge distribution can be treated as a set of concentric shells, like the layers of an onion or a pearl. Outside the sphere, the electric field is the same as that of a point charge, of equal total charge, at the center, since each shell produces such a field. Inside, the field is that produced by all layers interior to (at smaller r than) the test point, as if they were at the center. The layers at larger r contribute zero field within. For example, for a *uniform-density* ball of charge—our electron model—the charge interior to radius r, for $r \le a$, is proportional to r^3, while the field it produces falls off as $1/r^2$. Hence, the interior field varies linearly as r, while the exterior field varies as $1/r^2$, as the field of a point charge would. Specifically, the field of the uniform-density ball is

$$\mathbf{E}_{\text{ball}}(\mathbf{r}) = \begin{cases} \dfrac{Qr}{4\pi\epsilon_0 a^3} \, \hat{\boldsymbol{r}} \, , & \text{for } r \le a; \\[2ex] \dfrac{Q}{4\pi\epsilon_0 r^2} \, \hat{\boldsymbol{r}} \, , & \text{for } r \ge a. \end{cases} \qquad \text{(VII.15c)}$$

This field is continuous at the surface of the ball at $r = a$. The exterior field of any bounded spherical charge distribution is independent of its radius a; no measurement of the exterior field will indicate the value of that radius. The limit $r \to a$ of either result (VII.15b) or (VII.15c) yields the original Coulomb field (VII.7c) of a point charge. The Coulomb law follows as a consequence of the Gauss law and spherical symmetry, as promised.

One reason for writing the Coulomb constant as $1/(4\pi\epsilon_0)$ is clear from Eq. (VII.13): Thus, the Gauss law is simpler, without a factor of 4π. Results (VII.15b) and (VII.15c) show that the factor 4π in the Coulomb law is actually that in Eq. (VII.15a), i.e., the factor in the area $4\pi r^2$ of a sphere in three dimensions!

These results exactly parallel those given in Ch. IV Sec. A for the gravitational attraction of a spherical mass distribution. The gravitational results can be derived in exactly the same way, because there is a Gauss law for the Newtonian gravitational field $\mathbf{g}$. It takes the form

$$\Phi_{\mathbf{g}}(S_c) = -4\pi G M_{\text{in}}$$

i.e.,
$$\oint_{S_c} \mathbf{g} \cdot d\boldsymbol{\mathcal{A}} = -4\pi G M_{\text{in}} \, , \qquad \text{(VII.16)}$$

where M_{in} is the (active gravitational) mass enclosed within the closed surface S_c, and the form of the constant accords with that in force laws (IV.2), (IV.5), and (IV.6). This law can be motivated from those just as was done for the electric version, but again, this can be regarded as the more fundamental law.

In a similar manner, the Gauss law can be used in a cylindrically symmetric situation to calculate the electric field produced by an "infinite," i.e., very long, uniform line or cylinder of charge. Again the symmetry allows us to evaluate the electric flux through a judiciously chosen surface—here, a cylinder coaxial with the charge distribution—in terms of the local value of **E**. The Gauss law then links this with the charge. But the validity of the Gauss law itself does *not* depend on the symmetry.

Another useful example is the field of an "infinite," i.e., large, uniform, plane sheet of charge of *surface charge density* σ. In this case, the appropriate Gaussian surface is a right cylinder of cross-sectional area $\mathcal{A}$, extending equal distance z on either side of the sheet. The symmetry of the planar charge demands that the electric field be everywhere perpendicular to that plane, independent of position on the plane, and either away from the plane (for $\sigma > 0$) or toward the plane (for $\sigma < 0$) on both sides. This is illustrated in Fig. VII.7. The field takes the form $\mathbf{E} = E_\perp \hat{\boldsymbol{n}}$. Here, normal or perpendicular component $E_\perp$ is independent of position on the plane and the same value at any two points that are "mirror images" on opposite sides of the plane, and $\hat{\boldsymbol{n}}$ is the unit vector perpendicularly away from the plane on both sides. As **E** is tangent to the sides of the cylinder, the only flux through the surface is through the "lids," each contributing equally. The charge within the Gaussian surface is the surface charge intercepted by the cylinder. Hence, the Gauss law implies

$$2E_\perp \mathcal{A} = \frac{\sigma \mathcal{A}}{\epsilon_0} \ , \qquad\qquad\qquad \text{(VII.17a)}$$

giving electric field

$$\mathbf{E} = \frac{\sigma}{2\epsilon_0} \, \hat{\boldsymbol{n}} \ . \qquad\qquad\qquad \text{(VII.17b)}$$

Of course, **E** does not depend on $\mathcal{A}$, which is chosen arbitrarily. But it also does not depend on z; the field is independent of distance from the sheet of charge if its edges are too far away to be considered.

The Gauss law reveals important features of the behavior of electric charge placed on conducting objects. As has been described in Sec. B preceding, ordinary conducting materials contain enormous amounts of charge that can be moved in response to electric forces. This charge in turn modifies the electric-field distribution within the material. Eventually, the conductor must attain an equilibrium configuration in which the net force on the mobile charges in the conducting material is zero: The final electric field is $\mathbf{E} = \mathbf{0}$ in conducting material in static equilibrium. This implies that no applied charge can reside at any interior point of a conductor in equilibrium: Any such point could be surrounded by a tiny Gaussian surface entirely within conducting material on which the electric field is everywhere zero. Hence, the electric flux through the

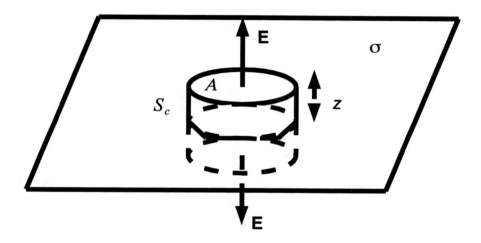

Figure VII.7: Gauss-law calculation of the electric field produced by an "infinite" uniform plane sheet of charge. The field **E** shown corresponds to positive surface charge density $\sigma > 0$.

surface is zero; by Gauss, there is no charge within.[30] If charge applied to a conductor cannot occupy the interior, it must reside on the outer surface like a layer of paint.[31] This accounts for the results of a famous experiment known as Faraday's Ice Pail. This was a metal pail with a metal lid. Faraday could lower a charged ball on a string into the pail through a small hole in the lid. But he found that once the ball touched the pail, it lost all its charge and encountered no electric forces within the pail.[32] All of the charge on the ball was transferred to the outside of the pail.

The electrical properties of cavities in conductors provide further applications of the Gauss law. Here, a *cavity* is not a hole through the surface of the conductor to the outside or a hole completely through the conductor like a doughnut hole, but an empty region within the conductor completely surrounded by conducting material.

- If no free charge is placed in the cavity, then the electric field there is the same as if the cavity were not present: $\mathbf{E} = \mathbf{0}$. The Gauss law, applied to a surface about any point on the cavity wall—part within the cavity and part within the conductor, with no electric field anywhere on the surface— implies that the charge density everywhere on the cavity wall is zero. Any charge on the conductor resides on its *outer* surfaces. Charges or electric fields outside the conductor exert no electrical influence within the cavity.

- If free charge q is placed anywhere in the cavity, then the Gauss law, applied to a "skin" that completely surrounds the cavity just within the surrounding conducting material—hence, with zero electric field everywhere on the "skin"—implies that the cavity wall must bear a total surface charge equal in magnitude and opposite in sign to that in the cavity: $Q_{\text{wall}} = -q$. The transfer of movable charge to (or away from) the cavity wall cannot create or destroy charge, so if the conductor was initially uncharged, then the outer surface must carry a total surface charge equal in magnitude and opposite in sign to Q_{wall}, i.e., $Q_{\text{outer}} = q$. Because the electric field in the conducting material, once static equilibrium is attained, remains zero, the *distribution* of the charge on the outer surface, and the resulting electric fields exterior to the conductor, are the same as if charge q had been put there in the first place—regardless of the shape or position of the cavity within the conductor or the position of the charge within the cavity. For example, if the outer surface of the conductor is a sphere, then the electric

---

[30] Of course, the atoms of which the material is made contain charges; this result applies on scales large enough that the material can be approximated as continuous.

[31] Again, on atomic scales, the charge occupies a layer, the density of which falls off exponentially with distance into the interior, on a length scale a few times the interatomic spacing. On macroscopic scales, then, the charge layer is effectively two-dimensional.

[32] This is not the same as the absence of electric field within a spherical shell of charge, per Eq. (VII.15b). That depends on spherical symmetry. The ice pail can be of any shape; the absence of electric forces within is due to the presence of movable charge in the conductor and the existence of charge of both signs.

field outside the conductor is given by

$$\mathbf{E}_{\text{outside}} = \frac{q}{4\pi\epsilon_0 r^2}\,\hat{r}\ , \tag{VII.18}$$

with r measured from, and unit vector $\hat{r}$ pointing away from, the center of the spherical surface. This field gives no indication of the shape or position of the cavity within the sphere or the position of charge q within the cavity—these need not conform to spherical symmetry. Only the value of q is manifest in the external field.

These results indicate that the surrounding conductor completely shields the cavity from external charges and fields,[33] but the reverse is not quite true: The conductor shields the exterior from every aspect of the cavity and its contents *except the total charge it contains*. This must appear as a feature of the exterior field; the Gauss law, applied to any exterior surface surrounding the conductor, demands this.

The charge on a conductor is distributed on its outer surface, but except for symmetric geometries such as a sphere, the distribution need not be uniform. The tools needed to describe the connection between the surface charge density and the surface geometry are developed in Sec. E to follow. But the connection between the surface charge density and the electric field just outside the surface follows from the Gauss law via a useful technique known as "pillbox integration." A traditional pillbox was a cylindrical container of height much less than its radius, with a hinged lid.[34] The Gauss law can be applied to such a cylinder partially buried in the conductor, with one lid just inside the surface and one just outside, as illustrated in Fig. VII.8. The height of the cylinder can be taken to be arbitrarily small, so that the only electric flux that need be considered is that through the lids. The electric field is zero on the lid in the conducting material. The field just outside is perpendicular to the surface—if it were not, the tangential component would move charges along the surface until equilibrium was re-established. The charge inside the pillbox is just that portion of the surface charge intercepted by the cylinder. If the pillbox is taken to be small enough that both charge density σ and field component $E_{\perp,\text{out}}$ can be regarded as constant across its area $\mathcal{A}$, then the Gauss law implies

$$E_{\perp,\text{out}}\mathcal{A} = \frac{\sigma\mathcal{A}}{\epsilon_0}\ , \tag{VII.19a}$$

or

$$\mathbf{E}_{\text{out}} = \frac{\sigma}{\epsilon_0}\,\hat{n}\ , \tag{VII.19b}$$

where $\hat{n}$ is the unit vector normal or perpendicular to the surface, everywhere directed out of the conductor. The electric field and the charge density are

---

[33]This is why a car is a useful shelter in a thunderstorm. Not because of the insulation of the tires—a lightning charge that has traveled through hundreds of meters of air is not stopped by a few centimeters of wet rubber—but because the charge is distributed over the exterior surface, the interior shielded.

[34]Cylindrical, partially buried, concrete gun emplacements used in the Second World War were called "pillboxes" because of their shape.

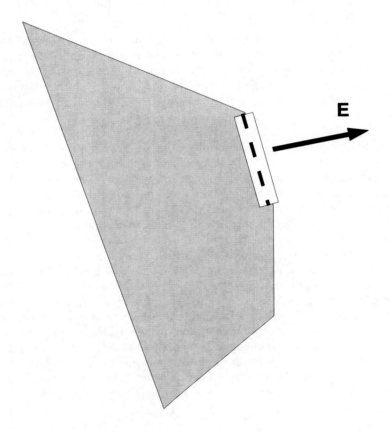

Figure VII.8: "Pillbox integration" at the surface of a conductor, seen in cross section.

proportional—where one is of large magnitude, so is the other. The field is outward-directed for positive σ, inward for negative.

This result differs from field (VII.17b) for an infinite plane sheet of charge. This can be understood by dividing the entire surface layer of charge on the outside of the conductor into a little "patch" in the neighborhood of any point, which can be treated as planar, and the entire "rest" of the layer. This is sketched in Fig. VII.9. Near enough to the surface, the patch can be regarded as a large planar sheet. It produces electric field $\mathbf{E}_{\text{patch}} = [\sigma/(2\epsilon_0)]\,\hat{\boldsymbol{n}}$ outside the conductor and $\mathbf{E}_{\text{patch}} = -[\sigma/(2\epsilon_0)]\,\hat{\boldsymbol{n}}$ inside—here $\hat{\boldsymbol{n}}$ points out of the conductor on both sides, not away from the surface layer. The rest of the charge on the conductor produces field $\mathbf{E}_{\text{rest}} = [\sigma/(2\epsilon_0)]\,\hat{\boldsymbol{n}}$ on both sides of the patch, as the field produced by the rest of the charge would be continuous there in the absence of the patch. The combination of these two contributions yields field $\mathbf{E}_{\text{out}} = (\sigma/\epsilon_0)\,\hat{\boldsymbol{n}}$ just outside the surface, as per Eq. (VII.19b), and zero field inside, as required.

The field $\mathbf{E}_{\text{rest}}$ exerts force on the patch of charge in the neighborhood of any point on the surface. Hence, the entire surface charge layer on a conductor experiences force per unit area

$$\boldsymbol{\Sigma} = \sigma \mathbf{E}_{\text{rest}}$$
$$= \frac{\sigma^2}{2\epsilon_0}\,\hat{\boldsymbol{n}}\,. \tag{VII.20}$$

As expected, this is outward for either sign of σ, as the charge of the surface layer repels itself. This is known as *electrostatic stress*. For example, a demonstration pith ball of radius $r = 1.0$ cm, with metallic coating, carrying charge $q = 10.$ nC has surface charge density

$$\sigma = \frac{q}{4\pi r^2}$$
$$= \frac{1.0 \times 10^{-8}\ \text{C}}{4\pi(1.0 \times 10^{-2}\ \text{m})^2}$$
$$= 8.0 \times 10^{-6}\ \text{C/m}^2\,. \tag{VII.21a}$$

It experiences electrostatic stress

$$\boldsymbol{\Sigma} = \frac{\sigma^2}{2\epsilon_0}\,\hat{\boldsymbol{r}}$$
$$= \frac{(7.96 \times 10^{-6}\ \text{C/m}^2)^2}{2[8.85 \times 10^{-12}\ \text{C}^2/(\text{N m}^2)]}\,\hat{\boldsymbol{r}}$$
$$= (3.6\ \text{Pa})\,\hat{\boldsymbol{r}}\,, \tag{VII.21b}$$

where an extra significant figure is used in the penultimate line to preserve the precision of the final result. This is a small outward pressure. But on much smaller scales, electrostatic stress can be very significant.

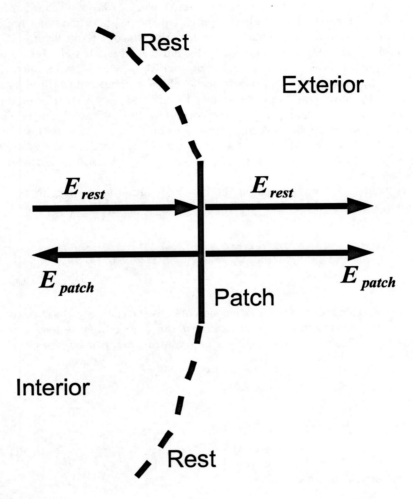

Figure VII.9: Charge layer and field contributions near the surface of a conductor, seen edge-on. The combined fields $\mathbf{E}_{\text{patch}}$ and $\mathbf{E}_{\text{rest}}$ yield field $\mathbf{E}_{\text{out}}$, from Eq. (VII.19b), outside the conductor and zero inside.

E   Electric potential energy and potential

Like Newtonian gravitation, and for the same reasons, the Coulomb force is a conservative force. The work done by this force can be tabulated by a potential-energy function similar in form to that in Eq. (IV.8): A charge q at position $\mathbf{r}$, subject to electric forces from source charges q_s at positions $\mathbf{r_s}$, has electric or electrostatic potential energy

$$U_e(\mathbf{r}) = \frac{q}{4\pi\epsilon_0} \sum_{\text{sources}} \frac{q_s}{|\mathbf{r} - \mathbf{r}_s|} + C \ . \qquad \text{(VII.22a)}$$

For an extended source treated as continuous, with *charge* density distribution $\rho(\mathbf{r}')$, this takes the form

$$U_e(\mathbf{r}) = \frac{q}{4\pi\epsilon_0} \int_{\text{source}} \frac{\rho(\mathbf{r}')}{|\mathbf{r} - \mathbf{r}'|} \, d\mathcal{V}' + C \ . \qquad \text{(VII.22b)}$$

As before, the constant C is chosen to set the zero of potential energy at a convenient location for the particular problem.

Electric potential

In electrostatics, it is particularly useful to define the *electric potential* field: the potential energy per unit receiver charge. This is the negative of the work done per unit receiver charge. That is, it is the negative of the integral of the force per unit receiver charge—the electric field—with respect to displacement:

$$V_e(\mathbf{r}) - V_e(\mathbf{r}_0) = -\int_{\mathbf{r}_0}^{\mathbf{r}} \mathbf{E} \cdot d\mathbf{s} \ . \qquad \text{(VII.23)}$$

This can be written in terms of the distribution of source charges as

$$\begin{aligned}
V_e(\mathbf{r}) &= \frac{1}{4\pi\epsilon_0} \sum_{\text{sources}} \frac{q_s}{|\mathbf{r} - \mathbf{r}_s|} + V_\infty \\
&= \frac{1}{4\pi\epsilon_0} \int_{\text{source}} \frac{\rho(\mathbf{r}')}{|\mathbf{r} - \mathbf{r}'|} \, d\mathcal{V}' + V_\infty \ ,
\end{aligned} \qquad \text{(VII.24)}$$

following from results (VII.22a) and (VII.22b), respectively. The constant V_∞ is the value $V_e(\mathbf{r})$ approaches far from a bounded source distribution.

Electric potential is measured in joules per coulomb, named *volts*:

$$1 \, \frac{\text{J}}{\text{C}} \equiv 1 \, \text{V} \qquad \text{(volt)}, \qquad \text{(VII.25)}$$

for Alessandro Volta.[35],[36] Hence, it is commonly called *voltage*. The volt also appears in the *energy* unit the *electron-volt* (eV), the energy of one elementary

---

[35] Alessandro Volta, 1745–1827

[36] This unit is an international collaboration: Joule was English, Coulomb French, and Volta Italian.

charge in a potential of one volt. This unit was used in Ch. V Problem 1 and in Eq. (VII.11a) in Sec. C of this chapter. To four significant figures, it is 1.000 eV $= 1.602 \times 10^{-19}$ J. This seems a small unit of energy, but it is the natural energy unit on atomic scales.

The potential distribution produced by a dipole source, for example, follows from the first form in Eq. (VII.24). In the notation of Eq. (VII.8) and Fig. VII.3, this becomes

$$
\begin{aligned}
V_e(\mathbf{r}) &= \frac{1}{4\pi\epsilon_0} \left(\frac{q}{|\mathbf{r} - \frac{a}{2}\,\hat{\mathbf{z}}|} - \frac{q}{|\mathbf{r} + \frac{a}{2}\,\hat{\mathbf{z}}|} \right) \\
&= \frac{q}{4\pi\epsilon_0} \left(\frac{1}{\left(r^2 - ar\,\cos\theta + \frac{a^2}{4}\right)^{1/2}} - \frac{1}{\left(r^2 + ar\,\cos\theta + \frac{a^2}{4}\right)^{1/2}} \right) \\
&= \frac{q}{4\pi\epsilon_0 r} \left\{ \left[1 + \tfrac{1}{2}\tfrac{a}{r}\cos\theta + O\left(\tfrac{a^2}{r^2}\right) \right] - \left[1 - \tfrac{1}{2}\tfrac{a}{r}\cos\theta + O\left(\tfrac{a^2}{r^2}\right) \right] \right\} \\
&= \frac{1}{4\pi\epsilon_0} \frac{\mathbf{p} \cdot \hat{\mathbf{r}}}{r^2} \left[1 + O\left(\tfrac{a}{r}\right) \right] \,,
\end{aligned}
$$

$$(VII.26)$$

the last two expressions in the "point dipole" limit $r \gg a$. Like the dipole electric field, this residual potential falls off with distance r faster than the naked Coulomb potential—here as $1/r^2$ rather than $1/r$.

The electrostatic potential for a bounded, spherically symmetric source distribution, outside the distribution, is the same as that for a point charge, since the exterior electric field is the same. For a thin spherical shell, the absence of electric field inside does not mean the potential is zero there. Rather, in accord with Eq. (VII.23), it means the potential is *constant* throughout the interior. Hence, it takes the form

$$
V_{\text{shell}}(\mathbf{r}) = \begin{cases} \dfrac{Q}{4\pi\epsilon_0 a} \,, & \text{for } r \leq a; \\[2ex] \dfrac{Q}{4\pi\epsilon_0 r} \,, & \text{for } r \geq a. \end{cases}
\qquad (VII.27a)
$$

This is in the notation of Eq. (VII.15b). The zero of potential is taken at $r \to +\infty$, i.e., the constant V_∞ is set to zero. For any other spherical distribution, the interior potential can be calculated via an integral of form (VII.23) over radius r, inward from the surface of the sphere. For example, for a uniform-density ball with electric field (VII.15c), the potential is

$$
V_{\text{ball}}(\mathbf{r}) = \begin{cases} \dfrac{Q}{8\pi\epsilon_0 a^3}\,(3a^2 - r^2) \,, & \text{for } r \leq a; \\[2ex] \dfrac{Q}{4\pi\epsilon_0 r} \,, & \text{for } r \geq a. \end{cases}
\qquad (VII.27b)
$$

Here $V_{\text{ball}} \to 0$ as $r \to +\infty$, but not at $r = 0$—the zero of potential can be selected only once in any calculation. In both cases, the potential is continuous at the surface $r = a$, even though the electric field is discontinuous there for the thin shell.

Electrostatic potential energy (VII.22) and potential (VII.24) are crucial in the study of atoms. For example, in the oversimplified model of the hydrogen atom evaluated in Eq. (VII.5), the potential produced by the proton at radius $a_0 = 5.29 \times 10^{-11}$ m is

$$
\begin{aligned}
V &= \frac{+e}{4\pi\epsilon_0 a_0} \\
&= \frac{(8.99 \times 10^9 \text{ N m}^2/\text{C}^2)(1.602 \times 10^{-19} \text{ C})}{5.29 \times 10^{-11} \text{ m}} \\
&= +27.2 \text{ V} ,
\end{aligned}
\tag{VII.28}
$$

where the zero of potential is at infinity, and the value of a_0 is the *Bohr radius* for the orbit of the electron in its ground state. This is neither a very large nor a very small number; the volt is a convenient unit at atomic scales. Since the electron has charge $-e$, this means the (classical) potential energy of the electron is -27.2 eV. This energy is also labeled -1.00 *hartree*[37] or -2.00 *rydberg*[38]; these are "nonstandard"' units useful in atomic physics and chemistry. The kinetic energy of the electron in its ground state is $-\frac{1}{2}$ times this value, so the total energy is half this value or -13.6 eV. The negative of this is the *ionization energy* of the atom, the energy that must be supplied to separate the electron from the proton. More accurate quantum-mechanical models of the hydrogen atom yield these same energy values.

Equipotential surfaces

A potential distribution can be visualized by drawing *equipotential surfaces*: Through any point, there passes a two-dimensional surface on which the potential function takes a constant value. A representative sample of such surfaces, say, at constant intervals ΔV, and usually drawn in cross section, illustrates the potential. Such a drawing resembles a contour map. (In fact, since a contour map illustrates ground elevation, and *gravitational* potential is proportional to elevation near the Earth's surface, a contour map is a gravitational equipotential diagram.) The features of the electric field can be inferred from an equipotential diagram: The field has no component tangent to an equipotential surface, as the work done on a charge moved along the surface is zero. That is, the equipotential surfaces are everywhere orthogonal to the field lines; the field **E** is everywhere perpendicular to the equipotentials. Also, the field magnitude $|\mathbf{E}|$ times the spacing between nearby equipotentials is the magnitude of the potential increment ΔV. Hence, for equipotentials drawn at fixed ΔV, the magnitude $|\mathbf{E}|$ is inversely proportional to the spacing. Where the equipotentials are closely spaced, the field magnitude is large; where the surfaces are widely spaced, the field is weak. For example, such an equipotential diagram

[37]Douglas Rayner Hartree, 1897–1958
[38]Johannes Robert Rydberg, 1854–1919

for the field of a point charge would consist of concentric spheres, with spacing proportional to the square of their radius. The equipotentials for an electric dipole are sketched in Fig. VII.10. These surfaces are orthogonal to the field lines of Fig. VII.5. As the two figures illustrate, equipotential contours have no directionality, while field lines do—the direction of the vector field **E**.

Since the electric field is zero in conducting material in static equilibrium, the potential must be constant throughout such material. Hence, the surface of a conductor is an equipotential surface. This implies that the electric field just outside the conductor is perpendicular to its surface, as surmised previously. It also enables us to describe the connection between the surface-charge density on a conductor and its shape. A rigorous derivation is beyond the scope of this text, but a somewhat contrived model will suffice: two spheres, of radii R and r, connected by a thin wire, far enough apart that the charge distribution on each sphere is not distorted by that on the other. The situation is that of Fig. VII.11. Electric charge placed on either or both spheres will redistribute itself via the wire until equilibrium is attained with the potential of the entire surface constant. If the wire is long enough, the potential on the surface of each sphere is dominated by the charge there. Hence, the charges Q and q on each sphere and their respective radii R and r are related via

$$\frac{Q}{R} = \frac{q}{r} \,, \qquad\qquad\text{(VII.29a)}$$

i.e.,

$$Q \propto R \,. \qquad\qquad\text{(VII.29b)}$$

The surface charge density on each sphere is its charge divided by $4\pi R^2$ or $4\pi r^2$, respectively, implying

$$\sigma \propto \frac{1}{R} \,, \qquad\qquad\text{(VII.29c)}$$

hence, per Eq. (VII.19b),

$$E_\perp \propto \frac{1}{R} \,. \qquad\qquad\text{(VII.29d)}$$

These last two results are general: The surface charge density and surface field component of a charge-bearing conductor are inversely proportional to its *radius of curvature*. Where the surface is sharply curved, as at an edge or point, surface density σ and field component $E_\perp$ are large; where the surface is flatter, both are diminished.

This accounts for the phenomenon known as *corona discharge*: At an edge or point on a charge-bearing conductor, the electric field can become large enough to ionize—rip the outer electrons from—nearby air molecules. The air becomes conducting, allowing the charge to bleed off the conductor at such points. As the molecules recombine or de-ionize, a visible glow is emitted, forming a *corona* or crown of light. Seen on the masts and yards of ships or the wings and tails of aircraft in stormy weather, this is known as *St. Elmo's*[39] *Fire.*

---

[39] Erasmus of Formiae (St. Elmo), d. 303 CE

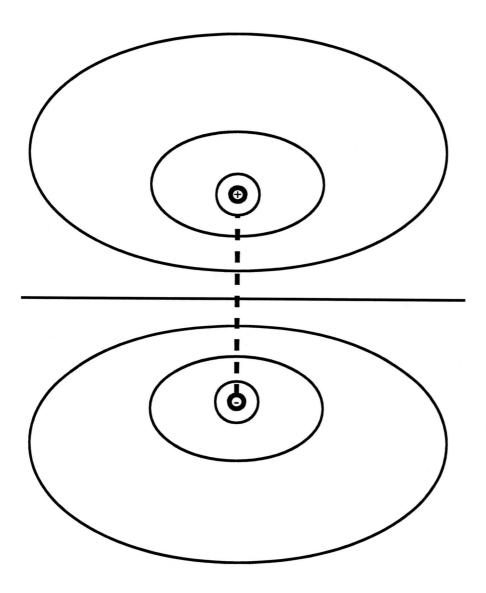

Figure VII.10: Sketch of equipotentials for a dipole source, as in Fig. VII.5. The shape of the surfaces, shown in cross section, is only approximate.

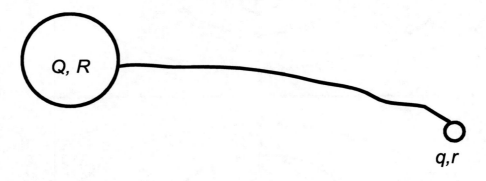

Figure VII.11: Two conducting spheres connected by a long wire. The spheres carry charges Q and q and have radii R and r, respectively. In static equilibrium, the entire configuration is an equipotential.

Corona discharge also underlies an important technology: the lightning rod, a sharp-tipped metal rod placed atop a structure and connected to the ground by a conducting cable. It is often stated, even in some advanced texts, that the purpose of the lightning rod is to *conduct* a lightning strike to the ground to protect the structure. This is not correct: A lightning rod that does this has failed and is likely to be destroyed, along with the structure. A lightning strike can carry one hundred thousand amperes[40] of current, driven by a potential difference of the order of one hundred thousand volts, sufficient to explode most conductors. The purpose of a lightning rod is to *prevent* a lightning strike: When a charged thundercloud passes overhead, an opposite charge collects in the ground beneath (which contains enough moisture to act as a conductor). If sufficient charge builds up, the air will ionize via a complicated process, allowing the charges to travel back and forth in a lightning strike. But corona discharge at the tip of the lightning rod, connected to the ground, can allow the charge to bleed off into the air and avert the strike if the buildup is not too quick. Benjamin Franklin is usually credited with this invention,[41] but in Japan, for example, wooden pagodas—some of which have stood for fourteen centuries, among the oldest wooden buildings in the world—are topped with metal spires connected to the ground by cables. And in the Bible, there are references to iron rods placed in the fields "as protection against lightning." So it is possible that this technology has emerged, perhaps by trial and error, at many times and in many places.

Calculating E from *V*

The electrostatic force is connected to potential energies (VII.22) in the manner described in general by Eq. (II.39) in Ch. II Sec. E preceding. Consequently, the electric field and potential—force and potential energy per unit charge—are similarly related by

$$\mathbf{E} = -\nabla V \quad = -\left(\frac{\partial V}{\partial x}\,\hat{\boldsymbol{x}} + \frac{\partial V}{\partial y}\,\hat{\boldsymbol{y}} + \frac{\partial V}{\partial z}\,\hat{\boldsymbol{z}} \right) , \qquad \text{(VII.30)}$$

where the first expression is general, the second appropriate in Cartesian (rectangular) coordinates. This is the converse of relation (VII.23). It implies that the direction of **E** is that in which *V* decreases fastest with distance, and the magnitude of **E** is the rate of decrease in that direction.

For example, a spherically symmetric, *linear* potential takes the form

$$V(\mathbf{r}) = kr$$
$$= k(x^2 + y^2 + z^2)^{1/2} , \qquad \text{(VII.31a)}$$

---

[40] André Marie Ampère, 1775–1836

[41] Franklin is also credited with recognizing the electrical nature of lightning via his famous kite experiment. It is apparently unclear whether Franklin actually performed the experiment or merely described it as an example. It is clear that several experimenters following Franklin were killed in the attempt. Flying a kite in a thunderstorm is an incredibly foolhardy thing to do; the reader is urged not to try it.

with k a constant of suitable dimensions. This yields electric field components

$$
\begin{aligned}
E_x &= -\frac{\partial V}{\partial x} \\
&= -k\frac{\partial}{\partial x}\,(x^2 + y^2 + z^2)^{1/2} \\
&= -\frac{kx}{(x^2 + y^2 + z^2)^{1/2}}\ , \qquad \text{et cetera,} \qquad \text{(VII.31b)}
\end{aligned}
$$

or electric field

$$
\mathbf{E}(\mathbf{r}) = -k\,\hat{r}\ . \qquad\qquad\qquad \text{(VII.31c)}
$$

The electric field is of uniform magnitude, everywhere radially inward for positive k. The charge distribution $\rho(\mathbf{r})$ that gives rise to this potential and field could be obtained from this by a suitable application of the Gauss law. Hence, given any of the three distributions $\rho(\mathbf{r})$, $\mathbf{E}(\mathbf{r})$, or $V(\mathbf{r})$, it is in principle possible to calculate the other two by integrating or differentiating as needed.

F   Classical radius of the electron

We have now assembled sufficient electrostatics to extract information from the classical electron model proposed at the beginning of this chapter. The connections among the electron mass and charge and the radius of the spherical model follow from the *electrostatic self-energy* of the electron. The self-energy of any charge distribution is defined as the amount of work required to "assemble" the distribution, starting from infinite charge separation. We may imagine building up the electron layer by layer like a pearl, from the first spot of charge at the center to the final electron radius. It is important to note, however, that this process is an artifice, useful only for the purpose of calculation: No one has ever assembled an electron from subconstituents or ever observed one disassembled in any way.

The work required to add a layer to the "partially built" electron, from infinite dispersal, is equal to the charge of the layer times the potential at the surface of the charge already assembled. With the potential given by Eq. (VII.27b) for a uniform-density sphere, this increment of work takes the form

$$
\begin{aligned}
dU &= \frac{1}{4\pi\epsilon_0}\left(\frac{-e(r^2/a^3)}{r}\right)\left(\frac{-e}{\frac{4}{3}\pi a^3}\right)4\pi r^2\,dr \\
&= \frac{1}{4\pi\epsilon_0}\,\frac{3e^2 r^4}{a^6}\,dr\ , \qquad\qquad\qquad \text{(VII.32a)}
\end{aligned}
$$

with a the ultimate radius of the electron and r the radius of the layer being added. The first two factors in the first expression are the potential at the surface of the "fraction" of the electron out to radius r. This is not the potential inside the *entire* final charge distribution at radius r—that would double-count the energy—but only that due to the "fraction" by itself. The last two factors are the charge density of the electron, assumed constant, times the volume of the layer at radius r of thickness dr. The total energy of the "assembled" electron, then, is

$$U = \int_0^a dU$$
$$= \frac{3e^2}{4\pi\epsilon_0 a^6} \int_0^a r^4 \, dr$$
$$= \tfrac{3}{5} \frac{e^2}{4\pi\epsilon_0 a} \; . \tag{VII.32b}$$

The factor $\tfrac{3}{5}$ is a feature of the model; the other factors are the only combination of electron charge, radius, and Coulomb constant with units of energy.

With Einsteinian mechanics in play, the energy of an electron is given by definition (I.96) from Ch. I Sec. H. An electron at rest, then, has energy $E_0 = m_e c^2$, with c the speed of light in vacuum. But an electron is a particle of charge—is this energy not simply the electrostatic self-energy just calculated? Identifying E_0 and U yields the *classical radius of the electron*[42]:

$$a = \tfrac{3}{5} \frac{e^2}{4\pi\epsilon_0 m_e c^2}$$
$$= \tfrac{3}{5} \frac{(8.9875517873681764 \times 10^9 \text{ N m}^2/\text{C}^2)(1.6021773 \times 10^{-19} \text{ C})^2}{(9.109390 \times 10^{-31} \text{ kg})(2.99792458 \times 10^8 \text{ m/s})^2}$$
$$= 1.690764 \times 10^{-15} \text{ m} \; . \tag{VII.33}$$

The precision of this result is artificial, only to show the precision to which the factors are known. The simplified model is not accurate enough to justify such precision. This radius is some five orders of magnitude smaller than the size scale associated with atoms and is comparable to that of atomic nuclei. This seems reasonable for a subconstituent of atoms. It appears that our simple classical model and basic electrostatic calculation have succeeded! The only shadow over this success is the fact that electrons do not *scatter* in high-energy experiments like spheres of this size, but rather can be treated as point-like. But we shall reserve further exploration of the limitations of the classical model to Ch. IX.

G Capacitance and field energy

Before leaving the subject of electrostatics, we must examine one more topic, as it will be needed later. The notion of *capacitance*—not to be confused with capaci*ty*—was introduced by Michael Faraday as a property of charge-bearing conductors.

---

[42]In many references the model-dependent factor $\tfrac{3}{5}$ is omitted, so the *classical radius of the electron* refers to $\tfrac{5}{3}$ of value (VII.33), or 2.817941×10^{-15} m.

Capacitance

The term *capacitance* is applied to a pair of conductors of any shape and arrangement not electrically connected to each other. If charge Q is transferred from one to the other so that one carries charge Q, the other $-Q$ (the combination remaining neutral), potential difference ΔV will arise between the conductors. The capacitance C of the pair is the ratio of these:

$$C \equiv \frac{Q}{\Delta V} , \tag{VII.34}$$

a positive quantity. In effect, capacitance measures the "energy cost" of storing charge—the lower the capacitance, the higher the potential difference for a given charge. A gravitational analog of this notion might be two storage tanks of the same volume but different shapes. When the tanks are filled, the potential energy of the stored liquid—i.e., the work required to fill the tank—is its weight times the height of its center of mass, per Eq. (III.23). The tank with the *lower* CM has the larger "gravitational capacitance." A most useful feature of electrical capacitance is that it is independent of Q or ΔV separately—their ratio depends only on the geometry of the conductors.

Electrical capacitance is measured in *farads*, i.e., coulombs per volt:

$$1\,\frac{C}{V} \equiv 1\text{ F} \quad \text{(farad)}. \tag{VII.35}$$

This is named for Michael Faraday,[43] the only scientist of sufficient distinction to have *two* SI units named for him. The other is the faraday of charge, as in Eq. (VII.4).

It is also possible to define the capacitance of a single conductor by taking the other to be the "sphere at infinity." The capacitance is the ratio of the charge on the conductor to its potential, with the zero of potential at infinity. For example, the capacitance of a conducting sphere of radius R is

$$\begin{aligned}
C_{\text{sph}} &= \frac{Q}{V}\\[1em]
&= \frac{Q}{\left(\dfrac{Q}{4\pi\epsilon_0 R}\right)}\\[1em]
&= 4\pi\epsilon_0 R , \tag{VII.36a}
\end{aligned}$$

simply a measure of its radius. As a conductor, the Earth has capacitance

$$\begin{aligned}
C_\oplus &= 4\pi\epsilon_0 R_\oplus\\
&= 4\pi[8.854 \times 10^{-12}\text{ C}^2/(\text{N\,m}^2)](6.371 \times 10^6\text{ m})\\
&= 0.7089\text{ mF} . \tag{VII.36b}
\end{aligned}$$

---

[43]This is another three-way international collaboration, as Faraday was English.

This suggests that the farad is a very large unit of capacitance. Also, a sphere of planetary size is not a particularly good design for a *capacitor*—a device constructed specifically for its capacitance.

Another important, basic conductor geometry is the *parallel-plate capacitor*, a pair of conducting plates separated by a gap. This is sketched in Fig. VII.12. With charge Q on one plate and $-Q$ on the other, electric field $\mathbf{E}$ exists between the plates and potential difference ΔV between them. In the simplest case, the plates are large enough that changes in the field at their edges (known as *fringing*) are negligible, and the space between them contains vacuum or air, an *air capacitor*.[44] The plates carry approximately uniform surface charge densities $\pm\sigma = \pm Q/\mathcal{A}$. The electric field between them, per Eqs. (VII.17b) or (VII.19b), is

$$\mathbf{E} = \frac{\sigma}{\epsilon_0}\,\hat{\boldsymbol{d}}$$

$$= \frac{Q}{\epsilon_0\mathcal{A}}\,\hat{\boldsymbol{d}}\,, \tag{VII.37a}$$

with $\hat{\boldsymbol{d}}$ the unit vector from the positive to the negative plate. Between the plates, away from the edges, the field is essentially uniform, and it is essentially zero outside. The potential difference between the plates, per Eq. (VII.23), is

$$\Delta V = |\mathbf{E}|d$$

$$= \frac{Qd}{\epsilon_0\mathcal{A}}\,. \tag{VII.37b}$$

The capacitance of the parallel-plate air capacitor, then, is

$$C_{\text{PP}} = \frac{Q}{\Delta V}$$

$$= \frac{Q}{\left(\dfrac{Qd}{\epsilon_0\mathcal{A}}\right)}$$

$$= \epsilon_0\,\frac{\mathcal{A}}{d}\,, \tag{VII.37c}$$

a simple result. In fact, it was to make the constant in this result simple that the Coulomb constant was written in form (VII.2a). This formula should not be confused with the *definition* of capacitance, Eq. (VII.34). This applies only to the parallel-plate geometry. Other geometries, such as concentric spheres and coaxial cylinders, are often used.

A parallel-plate air capacitor with plate separation $d = 1.00$ mm would have

---

[44]The tuners of old-fashioned radios contain an air capacitor consisting of interleaving metal plates. These can be rotated together or apart on a spindle to tune the radio. Modern radios, by contrast, have digital tuners.

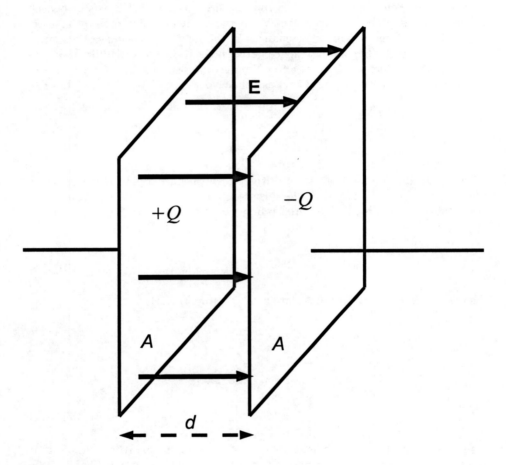

Figure VII.12: A parallel-plate capacitor. The plates have area $\mathcal{A}$ each and separation d, and carry charge Q and $-Q$, respectively.

capacitance per unit area

$$\frac{C}{A} = \frac{\epsilon_0}{d}$$
$$= \frac{8.854 \times 10^{-12} \text{ C}^2/(\text{N m}^2)}{1.00 \times 10^{-3} \text{ m}}$$
$$= 8.85 \text{ nF/m}^2 , \qquad \text{(VII.37d)}$$

the units of ϵ_0 being equivalent to F/m. This too suggests that the farad is a very large unit, and that this is not a particularly effective design. (As shall be seen, there are ways to improve it.) The small, disk-shaped capacitors in, say, a television set are parallel-plate capacitors with capacitances of the order picofarads (pF), while the cylindrical *electrolytic* capacitors typically have capacitances of tens to hundreds of microfarads[45] (μF). There are *ultracapacitors* with farad capacitances, though these require enhancements on this basic design.[46]

Like all the components we shall encounter in electric circuits in subsequent chapters, capacitors are often used in combination. There are two ways to combine any two components: in *parallel,* i.e., side by side, and in *series,* end to end. These are illustrated, for two capacitors, in Fig. VII.13. The standard symbol for a capacitor used in the figure is a stylized representation of a parallel-plate capacitor, though the actual component represented may or may not be of that geometry. The defining features of two components in parallel are that the potential difference across the combination is that across each component: $\Delta V = \Delta V_1 = \Delta V_2$, and the charge applied to the combination divides into the charges on each component without gain or loss: $Q = Q_1 + Q_2$. Dividing both sides of this by the common potential difference yields the combination rule for capacitances in parallel,

$$C_{\parallel} = C_1 + C_2 . \qquad \text{(VII.38)}$$

The capacitance of the parallel combination is greater than that of either component. Contrariwise, the defining features of components in series are that the charge through the combination passes unchanged through both components: $Q = Q_1 = Q_2$, while the potential difference across the combination is simply the sum of the differences across the individual components: $\Delta V = \Delta V_1 + \Delta V_2$. Dividing both sides of this by the common charge yields the combination rule for *inverse* capacitances in series:

$$\frac{1}{C_{\text{ser}}} = \frac{1}{C_1} + \frac{1}{C_2} . \qquad \text{(VII.39a)}$$

---

[45]The units actually printed on capacitors may cause confusion. Microfarads should be rendered μF and picofarads pF, although $\mu\mu$F is also used. But on older components made when the printing of Greek letters was problematic, the symbols MF for microfarads (*not* megafarads!) and MMF for micromicrofarads or picofarads can be found.

[46]One of the uses for such devices is in the triggers for nuclear weapons. The reader who shows too great an interest in them may attract unwanted attention.

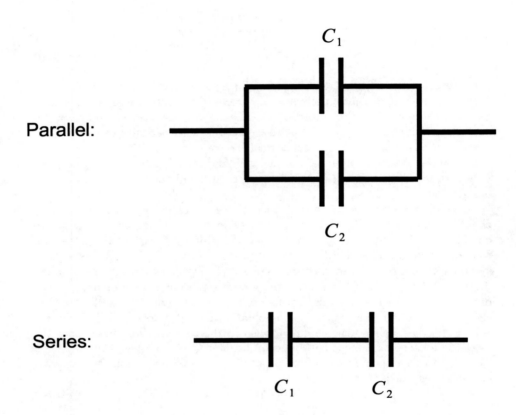

Figure VII.13: Two capacitors, in parallel and in series. The symbol used for the capacitors is standard notation for electric circuits.

This is equivalent to the combination rule

$$C_{\text{ser}} = \frac{C_1 C_2}{C_1 + C_2} \; . \qquad \qquad \text{(VII.39b)}$$

The capacitance of the series combination is less than that of either component. As will be seen in subsequent chapters, other electrical components satisfy similar, but not identical, combination rules; the reader must be careful to distinguish them.[47] The symbolic representations in Fig. VII.13 offer a mnemonic for remembering rules (VII.38) and (VII.39): The capacitors in parallel are, in effect, adding plate areas together, hence by Eq. (VII.37c) adding capacitances. Those in series are, in effect, adding plate spacings, since the two connected plates form an equipotential. Hence, the same formula implies that inverse capacitances are added in series.

Dielectrics

The electrical properties of *dielectric* or insulating materials can be very important. Such materials are not electrically inert. They contain positive and negative charge in amounts comparable to those of the conducting materials considered previously in Secs. C and D. The difference is that some of the charge in conductors can be made to move macroscopic distances, whereas the charge in a dielectric remains bound to its atoms and can be shifted only on atomic or molecular scales.

When a dielectric material is subjected to an electric field, the shifting of charge on atomic scales causes the atoms or molecules, while remaining electrically neutral, to acquire electric dipole moments—such as those that appear in Eqs. (VII.8) and (VII.26)—aligned with the field. Or, if the molecules already possess dipole moments—such as water molecules—then the torque (VII.10b) exerted by the field tends to align those dipoles with the field. In either case, the process is called *polarization*.[48] The polarized dielectric consists of layers of molecules with positive and negative ends aligned along the field. Inside the material, positive and negative ends are in proximity, and the material is neutral on macroscopic scales. But at boundary surfaces perpendicular to the applied field, the alignment produces surface layers of charge, called *bound charge* to distinguish it from *free charge*, which can be moved. For example, if the space between the capacitor plates in Fig. VII.12 was filled with a dielectric, the polarization of the material would create a positive bound-charge layer on the right side of the material and a negative layer on the left. These layers would produce an electric field directed to the left in the material, *reducing the field within the material* and (partially) *shielding the free charges on the capacitor plates*. This is an example of the law known to chemists as *Le Châtelier's*[49] *Principle of Spite*: A system in equilibrium—in this case, the dielectric material—subjected

---

[47]The surest way to do this is to rederive the rules from first principles when needed.

[48]This is entirely distinct from the polarization of transverse waves described in Ch. VI Sec. B. The two phenomena have in common only the choosing of a direction.

[49]Henry Louis Le Châtelier, 1850–1936

to a change in external conditions will respond in such a way as to diminish the change—here, the electric field to which the molecules are subjected.

This effect can be quantified: If $\mathcal{N}$ is the number of molecules per unit volume, a the thickness of a molecular layer, and $\pm q$ the charges found or induced on the ends of a molecule, then the effective (positive) bound-charge surface density is

$$
\begin{aligned}
\sigma_{\text{Pol}} &= \mathcal{N}aq \\
&= \mathcal{N}|\mathbf{p}| \\
&= |\mathbf{P}| \ ,
\end{aligned}
\tag{VII.40}
$$

where $\mathbf{p}$ is the dipole moment of a single molecule and $\mathbf{P}$ the *dipole moment per unit volume* or *polarization* of the material. The negative bound-charge surface density is just the negative of this. These layers produce a "counter" electric field within the material, in accord with Eq. (VII.37a), equal to $-\mathbf{P}/\epsilon_0$. The *net* electric field $\mathbf{E}$ within the material is

$$
\mathbf{E} = \mathbf{E}_0 - \frac{\mathbf{P}}{\epsilon_0} \ ,
\tag{VII.41}
$$

where $\mathbf{E}_0$ is the external electric field produced by the free charges on the capacitor plates.

That external field is represented by a rescaled version called the electric *displacement,* given by

$$
\mathbf{D} \equiv \epsilon_0 \mathbf{E} + \mathbf{P} \ .
\tag{VII.42}
$$

The name comes from James Clerk Maxwell's attempts to understand electric and magnetic forces as elastic forces, in this case as an elastic deformation of the vacuum. Maxwell's mechanical model did not prevail, but the name remains. In advanced treatments, displacement $\mathbf{D}$ and electric field $\mathbf{E}$ [as in Eqs. (VII.7)] are treated as independent vector fields: Free charges *produce* $\mathbf{D}$ and *feel* $\mathbf{E}$.

The stimulus-response relationship between $\mathbf{D}$ and $\mathbf{P}$, or the relationship between $\mathbf{D}$ and $\mathbf{E}$, need not be simple. However, for a broad class of materials, the former can be written

$$
\mathbf{P} = \left(1 - \frac{1}{\kappa}\right) \mathbf{D} \ ,
\tag{VII.43a}
$$

with *dielectric constant* $\kappa \geq 1$. Such materials are called *linear* because $\mathbf{P}$ scales with $\mathbf{D}$. And they are called *isotropic* because $\mathbf{P}$ is in the same direction as $\mathbf{D}$—the constant in this relation is a scalar rather than a tensor represented by a matrix—i.e., the material itself has no preferred direction for polarization. Linearity and isotropy are independent properties, and we shall encounter them subsequently in different contexts. The dielectric constant κ is defined only for linear, isotropic dielectrics. It is defined in this somewhat unwieldy form so that the electric field can be written

$$\mathbf{E} = \frac{\mathbf{D} - \mathbf{P}}{\epsilon_0}$$
$$= \frac{\mathbf{D}}{\kappa \epsilon_0}$$
$$= \frac{\mathbf{D}}{\epsilon} \,, \tag{VII.43b}$$

defining $\epsilon \equiv \kappa \epsilon_0$ as the *permittivity* of the dielectric material. With the electric field between the plates of a capacitor scaled down by κ for a given charge on the plates, the potential difference ΔV is likewise scaled down. Hence, the capacitance of a parallel-plate capacitor, say, filled with dielectric of constant κ is scaled up. It is given by

$$C_{\mathrm{PP}} = \kappa \epsilon_0 \frac{\mathcal{A}}{d}$$
$$= \epsilon \frac{\mathcal{A}}{d} \,, \tag{VII.44}$$

in comparison with result (VII.37c). The vacuum has dielectric constant unity. Values for some other materials include

$$\kappa_{\mathrm{air}} = 1.00054$$
$$\kappa_{\mathrm{Mylar}} = 3.1$$
$$\kappa_{\mathrm{H_2O}} = 80. \,, \tag{VII.45}$$

for air, Mylar, and water, respectively. Pure water is a dielectric or insulator—it becomes a conductor only when ionic materials are dissolved in it. The large value of $\kappa_{\mathrm{H_2O}}$ arises from the intrinsic polarity of water molecules [as used in Eqs. (VII.11)] and their freedom to align in the liquid state. For some purposes, conductors can be regarded as dielectrics with "very large" dielectric constants $\kappa \to +\infty$, as electric fields are completely suppressed in conducting materials in equilibrium.

The reader is doubtless wondering how the Gauss law is affected by the presence of dielectric materials. Applied on atomic scales, where all matter is free charges in vacuum, it is unchanged. But when matter is approximated as continuous, which charges are to be counted? One way to deal with the diminution of electric fields in dielectrics is to count both free and bound charge, in the form

$$\oint_{S_c} \mathbf{E} \cdot d\mathcal{A} = \frac{Q_{\mathrm{free}} + Q_{\mathrm{bound}}}{\epsilon_0} \,, \tag{VII.46a}$$

although this requires the calculation of bound charges on boundaries. With linear, isotropic dielectrics in play, it is possible to rescale the diminished electric fields and count only free charge, in the form

$$\oint_{S_c} \kappa \mathbf{E} \cdot d\mathcal{A} = \frac{Q_{\mathrm{free}}}{\epsilon_0} \,, \tag{VII.46b}$$

allowing for different values of κ on different portions of the Gaussian surface S_c. This and relation (VII.43b) suggest what is actually the most general form of the Gauss law,

$$\oint_{S_c} \mathbf{D} \cdot d\boldsymbol{\mathcal{A}} = Q_{\text{free}} \ . \qquad\qquad \text{(VII.46c)}$$

This is the form to be found in more advanced works.

Any dielectric can be converted into a conductor by treating it badly enough: A sufficiently large electric field will not only shift charges within the molecules of the material, but it will also tear some of them away, *ionizing* the molecules and liberating mobile charge carriers. The field magnitude at which this occurs for a given material—it's a matter of field strength, not potential—is called the *dielectric strength* or *breakdown field* E_{max} of the material. Lightning is perhaps the most spectacular example of dielectric breakdown in air. The spark one encounters on touching a doorknob after walking across a carpeted floor in dry weather is a smaller-scale example. Dielectric breakdown determines the capaci*ty* of a capacitor. When the charge or voltage exceeds a maximum value, the electric field between the conductors becomes large enough to break down the intervening dielectric, "shorting out" the conductors. Values of the dielectric strength for some materials include

$$E_{\text{max}}^{(\text{air})} \approx 3. \ \text{MV/m}$$
$$E_{\text{max}}^{(\text{Teflon})} \approx 60 \ \text{MV/m}$$
$$E_{\text{max}}^{(\text{vac})} \sim 3 \times 10^{18} \ \text{V/m} \ , \qquad\qquad \text{(VII.47)}$$

for dry air, Teflon, and vacuum, respectively. The precision of these values is not high, as breakdown can be sensitive to various conditions, such as the humidity of the air. The units shown here, V/m, are often used for electric fields; they are equivalent to N/C. The final value, for vacuum, seems out of place, as the vacuum is not a material. Dielectric breakdown of the vacuum is an Einsteinian and quantum phenomenon: A sufficiently strong electric field can create electron/positron (anti-electron) pairs out of the vacuum and separate them, allowing charge to flow. The field required for this is some twelve orders of magnitude stronger than the breakdown field of air. But this process has been observed in high-energy collisions of heavy atomic nuclei.

Dielectric material in a capacitor, then, can serve three purposes: First, it simply separates the conductors, sometimes with very small separations. Second, it can amplify the capacitance by a factor of its dielectric constant. Third, it can increase the capacity of the device by virtue of its dielectric strength. For example, in *electrolytic capacitors*, one conductor is metal foil, the other a conducting gel that reacts chemically with the metal to produce a very thin insulating layer. Typically, the whole structure is laminated with insulating material and rolled up, allowing a large area to be enclosed in a small (centimeter-size) container. Materials providing dielectric constants of the order 10^4 are used in the construction of ultracapacitors with farad capacitances.

Energy in capacitor and field

Capacitors are used to store electric charge and also to store energy. For example, modern camera flash units use a charged capacitor to supply the light energy of the flash. (Old-fashioned *flash bulbs,* with burning filaments, are becoming quite hard to find.) Portable cardiac defibrillators, now becoming commonplace in public buildings, use a charged capacitor to provide an emergency pulse of electric current and energy to the heart of a person suffering ventricular fibrillation to restore normal heart rhythm. These are rated by the energy they supply.

The energy stored in a capacitor is the work required to charge it. The first increment of charge can be moved from one conductor to the other "for free"; the rest must be moved "uphill" against the potential difference established by the charge on the conductors. The increment of external work required, i.e., of stored energy, is $dU = \Delta V \, dq$ for charge increment dq. Hence, the total energy stored with charge Q is

$$
\begin{aligned}
U &= \int_0^Q \Delta V \, dq \\
&= \int_0^Q \frac{q}{C} \, dq \\
&= \tfrac{1}{2} \frac{Q^2}{C} \\
&= \tfrac{1}{2} C V^2 \ ,
\end{aligned}
\tag{VII.48}
$$

from the definition (VII.34) of C. Here, $V = Q/C$ is the final value of the potential difference. For example, a capacitor with capacitance $C = 100.0 \ \mu\text{F}$ and maximum voltage $V = 150.0$ V can store energy

$$
\begin{aligned}
U &= \tfrac{1}{2} C V^2 \\
&= \tfrac{1}{2} (1.000 \times 10^{-4} \ \text{F})(1.500 \times 10^2 \ \text{V})^2 \\
&= 1.125 \ \text{J} \ .
\end{aligned}
\tag{VII.49}
$$

These are typical values for an off-the-shelf component.

Remarkably, the stored energy of Eq. (VII.48) can be associated with the separated charges or, equivalently, with the *electric field* within the capacitor. For example, if the capacitor were a parallel-plate capacitor with vacuum between the plates for simplicity (of calculation, not of construction), the energy could be written

$$
\begin{aligned}
U &= \tfrac{1}{2} C_{\text{PP}} (\Delta V)^2 \\
&= \tfrac{1}{2} \epsilon_0 \frac{\mathcal{A}}{d} (Ed)^2 \\
&= \tfrac{1}{2} \epsilon_0 E^2 (\mathcal{A}d) \ ,
\end{aligned}
\tag{VII.50}
$$

using formula (VII.37c). Since $\mathcal{A}d$ is the volume between the plates, this suggests that *field energy density*

$$
u_E \equiv \tfrac{1}{2} \epsilon_0 E^2
\tag{VII.51}
$$

should be associated with an electric field in vacuum.[50] Although shown here for a special case, this turns out to be a general result. For example, an electric field of magnitude equal to the breakdown field of dry air has energy density

$$
\begin{aligned}
u_E &= \tfrac{1}{2}\epsilon_0 E_{\text{max}}^{(\text{air})2} \\
&= \tfrac{1}{2}[8.854 \times 10^{-12}\ \text{C}^2/(\text{N}\,\text{m}^2)](3.\times 10^6\ \text{N/C})^2 \\
&= 40\ \text{J/m}^3\ .
\end{aligned}
\tag{VII.52}
$$

In any circumstance, it is important to count *either* the energy of the charges *or* the field energy, not both—that would be double-counting the same energy. In more advanced treatments, it is elegantly shown that the two measures of energy are equal.

The association of energy density with electric fields is more significant than it may appear. The electric field, like the gravitational field, was originally introduced to sidestep "action at a distance" questions, as described in Ch. IV Sec. A. But with the introduction of field energy, it begins to take on the attributes of an actual physical object or entity. The continuing expansion of the role of the field in physics will become apparent in subsequent chapters.

References

Electrostatics, electric fields, the Coulomb and Gauss laws, electric potential, conductors and dielectrics, and capacitance are the subjects of one or more chapters in all introductory physics texts. For example: Bauer and Westfall [BW14] devote four chapters to these topics; Fishbane, Gasiorowicz, and Thornton [FGT96] five; and Halliday, Resnick and Krane [HRK02] six. At the intermediate and advanced levels, these subjects are covered in specialized texts devoted to electricity and magnetism. Intermediate-level texts include Reitz, Milford, and Christy [RMC08] and Heald and Marion [HM94], the latter a companion volume to Thornton and Marion [TM04]. There are numerous advanced-level texts. For example, two volumes of the *Course of Theoretical Physics*, Landau and Lifshitz [LL80] and Landau, Pitaevskii, and Lifshitz [LPL84], treat electricity and magnetism. J. D. Jackson's *Classical Electrodynamics* [Jac98] has been the standard reference on the subject for many decades.

Problems

1. Even the land surface of the Earth contains enough moisture that the Earth can be regarded as a conductor, approximately a sphere of radius $R_\oplus = 6371.$ km. The electric field at the Earth's surface has magnitude 150. N/C and is directed vertically downward, i.e., radially inward.

---

[50]Field energy density in dielectric materials can be a complicated matter. Density $u_E = \tfrac{1}{2}\mathbf{E}\cdot\mathbf{D}$ is one expression used.

(a) Calculate the average surface charge density and total charge on the Earth.

(b) Calculate the electrostatic potential at the surface of the Earth, with the zero of potential at infinity. Why does this potential not pose a hazard to the occupants of the Earth's surface?

2. Suppose the Moon carried the same charge as that of the Earth from the previous problem. Looking up any necessary data, calculate the ratio of the electrostatic repulsion to the gravitational attraction of the two bodies. If measurements of sufficient precision were possible, what might be the effects of this repulsion on the size, shape, and stability of the Moon's orbit around the Earth?

3. Water molecules are electrically neutral, but they possess electric dipole moments of magnitude $p_{H_2O} = 6.2 \times 10^{-30}$ C m. In the field surrounding a point charge, the dipolar molecules will align with the field, but they will also experience a net force, as the field is not uniform. Calculate the magnitude and direction of the force on a suitably aligned water molecule a distance 1.0 cm from a point charge $q = \pm 10.$ nC, and compare this force with the *weight* of a water molecule. This effect can easily be demonstrated by bringing the tip of a charged rubber or glass rod, say, near a falling stream of water from a spigot.

4. A $^{238}_{92}$U nucleus can decay by emitting an α particle, leaving a $^{234}_{90}$Th daughter nucleus. The α particle is detected a great distance from the $^{234}_{90}$Th nucleus with kinetic energy 4.195 MeV. Assume that the α particle is emitted at rest at the surface of the $^{234}_{90}$Th nucleus. Treat the α particle as a point of charge $+2e$ and the $^{234}_{90}$Th nucleus as a uniform solid sphere of charge $+90e$. Assume the $^{234}_{90}$Th nucleus does not recoil. Estimate the radius of the $^{234}_{90}$Th nucleus.

5. **Method of Images I**: If a point charge is suspended by electrically neutral means above a large, flat, grounded—i.e., connected to a limitless source of charge at zero potential, as the Earth can be taken to be—surface or floor, charge of the opposite sign will pool on the floor beneath it. The electric field above the floor produced by the original and pooled charges will be exactly the same as would be produced by only the original charge plus a "mirror image" charge, equal in magnitude and opposite in sign, as far below the floor as the original is above it. (*Why* this is so is explained in intermediate-level and advanced texts.) Take the original charge to be positive charge q, suspended distance a above the floor. Calculate the surface charge density on the floor (as a function of distance from the point directly below the point charge), the total charge pooled on the floor, and the force exerted on the point charge toward the floor.

6. **Method of Images II**: If a point charge is suspended outside a grounded, conducting *sphere,* surface charge will pool on the sphere, attracted to

the point charge. The electric field outside the sphere produced by the original and pooled charges will be exactly the same as that of only the original plus a "mirror image" point charge of such value and location as to make the surface of the sphere an equipotential of potential zero. Taking the sphere of radius R with its center at the origin of coordinates and the original positive charge q at distance $a > R$ from the origin on the $+z$ axis, calculate the position and charge of the "image" (which is not real), the surface charge density on the sphere (which is) as a function of position, the total surface charge on the sphere, and the force exerted on the original charge toward the sphere. In the limit $R \to +\infty$, with $a - R$ finite, do these results reproduce those of the previous problem?

7. In principle, the electric field $\mathbf{E}(\mathbf{r})$ and potential $V_e(\mathbf{r})$ distributions can be obtained from the charge-density distribution $\rho(\mathbf{r})$ by successive integrations, per Eqs. (VII.7c) and (VII.23). Conversely, the field and charge density can be calculated from the potential distribution. (Hence, all of these quantities can be obtained from any one of them.) Find the electric field $\mathbf{E}(\mathbf{r})$ and charge density $\rho(\mathbf{r})$ associated with the spherically symmetric potential distribution

$$V_e(\mathbf{r}) = V_0 \exp\left(-\frac{r}{a}\right) ,$$

where V_0 and a are constants and $r = (x^2 + y^2 + z^2)^{1/2}$ is radial distance from the origin. What is the *total* charge associated with this potential?

8. Treating the Earth as a sphere of uniform *mass* density and a spherical electrical conductor carrying the charge of Problem 1, calculate and compare its electrostatic and gravitational self-energies. The gravitational self-energy can be obtained from a calculation analogous to that of Sec. E for the electron; the electrostatic self-energy can be obtained from a variant of that calculation or otherwise. Do the electrostatic forces associated with the charge the Earth carries exert much influence on its structure or dynamics?

9. Design a parallel-plate capacitor with capacitance $C = 47.0$ pF and maximum voltage $\Delta V_{max} = 160$ V. You may use conducting plates of any size and shape and Plexiglas sheet insulation of any desired size, shape, and thickness. The dielectric constant of Plexiglas is $\kappa = 3.40$, and its dielectric strength is $E_{max} = 40.0$ MV/m. Specify all dimensions of the capacitor. You may assume that fringing effects at the edges of the plates are negligible.

10. A capacitor of capacitance C_0 carries charge $\pm Q_0$ on its plates. A close-fitting slab of dielectric material with dielectric constant κ is inserted between the plates. The capacitor is electrically isolated so that the charge on the plates is undisturbed. Calculate and compare the energies stored in the capacitor before and after the insertion of the dielectric.

11. A capacitor of capacitance C_0 is connected to a battery that maintains constant voltage ΔV_0 between the plates. A close-fitting slab of dielectric material with dielectric constant κ is inserted between the plates without altering the connection to the battery. Calculate and compare the energies stored in the capacitor before and after the insertion of the dielectric. How much work is done by the battery throughout the insertion process?

12. A spherical soap bubble of radius r_0 is given an electric charge q. The resulting electrostatic stress causes the bubble to expand slightly to radius $r_0 + \delta r$. The bubble exists in equilibrium between the surface tension of the soap film, which tends to contract, and the air pressure differential between the interior and exterior of the bubble. Assume that the surface-tension contribution does not change: When the charged bubble expands, the interior pressure decreases, but the outward electrostatic stress makes up the difference. Assume that the air inside the bubble obeys the Ideal Gas Law, and that the soap film conducts heat sufficiently well so that its temperature does not change; assume also that the *amount* of air inside the bubble is unaltered.

    (a) Calculate the change in radius δr in terms of the original radius r_0, the charge q, and the original air pressure P_0 inside the bubble. Treat the changes as small, i.e., as differentials.

    (b) Dielectric breakdown in the air surrounding the bubble, i.e., corona discharge, limits the magnitude of charge q and the expansion δr. For radius $r_0 = 5.00$ cm and pressure $P_0 = 101.3$ kPa, and assuming the breakdown field of dry air, what is the maximum value of δr? Could this be measured in a laboratory experiment?

13. Using the methods of Sec. F, calculate the electrostatic self-energy of the *general*, spherically symmetric charge-density distribution $\rho(r)$, where r is radial distance from the origin. This distribution may or may not cut off at a finite radius. Express the self-energy in terms of definite integrals, paying careful attention to variables and limits of integration. Check that your result accords with Eq. (VII.32b) for the case of a uniform-density sphere of finite radius.

14. Calculate the total *electric-field energy* of the general, spherically symmetric charge distribution $\rho(r)$ of the previous problem. Again, the energy should be expressed in terms of definite integrals. Is this energy equal to that obtained in the previous problem?

Chapter VIII

The Nerve as Electrical Network

- *Analyze the behavior of a nerve-cell fiber (axon) as an electrical resistance network and determine the distribution of a time-independent signal on the fiber, the so-called electrotonus case.*

- *Analyze the axon as a resistive/capacitive network and derive a propagation equation for time-dependent signals.*

- *Examine the propagation of "action potentials" on the axon.*

- *Compare neural and semiconductor "logic circuitry."*

In the decades following the discoveries of Franklin, Coulomb, Cavendish, et al., some investigators built ever larger and more spectacular demonstrations of static electricity. But they were missing the boat. The real action in electrical science was in the study of *current* electricity—electricity in motion. And it was the harnessing of electric currents that would transform the world of the eighteenth century into that of the twentieth and twenty-first. This harnessing would also inspire fear that in its scientific quest, humanity was reaching for powers it should not possess. This notion—long predating the Atomic Age—inspired, for example, Mary Shelley's[1] famous novel *Frankenstein, or The New Prometheus*. While the novel does not explicitly detail the use of lightning to bring the monster to life, as depicted in later film adaptations, certainly lightning—hence, electricity—has long been identified with divine power, such as Shelley's protagonist was usurping. This again illustrates that science is an integral part of human culture, affecting and affected by other parts.

The example used here to illuminate the physics of electric currents and circuits is perhaps the most complex system in the known universe: the nervous system of animals, from the *Planaria* through *Homo sapiens sapiens* to the pinnacle of biological evolution on Earth—the Labrador Retriever. It is remarkable that features of this system can be understood through the application of basic laws governing electric charge in motion. That is, some aspects of the nervous system can be understood by modeling it with fairly simple electrical networks.

A   The neuron—axon, dendrites, and synapses

The nervous system—brain, spinal cord, and peripheral nerves—consists of nerve cells or *neurons*. Such a cell is illustrated in Fig. VIII.1. It consists of a cell body, containing a nucleus, cytoplasm, and the organelles of any cell, and a long extension called an *axon,* ending in *terminal fibers*. The protrusions of the cell body are called *dendrites* because of their resemblance in shape to teeth. The cell body is of the micron scale typical of cells, but the axon can be meters in length; a bundle of axons constitute a nerve. Nerve signals are transmitted down the axon, triggering the release of neurotransmitter molecules from the terminal fibers. These molecules cross gaps called *synapses* to bind to sites on the dendrites of adjacent neurons, thus sending signals through the nervous system. The arrangement of synapses in the reader's nervous system is the reader: Brain function, thought, and memory are all anchored in the architecture of neurons and synapses. The study of the structure and function of the brain and nervous system, i.e., neurophysiology or "brain science," is currently one of the hottest areas of scientific research.

The axon is a long, fluid-filled tube: A thin wall of lipid (fat) molecules surrounds a column of *axoplasm*. The whole arrangement is immersed in *interstitial fluid*; both fluids are similar in composition to seawater. In vertebrates, the axon is wrapped in layers of fatty tissue called *myelin,* like insulation around

---

[1] Mary Wollstonecraft Shelley, 1797–1851

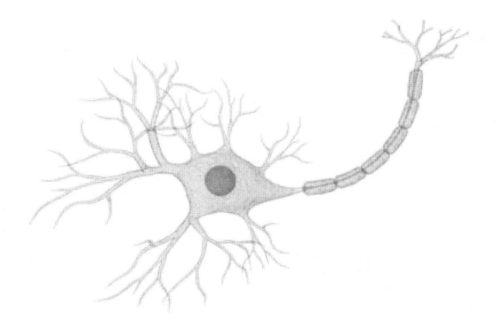

Figure VIII.1: Diagram of a neuron, showing the cell body with nucleus and dendrites, and the axon with myelin sheath and terminal fibers. Not necessarily to scale. (Illustration by Sebastian Koulitzki.)

a wire, produced by *Schwann*[2] *cells.* The myelin layer is interrupted by gaps called *nodes of Ranvier.*[3] The functions of these features will be examined later in this chapter.

Electric currents travel through the nervous system via the transport of chemical ions into, out of, and within the neurons. But these are no less electric currents than the movement of conduction electrons in metal wires. The same language and laws can be applied to both.

B   Electric current—charge in motion

The description of current electricity borrows heavily from fluid mechanics. In fact, the same sort of analysis used for electrical circuits can be applied to plumbing networks.

Current density

The most familiar and readily measured quantity used to describe electric charge in motion is *current,* but this is not the most fundamental quantity. That would be *current density*

$$\mathbf{j} = \rho_q \mathbf{v}_d$$
$$= nq\mathbf{v}_d \ , \tag{VIII.1}$$

where ρ_q is the volume *charge* density of mobile charge carriers (with the subscript to distinguish it from mass density), n is the volume *number* density of charge carriers, q is the carriers' individual charge, and $\mathbf{v}_d$ is the *drift velocity* or flow velocity of the charge carriers.[4] Current density $\mathbf{j}$ is a vector quantity. The fluid-mechanics counterpart of $\mathbf{j}$ is the mass flux $\rho\mathbf{v}$. However, j and $\mathbf{v}_d$ can be in opposite directions, unlike their fluid counterparts.

The SI unit of current density is coulombs per square meter per second or amperes per square meter, viz.,

$$1 \ \frac{C}{m^2 s} \equiv 1 \ \frac{A}{m^2} \ , \tag{VIII.2}$$

where the ampere of current is given by $1 \ A \equiv 1 \ C/s$. Strictly, the coulomb is *defined* as the charge transported by a current of one ampere in one second of time, i.e., by $1 \ C \equiv 1 A \, s$. The operational definition of the ampere—one of the seven fundamental units of SI—is based on the *magnetic* properties of electric currents and must wait until Ch. IX.

---

[2]Theodor Schwann, 1810–1882

[3]Louis Antoine Ranvier, 1835–1922

[4]Now the reader can see why this author eschews $\hat{\boldsymbol{\imath}}$, $\hat{\boldsymbol{\jmath}}$, and $\hat{\boldsymbol{k}}$ as the Cartesian unit vectors; the potential for confusion is too great.

Current

Electric current, then, is the *flux* of current density through any chosen surface. It is given by

$$I = \Phi_{\mathbf{j}}(S)$$
$$= \int_S \mathbf{j} \cdot d\mathcal{A} \, , \qquad \text{(VIII.3)}$$

where the flux of $\mathbf{j}$ through S is defined as the flux of any vector field is defined. It represents the rate of transport of charge through S. For example, the current in a wire is the flux through a cross section of the wire. Current is measured in amperes, i.e., coulombs per second. Like any flux of a vector field, this is a *scalar* quantity. As was illustrated in Fig. I.3, currents do not add as vectors do. The direction associated with the current is more properly associated with the surface S, the area of which *is* a vector.

Relations (VIII.1) and (VIII.3) imply that the direction of $\mathbf{j}$ and the direction associated with I are the directions that positive charge would flow *if the current were created by positive charge in motion*, even if the actual current is created by negative charges moving in the opposite direction. In this respect, Benjamin Franklin's convention for positive and negative charge is unfortunate: It is found to assign negative charge to the charge carriers in metal wires, so that in this ordinary situation, the current and the charge carriers move in opposite directions.[5]

Drift velocity $\mathbf{v}_d$

How fast is electricity? A rough estimate of the drift speed v_d can be made for a familiar situation. Suppose a current $I = 1.00$ A flows in copper wire of circular cross section, with radius[6] $r = 1.00$ mm. The charge carriers in this case are electrons. So relation (VIII.1) implies drift speed

$$v_d = \frac{j}{ne}$$
$$= \frac{I}{ne\mathcal{A}} \, , \qquad \text{(VIII.4a)}$$

where $\mathcal{A}$ is the cross-sectional area of the wire, if we approximate the current density as uniform across the wire. Each copper atom contributes its outermost electron to the conduction process, so the number density n of charge carriers

---

[5] In the 1960s and 1970s attempts were made to change the convention for current direction so that currents would be taken to be in the direction of electron flow. This necessitated replacing some "right-hand rules" with "left-hand rules." But by then, the body of literature on electricity was so vast, and the possibilites for confusion so great, that the attempts were abandoned.

[6] Wire is commonly rated by *gauge*, the number of strands laid side by side to make a one-inch width. The wire in the text example is close to 12-gauge wire in size, a common type. Small electrical devices may use 18- or 22-gauge wire. Household wiring may be 10-gauge wire, difficult to bend by hand. The current may come into the house in 2-gauge wire, which cannot be bent by hand.

is essentially the number density of copper atoms, i.e., the Avogadro number divided by the molar volume of copper. Hence, the drift speed can be estimated as

$$v_d = \frac{I}{ne(\pi r^2)}$$

$$= \frac{1.00 \text{ A}}{\left(\frac{1e}{\text{atom}}\right)\left(\frac{6.023\times10^{23} \text{ atom}}{63.5 \text{ g}}\right)\left(\frac{8.93\times10^6 \text{ g}}{\text{m}^3}\right)\left(\frac{1.602\times10^{-19} \text{ C}}{e}\right)\pi(1.00\times10^{-3} \text{ m})^2}$$

$$= 2.35\times10^{-5} \text{ m/s}$$

$$= 0.0235 \text{ mm/s} ,$$

$$\text{(VIII.4b)}$$

using common values for the molar mass and density of copper. Apparently, the current does not race through the wire: It takes more than forty seconds to cover a millimeter, *hours* to cover the meters from a wall switch, say, to an overhead light! It *oozes* through the wire slower than the proverbial molasses in January. Why, then, does the light come on as soon as the switch is closed? Because the electrons that flow through the light did not come from the switch—they were already at the light. When the switch is closed, an electric field configuration propagates through the wire to drive current through the light. That field configuration propagates at a speed comparable (not quite equal) to the speed of light in vacuum. It takes nanoseconds, not hours, to reach the light.

This commonly calculated result is somewhat misleading. The conduction electrons do not really ooze through the wire like molasses. We might approximate them as an ideal gas at room temperature. The reader can recall from the kinetic theory encountered in introductory chemistry that the average kinetic energy of the electrons is given by

$$\langle \tfrac{1}{2}m_e v^2 \rangle = \tfrac{3}{2} k_B T , \qquad \text{(VIII.5a)}$$

where m_e is the electron mass, k_B the Boltzmann constant, and T the absolute temperature. Hence, the average electron speed is

$$v_{\text{rms}} = \left(\frac{3k_B T}{m_e}\right)^{1/2}$$

$$= \left(\frac{3(1.381\times10^{-23} \text{ J/K})(293. \text{ K})}{9.109\times10^{-31} \text{ kg}}\right)^{1/2}$$

$$= 115. \text{ km/s} , \qquad \text{(VIII.5b)}$$

where room temperature is taken to be 20.°C = 293. K. (This average is the *root-mean-square* or rms value, the square root of the average of v^2.) The electrons are *screaming* through the metal, considerably faster than bullets or jet aircraft. Unperturbed, their directions are random, so there is no net movement of charge. When current flows, a *slight* (fractions of a part per billion) bias is introduced into the distribution of velocities, yielding a small *net* velocity $\mathbf{v}_d$. The effects of the current can be large because the amount of charge involved can be very large—many thousands of coulombs, as observed in Ch. VII.

Continuity equation

Current density satisfies an important identity: Its flux, i.e., the current, through any *closed* surface S_c is related to the charge Q_{in} within the surface by

$$\oint_{S_c} \mathbf{j} \cdot d\boldsymbol{\mathcal{A}} = -\frac{dQ_{\text{in}}}{dt} \ . \qquad \text{(VIII.6)}$$

This is called the *continuity equation*. It expresses the conservation of electric charge in fluid form. (The corresponding equation in fluid mechanics, with Newtonian mechanics in play, expresses the conservation of mass.) As with the Gauss law, the sign convention for $d\boldsymbol{\mathcal{A}}$ here is that the direction out of S_c is positive. Hence, if there is a net outflow of charge, the charge within must decrease at exactly that rate—charge can neither appear from nothing nor disappear. Of course, a net outflow can mean either positive charge flowing out or negative charge flowing in, and Q_{in} can decrease by becoming either less positive or more negative. But in any case, the rates must match, per continuity equation (VIII.6). The continuity equation for electric charge, i.e., the law of conservation of electric charge, is notable for its universality: It has never been observed to be violated under any circumstances.

C The Ohm "law"—a property of materials

The behavior of electric currents in materials is often—but not always—described by a relation named the Ohm[7] law. This is not strictly an unvarying law of nature; rather, it is a property of certain materials. In its fundamental form, it states that the current density $\mathbf{j}$ in a material subjected to electric field $\mathbf{E}$ is given by

$$\mathbf{j} = \sigma \mathbf{E} \ , \qquad \text{(VIII.7)}$$

where the *conductivity*[8] σ of the material is a scalar independent of $\mathbf{j}$ or $\mathbf{E}$. There is no violation here of the principle that $\mathbf{E}$ should vanish in conducting materials, as this material is not in static equilibrium. This is a stimulus-response relation, with $\mathbf{E}$ the stimulus and $\mathbf{j}$ the response. The Ohm law—specifically, the assertion that σ is a scalar independent of $\mathbf{j}$ or $\mathbf{E}$—states that the materials to which it applies are *linear, isotropic* conductors, in the same sense as these terms were applied to dielectric materials, e.g., in Eq. (VII.43a). Materials that behave this way are called *ohmic* materials.

There is an appearance of circularity here: The Ohm law applies to ohmic materials, which are defined as those materials that obey the Ohm law. The content of the Ohm law, then, is that *there are such materials*. In fact, most common conducting materials are approximately ohmic, at least over some range of $\mathbf{E}$ and $\mathbf{j}$ values. No material is perfectly ohmic under all circumstances. For example, the flow of current can change the temperature of the material, and

---

[7]Georg Simon Ohm, 1787–1854

[8]The notation here is conventional. Care must be taken not to confuse this with surface charge density σ; the two must be distinguished by context.

conductivities are temperature-dependent, so σ is not completely independent of $\mathbf{j}$ or $\mathbf{E}$.

Conductivity σ is a characteristic of materials. It has units equal to those of $\mathbf{j}$ divided by those of $\mathbf{E}$, viz.,

$$1\,\frac{\mathrm{A/m}^2}{\mathrm{V/m}} = \frac{1}{\Omega\,\mathrm{m}}$$
$$= 1\ \mathrm{mho/m}$$
$$= 1\ \mathrm{S/m}\ . \qquad\qquad\text{(VIII.8)}$$

Here the ohm (Ω) is the volt per ampere,[9] defined by $1\ \Omega = 1\ \mathrm{V/A}$. The mho ("ohm" spelled backwards) is the inverse ohm,[10] defined by $1\ \mathrm{mho} = 1\ \mathrm{A/V}$. But the siemens[11] (S) is the current SI designation for the inverse ohm, satisfying $1\ \mathrm{S} = 1\ \mathrm{A/V}$. Conductivity is remarkable for its range of values. For example, the values for silver—the best conductor of the elements—and sulfur (powdered yellow form), an excellent insulator, are

$$\sigma_{\mathrm{Ag}} \approx 6 \times 10^{7}\ \mathrm{S/m}$$
$$\sigma_{\mathrm{S}} \approx 1 \times 10^{-15}\ \mathrm{S/m}\ , \qquad\qquad\text{(VIII.9)}$$

ranging over almost 23 orders of magnitude.

The Ohm law appears at first glance to be at odds with Newton's Second Law of Motion. The former gives $\mathbf{j}$, essentially a flow *velocity,* as the response to $\mathbf{E}$, a force. But the latter indicates that the response to a force should be an *acceleration.* The resolution of this apparent paradox is that the Ohm law describes (dynamic) equilibria between two forces, the applied force represented by $\mathbf{E}$ and the dissipative force associated with interactions between the charge carriers and the atoms of the material. This does not appear explicitly as a force but is incorporated in the parameter σ. Hence, the Ohm law is akin to an expression for *terminal speed,* such as Eq. (I.58) or (I.67), but with a Stokes' law, i.e., linear, drag force as per Eq. (I.55b). If we were fish and formulated our physics in an environment in which drag was omnipresent and never negligible, our basic laws of motion might resemble the Ohm law.

A more familiar form of the Ohm law can be obtained by applying the fundamental form (VIII.7) to a simple situation. For a uniform conductor— or *resistor* if the feature of interest is its resistance to current flow—such as illustrated in Fig. VIII.2, both current density $\mathbf{j}$ and electric field $\mathbf{E}$ can be approximated as uniform through the material. The current is related to the magnitude of the current density by $I = j\mathcal{A}$, where $\mathcal{A}$ is the conductor cross section. The electric field is associated with a potential difference V along the length ℓ of the conductor given by $V = E\ell$. In terms of the more easily measured

---

[9]This is another three-way international collaboration.
[10]Yes, this is actually used in some older texts.
[11]Werner von Siemens, 1816–1892

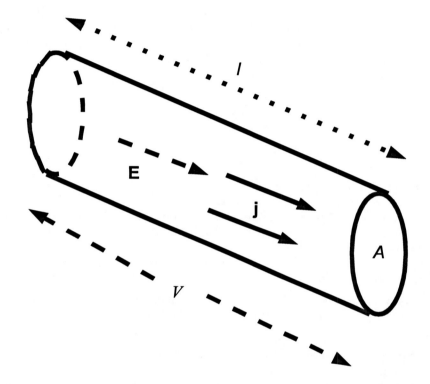

Figure VIII.2: Current flow in a uniform conductor or resistor of length ℓ and cross-sectional area $\mathcal{A}$. Electric field **E** drives current density **j**. The potential difference along the length of the conductor is V; the upper-left end is the high-potential end.

V and I, relation (VIII.7) takes the form

$$V = I \frac{\ell}{\sigma \mathcal{A}}$$

$$\equiv IR \, , \qquad\qquad\qquad \text{(VIII.10a)}$$

the form in which the Ohm law is most frequently encountered. Strictly, the last expression is the *definition* of *resistance*: $R \equiv V/I$. The Ohm law is the assertion that this is a constant independent of I or V, i.e., that a graph of V vs. I for the conductor is a straight line through the origin. *For this specific conductor geometry,* the resistance is given by

$$R = \frac{\ell}{\sigma \mathcal{A}}$$

$$= \rho \frac{\ell}{\mathcal{A}} \, . \qquad\qquad\qquad \text{(VIII.10b)}$$

The last expression defines the *resistivity* $\rho \equiv 1/\sigma$ of the conducting material.[12] This is a property of materials, e.g., the resistivity of copper, while resistance is a property of objects, e.g., the resistance of a meter of 12-gauge copper wire. Resistivity is measured in ohm-meters ($\Omega \, \mathrm{m}$), resistance in ohms.[13]

Ohmic materials come in four flavors, *not* labeled vanilla, chocolate, strawberry, and pistachio. Rather, ohmic materials can be grouped into four categories labeled, in order of decreasing resistivity or increasing conductivity:

- **Insulators**: poor conductors, usually nonmetallic materials. For example, fused quartz, the crystalline form of SiO_2, has resistivity $\rho_{\text{Quartz}} = 7.5 \times 10^{17} \, \Omega \, \mathrm{m}$.

- **Semiconductors**: Neither good insulators nor good conductors, these include semimetallic elements such as silicon and germanium. For example, elemental germanium has resistivity $\rho_{\text{Ge}} = 0.60 \, \Omega \, \mathrm{m}$. Semiconductors become *better* conductors at higher temperatures (satisfying $d\rho/dT < 0$) because increased thermal agitation of their atoms increases the supply of mobile charge carriers in these materials.

- **Conductors**: These include most metals and can have resistivities more than twenty orders of magnitude smaller than those of good insulators. For example, copper, the second-best conductor among the elements after silver, has resistivity $\rho_{\text{Cu}} = 1.72 \times 10^{-8} \, \Omega \, \mathrm{m}$. Typically, conductors conduct better at lower temperatures (i.e., satisfy $d\rho/dT > 0$) because increased thermal agitation of their atoms increases the scattering of moving charge carriers, increasing dissipation and resistance.

---

[12] Here again, the use of ρ for resistivity must be distinguished by context from the use of ρ for density.

[13] Likewise, while conductivity (measured in siemens per meter) is a property of materials, while *conductance*—the inverse of resistance, measured in siemens—is a property of objects.

- **Superconductors**: Certain metals and metal/metal-oxide alloys lose all resistivity ($\rho = 0$) below a critical temperature characteristic of each material (i.e., for $T < T_c$). The transition of the material into this superconducting state is a *first-order phase transition,* like the freezing of water into ice.

Insulators and conductors were known to Franklin and his contemporaries, who also recognized that good conductors of electricity were also good conductors of heat.[14] Semiconductors were identified in the nineteenth century, although their extraordinary usefulness did not come to light until the mid-twentieth century. Superconductors were discovered in the twentieth century.

The mechanisms that make semiconductors conduct better, and conductors worse, with increasing temperature—the liberation of more charge carriers on the one hand, and their increased scattering on the other—actually operate in both materials.[15] But in semiconductors, the supply of mobile carriers is sparse, so the effects of an increase overwhelm the effects of increased scattering. In conductors, the supply of mobile carriers is so great that the liberation of more is of little consequence—the effects of increased scattering dominate.

In fact, it is the supply of mobile charge carriers that determines whether a material is characterized as an insulator, semiconductor, or conductor. This, in turn, is determined by the configuration of quantum states available to electrons within the atomic structure of the material. Our understanding of this—the *band structure* of solids—is one of the triumphs of quantum mechanics. But it must remain the subject of a text beyond this one.

Superconductivity arises from the specific quantum behavior of electrons in these materials. The first superconductors were identified by Kamerlingh Onnes[16] in 1911. These were metals with critical temperatures a few kelvin above absolute zero. The need to maintain temperatures well below that of liquid nitrogen (77 K) has sharply restricted the large-scale application of superconducting technology. In the 1980s complex metal/metal-oxide mixtures were identified with critical temperatures over 100 K and called *high-T_c superconductors.* This set off a race to find materials with ever-higher T_c values, perhaps even within range of ordinary refrigeration techniques.[17] In the intervening decades, however, the fabrication of superconducting devices from these materials has remained elusive. They remain a subject of cutting-edge research rather than a technological breakthrough.

---

[14]They did not know *why*: Because the same mobile electrons that transport charge in a conductor also transport heat.

[15]The *temperature coefficient of resistivity* $\alpha \equiv (1/\rho)(d\rho/dT)$ is typically a few parts per thousand per kelvin (per °C), positive or negative, for conductors or semiconductors, respectively.

[16]Heike Kamerlingh Onnes, 1853–1926

[17]At one time, a record $T_c = 242$ K was claimed, though this may not have stood the test of time.

D   Power (energy, work) and electric currents

That electric currents transport energy is well-known. If current flows, the electric field **E** driving the current does work at rate

$$P = -\Delta V \frac{dQ}{dt}$$
$$= V_{\text{drop}} I \, , \qquad\qquad\qquad \text{(VIII.11a)}$$

where the positive quantity V_{drop} is the magnitude of the change in potential associated with the electric field and the distance the charge is transported, while the current I is the charge flow rate dQ/dt. For any resistor, ohmic or otherwise, this takes the forms

$$P = I^2 R$$
$$= \frac{V_{\text{drop}}^2}{R} \, , \qquad\qquad\qquad \text{(VIII.11b)}$$

as follow from the definition of resistance (VIII.10a). Hence, given any two of the quantities P, V_{drop}, I, and R, the others can be calculated.

The power P expended on current transported through a resistor is dissipated in its conducting material as heat. This process is named *Ohmic heating* or *Joule heating*—especially when it is wanted, as in an electric toaster. Or it can be called *$I^2 R$ loss* when it is not wanted, as in power-line transmission.

For example, a household 60-watt incandescent light bulb[18] is operated at a potential difference of 110. V. These are actually the root-mean-square (rms) values of sinusoidally varying *alternating-current* (AC) quantities, but for the purposes of this example, they can be treated as steady *direct-current* (DC) quantities.[19] The current through the bulb is given by

$$I = \frac{P}{V_{\text{drop}}}$$
$$= \frac{60. \text{ W}}{110. \text{ V}}$$
$$= 0.55 \text{ A} \, , \qquad\qquad\qquad \text{(VIII.12a)}$$

a respectable current.[20] The resistance of the light bulb is

$$R = \frac{V_{\text{drop}}^2}{P}$$
$$= \frac{(110. \text{ V})^2}{60. \text{ W}}$$
$$= 200 \ \Omega \, , \qquad\qquad\qquad \text{(VIII.12b)}$$

---

[18]This may be a vanishing technology.

[19]In fact, for an ordinary light bulb, such treatment is almost exactly correct.

[20]A steady current of 75 *milli*amperes through the chest will stop a human heart. Household currents of the order of amperes and larger currents command respect.

calculated from the original data, rather than the previous result, as a precaution against possible error. This is the resistance of the light bulb *in operation*. Measurement of a 60. W light bulb with an ohmmeter would yield a signficantly smaller value. The operating temperature of the bulb filament, some thousands of kelvin, is sufficiently higher than room temperature that the change in filament resistance is noticeable. Operated between room temperature and its normal operating temperature, then, an incandescent light bulb is *not* an ohmic resistor.

E DC Circuits: closed paths for current

Circuit theory is an extensively developed subject in its own right, with numerous textbooks and entire courses dedicated to it. Here, we shall need only the basics, the principles governing *direct-current* (DC) circuits—circuits in which the currents flow in one direction.

In general, a circuit is a closed path for current flow, which may consist of one or more loops. The path must be closed: On an open path, the accumulation of charge at the "ends" would give rise to electrostatic forces that would stop the flow. A simple closed path is of no interest in itself. Charge flows from a point or points of high potential to points of lower potential,[21] but it must then be raised back to high potential to maintain the flow. Devices that transport charge in this way are sources of *electromotive force* or *EMF*.

Electromotive force (EMF) devices or sources

EMF devices or *EMF sources* transport charge, formally, from lower to higher potential—or most commonly, transport electrons in the other direction. The term *electromotive force* is widespread and historic, but it is a misnomer: It is a potential difference measured in volts, not a force measured in newtons. The chemical cells and batteries that power countless devices are rated by the potential or voltage they maintain, not by the force they exert.

The transport of charge can be accomplished in many ways: mechanical, chemical, photoelectric, thermoelectric, et cetera. For example, in a Van de Graaff[22] generator, sketched in Fig. VIII.3, charge is sprayed (via corona discharge) onto a moving belt, which transports it into a hollow spherical conductor at the top of an insulating column. The charge is transferred to the sphere, where it moves to the outer surface. A tabletop demonstration model can raise the potential of the sphere to tens of thousands of volts. Larger versions, meters in size, can produce megavolt potentials; these are used to accelerate charged particles to high speeds. Or the forces needed to move charge can be exerted on molecular scales, via chemical reactions, in electrochemical cells. Or the energy of light waves can be used to move charge in photoelectric

---

[21]Electrons in conductors move from low to high potential, the current flowing the other way.

[22]Robert Jemison Van de Graaff, 1901–1967

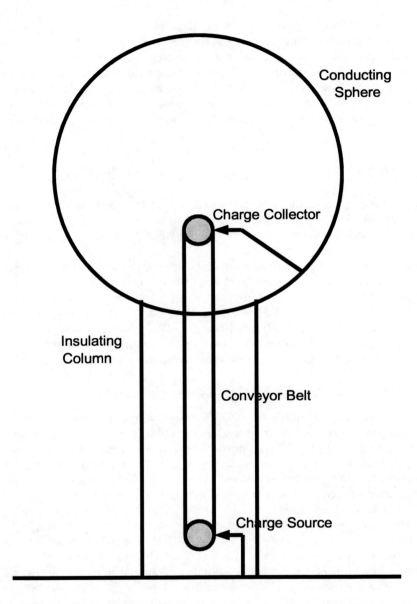

Figure VIII.3: Sketch of a Van de Graaff generator. Electric charge is sprayed onto the moving belt (on pulleys turned by a motor), which transports it mechanically into the spherical conductor, where it moves to the outer surface. The sphere is raised to very high electrostatic potential.

or photovoltaic cells. Or thermal energy can be used: If a junction of dissimilar metals is heated, charge is shifted and a potential difference is created across the junction. Such a device is called a *thermocouple*. These are integral to the *Radioisotope Thermal Generators* (RTGs) used to power spacecraft on deep-space missions, where solar energy is insufficient. A suitable radioisotope sample, such as plutonium—essentially, a self-heating rock, which will maintain a temperature of some 800 K for many years—is connected to thermocouples to provide EMF for the spacecraft systems.

Electrochemical cells and batteries, in many different forms, are ubiquitous in modern technology—they power everything from heart pacemakers to jet aircraft systems. The distinction between a *cell* and a *battery* here is the same as the distinction between a *gun* and a *battery* in artillery: A cell is one unit, a battery an array of cells. The AA "batteries" sold in stores are actually cells; an automobile battery consists, typically, of six connected chambers, each of which is a cell. Like the capacitors of Ch. VII Sec. G previously, the cells of a battery can be connected in parallel or in series. In parallel, the potential difference of the battery is that of an individual cell while their charge contributions add; in series, the potentials of the cells add while a common charge passes through all. The potential difference maintained by a cell is determined by the ionization potentials of the reactants in the cell, i.e., by the chemistry chosen. For example, in a typical car battery, metallic lead is ionized and goes into solution as lead sulfate. This reaction sustains a potential difference of 2.0 V. Six cells in series constitute a 12-volt battery.

An *ideal* EMF device maintains a fixed potential difference while supplying an arbitrary amount of current. *Real* batteries or other devices contain finite amounts of matter and can supply only a limited current or transport a limited amount of charge. These limitations are incorporated into the analysis of circuits by assigning an *internal resistance* R_i to the EMF source. This is illustrated in Fig. VIII.4, which introduces some standard notation for circuit diagrams. The symbol for the EMF source is a stylized representation of a wet-chemical battery or *voltaic pile,* although the actual device may be of any type. The sawtooth symbols represent resistances, both the internal resistance of the EMF source and the load. The former is a convenient representation; it is not a component that can be detached from any real EMF source. The latter may be any device with electrical resistance, such as a light bulb or heating element, or it may be a component specifically constructed to provide resistance. Straight-line segments represent conducting paths of negligible resistance; hence, these are equipotentials. The switch, represented by a standard symbol, is properly installed: immediately adjacent to the power supply on the high-potential side, so that when the switch is open, the rest of the circuit is at low potential (a safety feature). It may occur to the reader that the switch is a triviality in such a simple circuit, but let the reader be warned: *Do not scorn the humble switch.* The triangle symbol at the bottom of the diagrams indicates the *common voltage.* It identifies the reference voltage level for the entire circuit. The chassis of an electrical device often provides such a reference.

With switch S open, no current flows in the circuit. There is no voltage

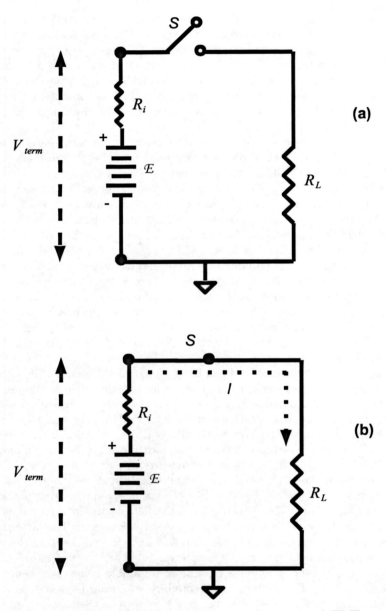

Figure VIII.4: A one-loop resistive circuit, containing a real EMF source, a switch, and a load resistor R_L. In (a), switch S is open and no current flows; in (b), the switch is closed and current flows through the load.

drop across the internal resistance R_i. The potential difference between the terminals of the EMF source is the native EMF $\mathcal{E}$ of the device, determined by its chemistry or other mechanism. When the switch is closed, current I flows. The potential difference between the terminals is

$$V_{\text{term}} = \mathcal{E} - IR_i \ , \tag{VIII.13a}$$

reduced by the voltage drop across R_i. Per Eq. (VIII.10a), voltage V_{term} must equal IR_L. This determines the current, viz.,

$$I = \frac{\mathcal{E}}{R_i + R_L} \ . \tag{VIII.13b}$$

Result (VIII.13a) shows that the internal resistance R_i of the EMF device can be determined as the zero-current EMF $\mathcal{E}$ (there are ways to measure the potential of the device without drawing current from it) divided by its maximum current, the current it produces when *shorted* ($V_{\text{term}} = 0$). For an ordinary AA cell, say, this might be a few tenths of an ohm, while an automobile battery, which might drive "600 cold-cranking amps[23]" at 12 volts,[24] would have internal resistance $R_i = 0.020 \ \Omega$. As a cell or battery wears out, its intrinsic EMF does not change, but its capacity to drive current decreases. This is manifest as an *increase* in its internal resistance.

For a circuit with resistances much larger than the internal resistances of its EMF sources, e.g., for $R_L \gg R_i$ here, the latter can be neglected and the sources approximated as ideal. However, it is a famous result—which the reader can easily derive—that the *power* dissipated in the load resistance of a circuit like that in Fig. VIII.4 is a maximum when the load resistance is equal to the internal resistance of the power source ($R_L = R_i$). This is an example of a general phenomenon known as *impedance matching*.

Resistors

As indicated, resistors in circuits can be devices, the principal electric characteristic of which is resistance, or they can be devices simply designed to contribute resistance (for example, to limit current in a circuit). Purpose-made resistors can take many forms, from large devices to microscopic portions of integrated-circuit (IC) chips. A frequently encountered design is a cylinder, often of powdered carbon, the dimensions of which determine the resistance. The cylinder has a wire lead at each end and is enclosed in an insulating casing with three or four colored stripes painted or printed on it. The first two stripes specify significant figures, the third a power of ten, and the fourth the precision of the resistance value. For the first three stripes, the digits 0 through 9 are represented by the colors black, brown, red, orange, yellow, green, blue, violet, gray,

---

[23]Delco$^{(TM)}$ batteries used to be advertised this way.

[24]The common lead-acid automobile battery is a formidable device, deserving of respect. It offers both chemical and electrical hazards to any who handle it: It is filled with sulfuric acid and can evolve explosive hydrogen gas; it can easily drive lethal electric currents. Modern "maintenance-free" batteries are sealed and should stay that way.

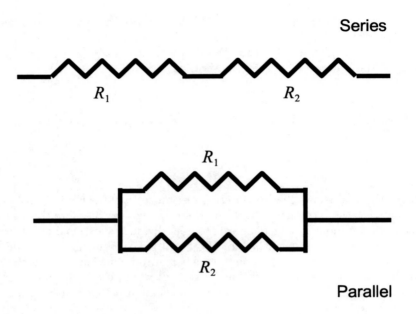

Figure VIII.5: Resistors in series and in parallel.

and white, in order. For the fourth stripe, gold denotes 5% tolerance, silver 10%, and no fourth stripe 20%. (Resistors manufactured to greater precision usually have the tolerance, e.g., 1% or 2%, printed on the resistor.) For example, red-red-yellow-silver represents $(22 \pm 2) \times 10^4 \; \Omega = (220 \pm 20)$ kΩ. Resistors of this design are frequently millimeters in diameter and a centimeter or so in length. They are found in a great variety of electrical and electronic devices.

Like pairs of capacitors, pairs of resistors can be connected in series (end to end) or in parallel (side by side), as illustrated in Fig. VIII.5. In the former, the same current passes through both, while the potential difference across the combination is simply the sum of those across the individual components. Dividing these potential differences by the common current and using the definition of resistance yields the rule

$$R_{\text{ser}} = R_1 + R_2 \qquad\qquad\text{(VIII.14a)}$$

for resistors in series. Resistors in parallel encounter a common potential difference, while the current through the combination is the sum of the individual currents. (No current is lost or gained at the junctions.) Dividing the currents by the common potential difference yields the rule

$$\frac{1}{R_\parallel} = \frac{1}{R_1} + \frac{1}{R_2} \ , \qquad\qquad \text{(VIII.14b)}$$

or equivalently,

$$R_\parallel = \frac{R_1 R_2}{R_1 + R_2} \ , \qquad\qquad \text{(VIII.14c)}$$

for resistors in parallel. Result (VIII.10b) for the uniform conductor provides a means for remembering these rules: Resistors in series "add lengths," i.e., resistances; resistors in parallel "add cross-sectional areas," so *inverse* resistances add. These are the same rules as Eqs. (VII.38) and (VII.39) for capacitors, but for the opposite cases; care must be taken to avoid confusion.[25]

Kirchhoff rules

A circuit or network can be *solved,* i.e., the performance of every component determined, via the application of two simple principles called the *Kirchhoff*[26] *rules.* The *loop rule* states that for any closed path or loop within a circuit, the algebraic sum of potential differences ΔV_i from point to point around the loop must be zero, i.e., the potential differences satisfy

$$\sum_{\text{Loop}} \Delta V_i = 0 \ . \qquad\qquad \text{(VIII.15a)}$$

This is just the assertion that the potential V is well-defined as a function of position in the circuit, i.e., that the Law of Conservation of Energy is in play. The *junction rule* states that at any junction of current paths in the circuit, the algebraic sum of currents out of the junction—i.e., with outbound currents positive and inbound currents negative (or equivalently, the sum of currents in, with inbound currents postive and outbound negative)—must be zero, must satisfy

$$\sum_{\text{Jct}} I_{\text{out}} = 0 \ . \qquad\qquad \text{(VIII.15b)}$$

This is just a discrete version of the continuity equation (VIII.6), the Law of Conservation of Electric Charge, asserting that charge neither piles up nor drains away at the junction. Strictly, the Kirchhoff rules apply to *steady-state* circuits and networks, for which this assertion is exact. However, as shall be seen, there are "dodges" that allow the rules to be applied even to more general circuits.

Kirchhoff and his colleague Bunsen[27] are titans of science. They discovered the science of *spectroscopy*: the determination of the composition of a material from the frequencies of light it emits, absorbs, or reflects. Few discoveries in all

---

[25] Alternatively, the combination rules could be written for *conductances,* i.e., inverse resistances. Then the rules would match those for capacitances exactly.

[26] Gustav Robert Kirchhoff, 1824–1887

[27] Robert Wilhelm Bunsen, 1811–1899

of human intellectual history have had more wide-ranging impact. Spectroscopy is central to physics at every scale, to chemistry, to biology, and to astronomy—it is the only means by which we know the composition of the stars, as well as their temperatures, sizes, and velocities. Yet Kirchhoff's name is remembered for these two elementary applications of fundamental conservation laws, while Bunsen's is remembered for a pipe with a valve attached—the Bunsen burner. Fame is a fickle mistress indeed.

To avoid "going around in circles" when analyzing multiloop circuits—circuits that cannot be reduced to a single loop—it is best to follow a systematic procedure. One such procedure is:

- Simplify all component combinations, i.e., replace all series and parallel combinations with single equivalent components, and utilize any available symmetry to reduce the extent of the problem.

- Divide the resulting circuit into independent loops, assigning a current to each loop. There must be at least one such current through every component of the circuit, but no loop should consist of the superposition of two or more others—that will only introduce extraneous variables and redundant equations.

- Apply the loop (and junction) rules to each of the independent loops. In simpler cases, it is possible to incorporate the junction rules into the choice of currents through the components. For more complex circuits, it may be necessary to assign an independent current to each segment of the circuit and impose the junction rule explicitly for each junction, along with the loop rule for each chosen loop. In any case, the result is a system of n equations for n independent currents. Mathematically, this is a "guaranteed win," at least in principle.

- Solve the system of simultaneous equations for the unknown currents.

Once the current through any component is known, any other feature can be determined—voltage, power, et cetera.

As an example, we can consider the two-loop circuit of Fig. VIII.6. This consists of two EMF sources V_1 and V_2, approximated as ideal, and three resistors R_1, R_2, and R_3. Switches in this circuit are assumed closed and not shown. The common-potential symbol in this figure is the more traditional symbol for *ground*[28] potential. To analyze this circuit, the procedure given here should be followed with some care.

- First, resistors R_1 and R_2 are *not* in parallel despite their positions in the diagram: Parallel components must encounter a common potential difference, but the presence of source V_2 means that is not the case here. Likewise, resistors R_1 and R_3 are not in series, because all the current

---

[28] For the wiring of a house, for example, this connection is made to an actual iron pipe driven into the actual ground. The surface of the Earth sets the common potential for all the circuitry.

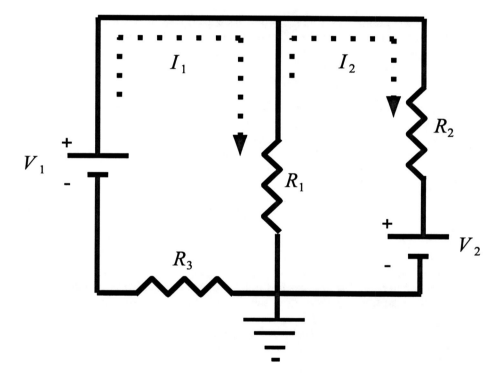

Figure VIII.6: A two-loop resistive circuit or network. For simplicity, EMF sources are taken to be ideal, and all switches are taken to be closed.

through R_1 need not flow through R_3. There are no combinations to simplify, and no symmetry to apply, in this case.

- Second, currents I_1 and I_2 are assigned to the left-hand and right-hand loops, as shown. A third loop, around the outside of the circuit, is superfluous, as this would be the superposition of the left- and right-hand loops. How do we know in advance that the currents flow in the loops in the directions shown? We do not, nor do we care; the directions are assigned as shown just to be systematic. If we have "guessed" wrong, the algebra will take care of it, assigning a negative value to the corresponding current.

- Third, we first apply the Kirchhoff loop rule to the left-hand loop. We start, say, in the "southwest" corner and proceed clockwise. Taking care with the *signs* of the changes in V, we obtain

$$V_1 - (I_1 - I_2)R_1 - I_1 R_3 = 0 . \qquad \text{(VIII.16a)}$$

The second term on the left is the correct voltage drop in the downward direction through R_1, since the net current in that direction is $I_1 - I_2$. This assignment automatically incorporates the Kirchhoff junction rule at the junctions top center and bottom center of the diagram. Applying the loop rule to the right-hand loop, beginning at the northeast corner and proceeding clockwise, yields

$$-I_2 R_2 - V_2 - (I_2 - I_1)R_1 = 0 . \qquad \text{(VIII.16b)}$$

Here the second term on the left has a negative sign because EMF source V_2 is crossed from the high to the low side. The third term is the voltage drop across R_1 in the *upward* direction. For both equations, the change in voltage across the line segments—taken to be wires of negligible resistance—is zero. The two equations can be rearranged into the system

$$\begin{aligned} I_1(R_1 + R_3) - I_2 R_1 &= V_1 \\ -I_1 R_1 + I_2(R_1 + R_2) &= -V_2 , \end{aligned} \qquad \text{(VIII.16c)}$$

two simultaneous linear equations for the currents I_1 and I_2.

- Finally, this system can be solved by any convenient method. We obtain

$$\begin{aligned} I_1 &= \frac{V_1(R_1 + R_2) - V_2 R_1}{R_1 R_2 + R_1 R_3 + R_2 R_3} \\ I_2 &= \frac{V_1 R_1 - V_2(R_1 + R_3)}{R_1 R_2 + R_1 R_3 + R_2 R_3} . \end{aligned} \qquad \text{(VIII.16d)}$$

Any feature of any component in the circuit can be calculated from these currents.

For example, with $V_1 = V_2 = 12.0$ V and $R_1 = R_2 = R_3 = 100.$ Ω, the two currents take values

$$I_1 = \frac{V_1}{3R_1}$$
$$= \frac{12.0 \text{ V}}{3(100. \text{ }\Omega)}$$
$$= 40.0 \text{ mA} \qquad \text{(VIII.17a)}$$

and

$$I_2 = \frac{-V_1}{3R_1}$$
$$= \frac{-12.0 \text{ V}}{3(100. \text{ }\Omega)}$$
$$= -40.0 \text{ mA} . \qquad \text{(VIII.17b)}$$

As mentioned, the results indicate that for these component values, the current in the right-hand loop is *not* as drawn in Fig. VIII.6. Rather, current I_2 flows counterclockwise, so that the total current through resistor R_1 is 80.0 mA. The algebra will always take care of this as long as the Kirchhoff-rule equations are set up correctly.

A complete circuit diagram for one of the workstations on which this text was composed, drawn to the scale of Fig. VIII.6, might cover a sizable portion of North America. Nonetheless, the circuitry could be analyzed, at least in principle, by applying the methods illustrated here.

F   Axon as resistance ladder: electrotonus case

We now have tools sufficient to begin the analysis of the axon as an electrical network. The simplest situation with which to start is that of a time-independent, steady-state voltage on the axon. Neurophysiologists label this the *electrotonus* case.

The structure of the unmyelinated axon is simple. It is a thin-walled tube: The wall consists of lipid molecules and is about 5 nm thick. The diameter of the tube can vary from order 1 μm in vertebrates to 500 μm—half a millimeter!—in, for example, the giant axon of the squid. (This is the basis of many experiments, as it can be seen and manipulated without a microscope.) The *axoplasm* that fills the tube is similar to seawater but with a deficit of Na^+ ions relative to equilibrium concentration, a slight excess of K^+ ions, and Cl^- ions in equilibrium. In the resting state of the neuron, the axoplasm is at potential $V_{rest} \sim -70$ mV relative to the surrounding medium. That medium, the *interstitial fluid,* has an excess of Na^+ ions, a slight deficit of K^+ ions, and Cl^- ions in equilibrium. It is at common potential $V_{rest} = 0$.

The electrical analysis of the axon illustrates an important technique for dealing with continuous systems: We imagine the axon subdivided into infinitesimal segments, determine the change from one segment to the next, and integrate the resulting differential equation. In each segment, of infinitesimal

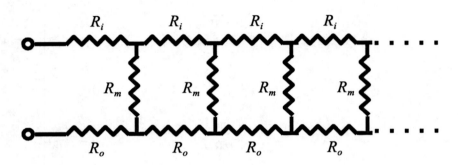

Figure VIII.7: Resistance-ladder model for the axon in the electrotonus (steady-state) case.

length dx, current can flow (via ion transport) down the axoplasm, down the interstitial fluid outside the axon, or through the axon membrane. For the first two currents, the resistance is proportional to the length of the segment, viz.,

$$R_i = r_i\,dx \qquad\qquad \text{(VIII.18a)}$$

and

$$R_o = r_o\,dx \qquad\qquad \text{(VIII.18b)}$$

for the "inside" and "outside" currents, respectively. Here, r_i and r_o, the inside and outside resistances per unit length, are characteristics that can be measured in the laboratory. For the current through the membrane, the cross-sectional area through which the current flows is proportional to dx. Per Eq. (VIII.10b), the resistance is *inversely* proportional to this. It can be written

$$R_m = \frac{1}{c_m\,dx}\;, \qquad\qquad \text{(VIII.18c)}$$

with c_m the *conductance* per unit length of the axon membrane. The axon can be modeled as the "infinite" resistance ladder depicted in Fig. VIII.7, with inside, outside, and membrane resistances for each segment as indicated.

A Kirchhoff-rule analysis of this model following the procedure outlined in the preceding section would yield an infinite sequence of coupled equations for the currents in each section of the ladder. The mathematics for treating such systems exists; the reader is invited to try this approach. This author will use something sneakier, based on a "recursion" approach: the recognition that an infinite system, minus one piece, is the same infinite system. For example, the apparently impenetrable equation

$$x^{x^{x^{x^{.^{.^{.}}}}}} = 2 \qquad\qquad (VIII.19)$$

is equivalent to $x^2 = 2$, with solution $x = \sqrt{2}$, as the exponent of the first x is the original expression. (This solution is, in fact, correct.) Likewise, the resistance ladder can be considered resistances R_i and R_o in series with the parallel combination of resistance R_m and *the original ladder*. Hence, the resistance R of the entire ladder satisfies

$$R = R_i + R_o + \frac{R_m R}{R_m + R} , \qquad\qquad (VIII.20a)$$

or

$$R^2 - (R_i + R_o)R - (R_i + R_o)R_m = 0 . \qquad\qquad (VIII.20b)$$

The positive (i.e., physical) solution of this quadratic equation is

$$\begin{aligned}
R &= \tfrac{1}{2}\left\{ R_i + R_o + \left[(R_i + R_o)^2 + 4(R_i + R_o)R_m \right]^{1/2} \right\} \\
&= \tfrac{1}{2}\left\{ (r_i + r_o)\,dx + \left[(r_i + r_o)^2\,dx^2 + \frac{4(r_i + r_o)}{c_m} \right]^{1/2} \right\} \\
&= \left(\frac{r_i + r_o}{c_m} \right)^{1/2} , \qquad\qquad (VIII.20c)
\end{aligned}$$

where the second expression is obtained by using forms (VIII.18) for the resistances and the last in the limit $dx \to 0$, i.e, by discarding infinitesimals as negligible in comparison to the finite term. This yields a finite, measurable value for the resistance of the axon.

The voltage distribution $V(x)$ along the axon can be obtained by considering successive "rungs" at positions x and $x + dx$, say. All the current that flows into the leg of the ladder to the right of position x (and flows back through the leg on the other side) must enter the parallel combination of the rung at $x + dx$ and the rest of the ladder. Since, by recursion, all the ladder to the right of position x is equivalent to all the ladder to the right of $x + dx$, the equality of these currents can be expressed by

$$\frac{V(x)}{R} = V(x + dx) \left(\frac{1}{R_m} + \frac{1}{R} \right) , \qquad\qquad (VIII.21a)$$

where expression (VIII.14b) is used on the right-hand side for the inverse of the resistance of the parallel combination of rung and remaining ladder. This form makes it clear that the current that flows in the leg to the right of x must either flow through the membrane at $x + dx$ or continue down the ladder. That is, this is simply the Kirchhoff junction rule (VIII.15b) applied to the junction at position $x+dx$. Defining differential dV via $V(x+dx) = V(x)+dV$, rearranging, and using results (VIII.18c) and (VIII.20c) yields

$$dV = -V\,\frac{R}{R_m}$$
$$= -V\,[(r_i + r_o)c_m]^{1/2}\,dx \ , \qquad\qquad (VIII.21b)$$

where the second-order differential term $dV\,dx$ on the right-hand side is neglected in comparison with the first-order terms. This is a differential equation as they were originally written. It is an application of the Most Important Differential Equation in the Universe [identified in the footnote following Eqs. (V.2)]: It describes a quantity that changes at a rate proportional to itself. It implies

$$\int_{V_0}^{V(x)} \frac{dV'}{V'} = -[(r_i + r_o)c_m]^{1/2} \int_0^x dx' \ , \qquad\qquad (VIII.22a)$$

or

$$\ln \frac{V(x)}{V_0} = -[(r_i + r_o)c_m]^{1/2}\,x \ , \qquad\qquad (VIII.22b)$$

or

$$V(x) = V_0\,\exp\left\{-[(r_i + r_o)c_m]^{1/2}\,x\right\} \ . \qquad\qquad (VIII.22c)$$

A static voltage V_0 applied at the neuron cell body decays exponentially down the length of the axon.

The parameters of this system are known from laboratory measurements, e.g., of squid giant axons, going back to the 1940s. Typical values are

$$r_i = 6.37 \times 10^9 \ \Omega/\text{m}$$
$$r_o = 6.37 \times 10^9 \ \Omega/\text{m}$$
$$c_m = 1.25 \times 10^{-4} \ \text{S/m} \ . \qquad\qquad (VIII.23)$$

Hence, the electrotonus voltage distribution can be written

$$V(x) = V_0\,\exp\left\{-[2(1.25 \times 10^{-4} \ \text{S/m})(6.37 \times 10^9 \ \Omega/\text{m})]^{1/2}\,x\right\}$$
$$= V_0\,\exp[-(1.26 \times 10^3 \ \text{m}^{-1})\,x]$$
$$= V_0\,\exp[-x/(0.792 \ \text{mm})] \ . \qquad\qquad (VIII.24)$$

This last form shows that the voltage across the axon membrane—strictly, the *signal,* i.e., the difference between that and the resting potential difference—is reduced by factor $e^{-1} = 0.368$ after less than 0.8 mm distance. In eight millimeters distance, it is more than twenty-two thousand times smaller than its initial value. Yet axons and nerves can be centimeters, even meters in length. Failure? Not yet: At least the resistance-ladder model has shown that *neurons do not function via static or quasi-static voltage signals.*

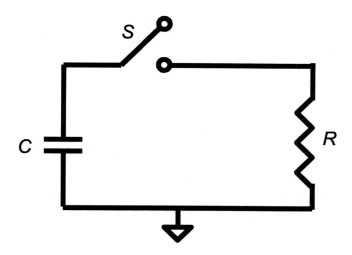

Figure VIII.8: RC circuit for discharging a capacitor.

G   RC Circuits: time-dependent currents

To model time-dependent signals on the axon, we must allow for the build-up or draining away of charge on the axon membrane. That is, we must incorporate the capacitance of the axon segments, as well as their resistances. Circuits in which both resistance and capacitance are in play are called RC circuits. (A third player will be introduced in Ch. X.) The simplest such circuit, with discrete components, is shown in Fig. VIII.8. Capacitor C is charged. When switch S is closed, current flows clockwise through resistor R, allowing the capacitor to discharge. Applying the Kirchhoff loop rule,[29] starting at the southwest corner and proceeding clockwise, yields

$$\frac{Q}{C} - IR = 0 \ . \qquad \text{(VIII.25a)}$$

But the current I is the rate at which the charge comes off the capacitor; it satisfies $I = -dQ/dt$. Hence, this is the differential equation

$$R\frac{dQ}{dt} + \frac{Q}{C} = 0 \ , \qquad \text{(VIII.25b)}$$

---

[29]This is not a steady-state circuit. But energy conservation, i.e., the loop rule, is still in play. And as long as we treat the capacitor as a unit and do not ask what is happening on its individual plates, even the junction rule could still be used if needed.

or

$$\frac{dQ}{dt} = -\frac{Q}{RC} \, . \qquad\qquad (\text{VIII.25c})$$

It is not necessary to repeat the integration Eqs. (VIII.22). We need only recognize that this differential equation is of the same form as Eq. (VIII.21b): It describes a quantity changing at a rate proportional to itself. Hence, its solution is of exponential form, viz.,

$$Q(t) = Q_0 \exp\left(-\frac{t}{RC}\right)$$

$$I(t) = \frac{Q_0}{RC} \exp\left(-\frac{t}{RC}\right) \, , \qquad\qquad (\text{VIII.26})$$

where Q_0 is the charge on the capacitor at $t = 0$ when switch S is closed. The capacitor does not discharge instantaneously. The charge decreases to $e^{-1} = 36.8\%$ of its initial value in the first RC of time, 36.8% of that or 13.5% of Q_0 after the next RC, and so on. The time scale RC is called the RC *time constant* of the circuit. For example, for a 100. kΩ resistor and a 100. μF capacitor (both readily available), it is

$$\tau_{RC} \equiv RC$$

$$= (1.00 \times 10^4 \ \Omega)(1.00 \times 10^{-4} \ \text{F}) \quad = 1.00 \ \text{s} \, ; \qquad (\text{VIII.27})$$

the reader will wish to confirm that the ohm-farad (Ω F) is indeed equivalent to the second. In this case the exponential decay of the charge Q (i.e., of the voltage across the capacitor, Q/C) or the current I could easily be observed on a suitable meter.

How did the capacitor get charged in the first place? Perhaps by means of the circuit illustrated in Fig. VIII.9. If switch S is closed at time $t = 0$, say, then current I flows in a clockwise direction. Positive charge Q accumulates on the upper plate of the capacitor and flows off the lower plate to complete the circuit. (Or rather, electrons accumulate on the lower plate and flow off the upper at the same rate.) Applying the Kirchhoff loop rule, starting in the southwest corner and proceeding clockwise, yields

$$V - IR - \frac{Q}{C} = 0 \, , \qquad\qquad (\text{VIII.28a})$$

or

$$R\frac{dQ}{dt} + \frac{Q}{C} = V \, , \qquad\qquad (\text{VIII.28b})$$

since here the current is the rate at which the charge *increases*, satisfying $I = +dQ/dt$. This is similar to Eq. (VIII.25b), but that equation is *homogeneous* (every term contains a Q to the first power), while this includes the *inhomogeneous* term V. There are a variety of techniques for solving this equation, treated in introductory differential-equations courses. But for this text, it is most straightforward simply to rearrange the equation into the form

$$\frac{d}{dt}(CV - Q) = -\frac{1}{RC}(CV - Q) \, . \qquad\qquad (\text{VIII.28c})$$

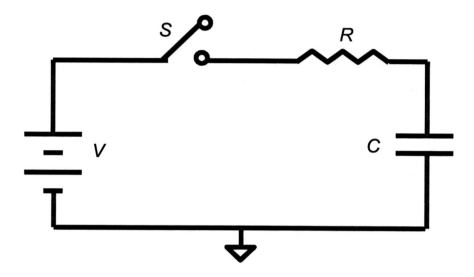

Figure VIII.9: Series RC circuit for charging a capacitor.

As C and V are constants, the right-hand side is just $-dQ/dt$. This equation also describes a quantity changing at a rate proportional to itself. Its solution is

$$CV - Q(t) = CV \exp\left(-\frac{t}{RC}\right) , \qquad \text{(VIII.29a)}$$

using $Q(0) = 0$, or

$$Q(t) = CV \left[1 - \exp\left(-\frac{t}{RC}\right)\right]$$

$$I(t) = \frac{V}{R} \exp\left(-\frac{t}{RC}\right) . \qquad \text{(VIII.29b)}$$

The capacitor does not charge instantaneously either. Charge Q reaches $1 - e^{-1} = 63.2\%$ of the value CV after time RC, increases 63.2% of the rest of the way, or to 86.5% of CV, after another RC, and so on. It approaches charge CV asymptotically; after ten time constants RC it attains 99.9955% of that value. Meanwhile, the current I starts at value V/R and decreases exponentially toward zero. These results illustrate a useful general feature of RC circuits and networks: When the switches in such circuits are first closed, the uncharged capacitors can be treated as closed switches or conducting wire segments; if the asymptotic $t \to +\infty$ behavior of the circuit is sought, the fully charged capacitors can be treated as open switches.

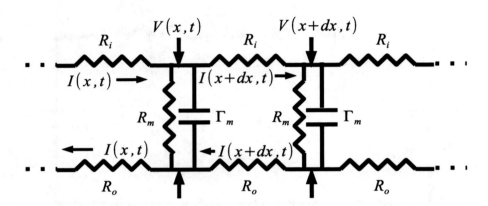

Figure VIII.10: Resistive-capacitive-ladder model of the axon.

H   Telegraphers' Equation: time-varying signals

For analyzing time-dependent signals on the axon, we can model it with the resistive/capacitive ladder illustrated in Fig. VIII.10. Each infinitesimal segment of the membrane can be treated as a parallel-plate capacitor with "plate" area, hence, capacitance, proportional to segment length dx. That is, the segment capacitance is given by

$$\Gamma_m = \gamma_m \, dx \ , \tag{VIII.30}$$

where membrane capacitance per unit length γ_m is a measurable property. The axon is a continuous or *distributed* system; this ladder model is a *lumped-component* approximation. Applying the Kirchhoff rules to the section of the ladder between positions x and $x + dx$, as shown, yields

$$I(x,t) = \frac{V(x,t)}{R_m} + \Gamma_m \frac{\partial V(x,t)}{\partial t} + I(x + dx, t) \tag{VIII.31a}$$

from the junction rule (the second term on the right is the current flowing into the membrane capacitance, as $\Gamma_m V$ is the charge there), and

$$V(x,t) = R_i I(x + dx, t) + V(x + dx, t) + R_o I(x + dx, t) \tag{VIII.31b}$$

from the loop rule. We can introduce differentials

$$I(x + dx, t) - I(x, t) = \frac{\partial I}{\partial x} \, dx$$

$$V(x + dx, t) - V(x, t) = \frac{\partial V}{\partial x} \, dx \; , \qquad \text{(VIII.32)}$$

where partial derivatives are needed because both the current and voltage distributions are now functions of x and t. Using forms (VIII.18) and (VIII.30), discarding second-order differentials, and rearranging yields

$$\frac{\partial I}{\partial x} = -c_m V - \gamma_m \frac{\partial V}{\partial t} \qquad \text{(VIII.33a)}$$

and

$$\frac{\partial V}{\partial x} = -(r_i + r_0)I \; . \qquad \text{(VIII.33b)}$$

These coupled, first-order partial differential equations can be combined into a single, second-order PDE for the voltage distribution V by differentiating both sides of Eq. (VIII.33b) with respect to x, then substituting $\partial I/\partial x$ from Eq. (VIII.33a). The result is

$$\frac{\partial^2 V}{\partial x^2} - (r_i + r_o)c_m V - (r_i + r_o)\gamma_m \frac{\partial V}{\partial t} = 0 \; , \qquad \text{(VIII.33c)}$$

a form of a famous partial differential equation known as the *Telegraphers' Equation*. Our hearts soar—we recall from Ch. VI that the wave equation for any physical system can tell us everything about signal propagation on that system, at least in principle. Alas and alack, this is *not* a wave equation. According to the criteria following Eq. (VI.9), this is a *parabolic* second-order PDE, akin to the equations for diffusion or heat flow. The speed and precision of nerve signals are unlikely to be described by solutions of this equation. Although the basic network model of the axon is amenable to analysis, it is insufficient.

I Action potentials on the axon

The Telegraphers' Equation was first derived not for nerve fibers, but for transoceanic cables laid along the sea floor to carry first intercontinental telegraph messages (hence the name), and later telephone signals.[30] The diffusive effects described by the Telegraphers' Equation were corrected by installing *repeaters* at intervals along the cables—transceivers that received incoming signals, cleaned and amplified them, and retransmitted them. The cable system was thus an *active* network, the electrical properties of which changed in response to the signals it carried. Remarkably, the axon works in a similar way.

The signals that are transmitted down the axon and communicated to neighboring neurons are called *action potentials*. They consist of voltage changes

---

[30]These networks still exist, but they have been superseded first by communiation satellites, then by fiber-optic cables.

(from the resting state) above a certain threshold, large enough to *depolarize* the axon membrane at a location. As a result, the permeability of the membrane increases. This allows Na^+ ions from the surrounding fluid to flow into the axoplasm, sustaining the elevated voltage, depolarizing neighboring areas, and preventing the exponential decay described by Eqs. (VIII.22c) and (VIII.24). This is termed *regeneration* of the action potential. Eventually, K^+ outflow[31] from the axoplasm *repolarizes* the membrane, and the normal ion concentrations of the axon are restored, as the pulse propagates away. A more detailed description of these processes requires more physical chemistry than would be appropriate here.

This mechanism explains the role of the *myelin* that wraps around the axons in vertebrates. Like insulation on wire, it decreases the conductivity of the axon membrane. Signal attenuation is less, and propagation is faster, on myelinated sections of axon. The ion transport necessary for regeneration of action potentials takes place at the exposed *nodes of Ranvier* between the sections of myelin. Some anaesthetic drugs are believed to work by binding to and blocking these nodes, suppressing the regeneration process so that nerve signals damp out before reaching their destinations.

Invertebrates like the squid have just as much need for fast neural responses as vertebrates do—a squid must evade sperm whales that are hunting in the ocean depths and coordinate eight legs and two tentacles. Lacking myelin, their nervous systems compensate with much larger-diameter axons. Hence, vertebrates can pack much greater numbers of nerve fibers in the same space. The image of a fly's eye, for example, may contain only a few hundred pixels, compared to the order of a hundred million for a human eye.[32]

J   Semiconductor devices

It is a strangeness, at least to this author, that introductory texts even in the twenty-first century do not include semiconductor electronics in their discussions of electric circuits. Some contain this subject in later chapters on modern solid-state physics, others not at all.[33] Yet these devices are central to a technological revolution that has wrought far-reaching changes to many aspects of human life, all over the world, over the last half century. Moreover, their basic operation is not difficult to understand.

Semiconductor materials

Semiconductor materials were described in the previous chapter: They are neither good conductors nor good insulators. Pure or *intrinsic* semiconductors,

---

[31]Both sodium and potassium salts or electrolytes are fundamental nutrients, necessary *inter alia* for nerve function.

[32]This may be part of the reason the human is the one wielding the swatter.

[33]At least one fairly recent textbook still described vacuum tubes, but not transistors—although many current readers may never see a vacuum tube.

such as elemental silicon and germanium, have very few mobile charge carriers. Their electrical properties can be altered substantially by adding small amounts of impurities. This process is known as *doping,* and doped semiconductors are called *extrinsic.* "Small amounts" means just that—typically one impurity atom per hundred million atoms of the intrinsic material.[34] Extrinisic semiconductors are of two varieties: *Dopant* atoms such as P, As, Sb, et cetera release additional mobile electrons into the material, creating *n-type* semiconductors. Atoms such as Al, B, In, et cetera seize electrons from neighboring atoms, which then attract electrons from their neighbors. The *absence* of an electron from an atom behaves as a mobile positive charge carrier[35]; these are called *holes.*[36] Materials with a surplus of holes are *p-type* semiconductors. The materials remain electrically neutral overall—it is the supply of *mobile* charge carriers that is manipulated. The manipulations have great effect because the supply of mobile carriers in the intrinsic material is so small compared to conducting materials.

The *pn* junction

The simplest device that can be constructed from *p*- and *n*-type semiconductors is the *pn* junction, sketched in Fig. VIII.11 with its electrical connections. Where the two types of material meet, surplus holes and electrons diffuse across the junction and combine. This removes them from the mobile charge-carrier supply, so that in the immediate area of the junction the material behaves as the intrinsic semiconductor. This area, indicated in the figure, is called the *depletion region* or *depletion layer.* As holes are more prevalent in the *p*-type material and free electrons in the *n*-type, random motion causes a net diffusion of holes across the junction to the right, as shown, and electrons to the left. These flows constitute a *diffusion current* from the *p*-side to the *n*-side across the junction.

If no external potential difference is applied to the junction, it is said to be *unbiased.* In such case, the *n*-side of the device acquires positive charge and the *p*-side negative. (Care must be taken to avoid confusion here.) The device is "self-polarizing." Hence, both holes and electrons must diffuse "uphill" across the junction, as the *n*-side attains higher potential, the *p*-side lower. This limits the diffusion current. It also drives a countervailing *intrinsic current* in the other direction, limited by the supply of charge carriers in the depletion region. The two currents attain a dynamic equilibrium, with equal currents—typically, of the order nanoamperes—flowing in both directions.

The junction is *forward biased* if an external potential V is applied that raises the potential of the *p*-side and/or lowers that of the *n*-side. This reduces the height of the "hill" that the diffusing charge carriers must climb. Hence, a much larger fraction of them have the energy for the climb. The population

---

[34] Although semiconductors were known in the nineteenth century, the semiconductor revolution had to await the development of techniques for fabricating these materials at very high purities and controlling their composition with very high precision.

[35] For example, in a classroom full of computers with roller-ball mice, a student whose mouse is missing a ball can surreptitiously take one from a neighbor. The neighboring student can do the same, and so on. The *absence* of the ball thus becomes a moving object.

[36] In Einsteinian quantum mechanics, these are *antiparticles.*

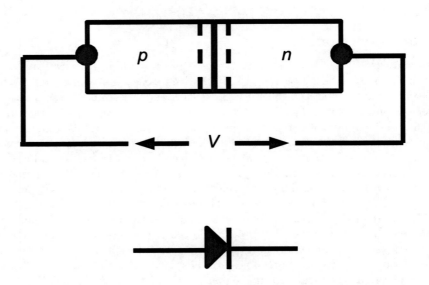

Figure VIII.11: A *pn* junction and the symbol for a diode. The area between the dashed lines around the junction is the *depletion region*.

distribution is exponential in energy, proportional to $\exp[-E/(k_B T)]$, with k_B the Boltzmann constant and T the absolute temperature—like the molecules of a gas. So lowering the hill increases the diffusion current by orders of magnitude, e.g, to milliamperes. The intrinsic current is decreased slightly. The result is a substantial current flow from p-side to n-side across the junction.

If a potential difference in the other direction is applied, raising the potential of the n-side and/or lowering that of the p-side, the junction is *reverse biased*. The hill faced by the diffusing charge carriers is raised; the number with sufficient energy for the climb is reduced exponentially. The diffusion current is choked off. The intrinsic current is increased slightly, but it remains limited by the supply of intrinsic charge carriers. The result is a slight—e.g., of the order nanoamperes—current from the n-side to the p-side of the junction.

The behavior of a pn junction under forward and reverse bias is illustrated by the *Shockley*[37] *equation*. Strictly, this applies to a perfect pn junction with no boundaries; it is approximate for real junctions of finite size. It gives the current across the junction as function

$$I(V) = I_0 \left[\exp\left(\frac{eV}{k_B T} \right) - 1 \right] , \qquad \text{(VIII.34)}$$

of the bias potential V, positive for forward bias, negative for reverse. Here, I_0 is an intrinsic current value, e the elementary charge, and T the absolute temperature of the device. This function is graphed in Fig. VIII.12. Essentially, the pn junction is a *one-way conductor*, carrying milliamperes of current in the forward direction, only nanoamperes in the reverse. This contrasts with ordinary resistors, which conduct or resist equally in either direction. What is the use of a one-way conductor?

The diode

A circuit component that functions as a one-way conductor is called a *diode*, although the name only indicates a component with two connecting leads. Its symbol in circuit diagrams is shown in Fig. VIII.11; the arrow indicates the direction the diode allows current to flow. Formerly, diodes were one type of *vacuum tube*: an evacuated glass bulb on a base containing its electrical connections, enclosing an incandescent filament to provide a beam of electrons, and metal grids and plates to control the current. Now most diodes are *pn*-junction diodes. A discrete diode is typically cylindrical, perhaps 2 mm in diameter and 1 cm long, enclosed in an insulating casing embossed with the diode symbol (to distinguish it from a resistor). But diodes can also consist of microscopic layers of semiconductor etched on an integrated-circuit (IC) chip, which might contain hundreds of thousands or millions of individual components on a wafer centimeters in size. The pn-junction diode is much more compact, faster, and less power-consuming than its vacuum-tube predecessor. Nowadays,

---

[37]William Bradford Shockley Jr., 1910–1989

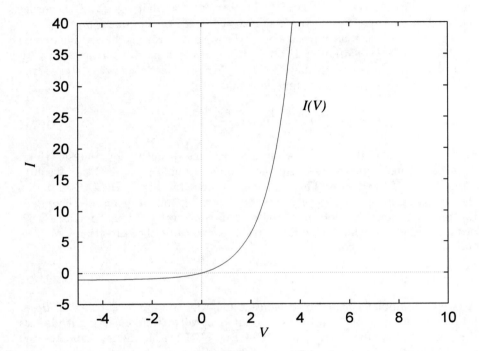

Figure VIII.12: The *I-V* curve for a *pn* junction satisfying the Shockley equation. The scales are arbitrary, chosen to show the nature of the curve.

the latter is likely to be found only in high-power or low-noise devices, where its larger size is a virtue, or in antique devices kept in service by aficionados.

A diode in a circuit can serve as a *rectifier,* converting oscillating *alternating current* (AC) to direct current (DC) flowing in a single direction. A single diode would allow current flow during only half an AC cycle, cutting off the current during the other half. Actual rectifiers contain two or more diodes, connected to allow current to flow in the same direction during both halves of the cycle, along with filter circuits to smooth out the output current. A diode can also be regarded as a switch, closed for current in one direction, open for current in the other—a switch operated by the very signal it switches. An ideal switch has zero resistance when closed, infinite resistance when open, and changes instantaneously between the two. A typical *pn*-junction diode might have resistance $R_+ \sim 10^{-7}\ \Omega$ for current in the forward direction, $R_- \sim 10^9 J\ \Omega$ in the reverse, a fair approximation to the ideal. Its small size contributes speed and allows large numbers of them to be grouped together. The reader is again cautioned *not to scorn the humble switch*; the significance of large numbers of fast switches will be considered subsequently.

The transistor

Despite its utility, the *pn*-junction diode does not represent a world-transforming revolution. The device that does is the *transistor,* so called because it *trans*fers charge across a res*istor*—i.e., it is a variable resistor. It was invented in 1947 at Bell Labs, now Lucent Technologies, perhaps the greatest industrial research facility ever run. (For example, the Cosmic Microwave Background Radiation—the afterglow of the Big Bang—was discovered in 1965 by Penzias[38] and Wilson,[39] working on microwave antennae at Bell Labs.) The transistor was invented by John Bardeen,[40] Walter Brattain,[41] and William Shockley, who won the Nobel[42] Prize in 1956 for their work. Bardeen would share the 1972 Physics Nobel with Cooper[43] and Schreiffer[44] for their BCS model of superconductivity, making John Bardeen the only person to date to win the Nobel Prize *in physics* twice.

Modern readers may not credit accounts that this author might give of pre-transistor electronics. For example, there was a time when if one was watching television and did not like what was on, it was necessary *to stand up and walk over to the set* to turn it off or change the channel.[45] Moreover, if one wished to watch television at 8:00 p.m., it was necessary to turn on the set at 7:45 p.m. and let it "warm up," because vacuum-tube filaments needed time to heat up, and

---

[38] Arno Allan Penzias, 1933–

[39] Robert Woodrow Wilson, 1936–

[40] John Bardeen, 1908–1991

[41] Walter Houser Brattain, 1902–1987

[42] Alfred Bernhard Nobel, 1833–1896

[43] Leon N. Cooper, 1930–

[44] John Robert Schieffer, 1931–

[45] There was also a time when a dog, entering the room with tags jingling, could change the TV channel or switch it off.

the components needed time to reach their operating parameters. Bardeen and his team staggered their colleagues by constructing an audio amplifier—Bardeen's Box—which *worked as soon as it was turned on.* It was built, of course, around transistors and other semiconductor components, termed collectively *solid-state circuitry.*

There are various designs for transistors, producing the same effects by different means. One design is the Bipolar Junction Transistor or BJT,[46] a "sandwich" of *n*- and *p*-type semiconductors such as is sketched in Fig. VIII.13. The scales in the figure are for clarity, not accuracy: The base layer (the *p*-layer in the figure) may be only microns in thickness. And the transistor itself is not the size of a tomato-paste can. Ordinary discrete transistors are the size of pencil erasers. The largest transistors, used in power supplies, might be two or three centimeters in diameter. At the other end of the scale, transistors can be microscopic segments of semiconductor material layered onto IC chips containing many thousands of them. One of the goals of cutting-edge research in this field is a transistor that works by transferring *a single electron* from one side to the other. That should be the limit of miniaturization.

The transistor shown in Fig. VIII.13 is an *npn* transistor. A *pnp* transistor is made with a *base* layer of *n*-type semiconductor sandwiched between *p*-type *emitter* and *collector* segments—"emitter" and "collector" of charge carriers in either case. The two types work the same way—the only difference between them is the sign of the charge carriers transferred, hence, the direction of current through the device. The symbol shown in the figure represents an *npn* transistor, whether of BJT design or any other. The symbol for a *pnp* transistor is the same, except that the arrow points the other way. The arrow, always drawn between the base and emitter leads of the transistor, indicates the direction of current flow, following the traditional convention. One can remember that the symbol in the figure represents an *npn* transistor because the arrow is *N*ot *P*ointing i*N*.

With three leads—its vacuum-tube counterpart is called a *triode* for this reason—a transistor must be connected in a circuit of at least two loops. There are several ways to do this; one example is the *common-emitter* circuit of Fig. VIII.14, in which the emitter lead of the transistor is common to both loops of the circuit. When the base potential V_B is low, the emitter-base *pn* junction is reverse biased, and negligible current flows in the emitter-collector loop. If V_B is raised so that the emitter-base junction becomes forward biased, charge carriers—in this case electrons—flow from the emitter into the base. The base-collector junction is still reverse biased, i.e., the collector is at higher potential. But for the electrons flooding into the base layer, higher potential is "downhill"; they continue into the collector and current flows through the transistor. The base current I_B serves to replace electron-hole pairs that combine in the base layer and are lost. The Kirchhoff junction rule requires $I_E = I_B + I_C$, but I_B is typically about 1% of the emitter current I_E or the collector current I_C, which

---

[46]Perhaps because of extensive connections between the military and scientific research since the 1940s, acronyms are used extensively in this field. For example, a field-effect transistor, which uses electric fields to control current flow, is an FET. If made from metal-oxide semiconductors, it is a MOSFET. Once again, the alphabet soup flows like wine.

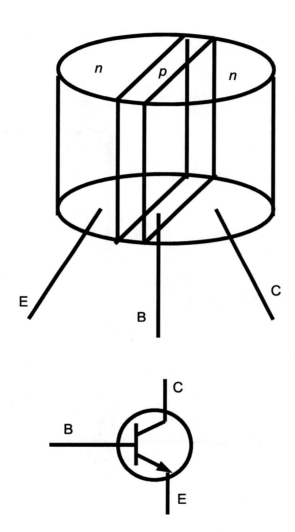

Figure VIII.13: An *npn* Bipolar Junction Transistor (BJT) and its symbol, showing the emitter (E), base (B), and collector (E) leads.

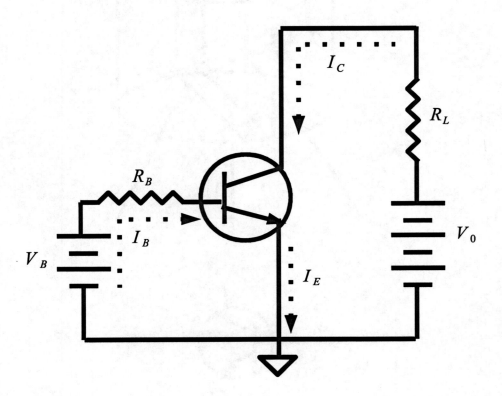

Figure VIII.14: An *npn* transistor in a common-emitter circuit.

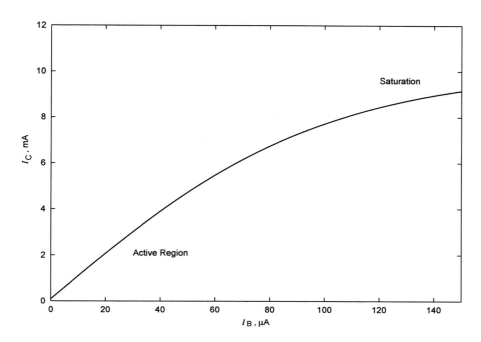

Figure VIII.15: Transistor as amplifier.

are essentially equal. As V_B is raised, I_E and I_C increase until the base potential is high enough that the base-collector junction becomes forward biased. Then a "backwash" of charge carriers from the collector causes the currents to level off, a condition known as *saturation*. The saturation current in the collector-emitter loop is set by the voltage V_0 and the load resistor R_L.

A graph of collector current I_C versus base current I_B for such a circuit is shown in Fig. VIII.15. The scales shown are realistic current values for a typical transistor in a typical circuit. The range of base currents over which the curve is linear, i.e., the collector current is essentially proportional to the base current, is called the *active region*. If the transistor is operated in this range of currents, changes in I_B are reproduced in I_C, but on a much larger scale. This is the essence of *amplification*: A small signal in I_B controls a much larger version in I_C, like a small motion in a valve controlling a huge flow of water.[47] Practical amplifiers typically do not use a single transistor—they have multiple stages and feedback circuitry for stability—but they nonetheless rely on this basic function. The slope of the I_C-I_B curve in the active region, i.e., the amplification factor, is called the *dc current gain* or *dc β* of the transistor. Values $\beta \sim 100$, such as illustrated in the figure, are typical of readily available transistors.

---

[47]Vacuum-tube triodes are also called *valves* for this reason.

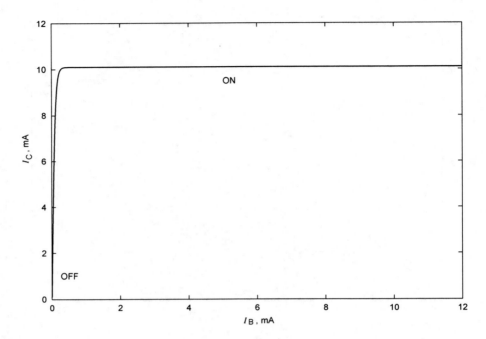

Figure VIII.16: Transistor as switch.

The reader may object that Fig. VIII.15 is not honest, as the two variables are graphed on different scales. Fair enough. A graph of I_C versus I_B on equal scales, for the same typical transistor in the same typical circuit, is shown in Fig. VIII.16. Operated through this range of base currents, the transistor either blocks the emitter/collector current (OFF) or allows its full saturation value to flow (ON). It makes a fast transition from a state of near-infinite resistance to one of near-zero resistance: It functions as a switch. *This* is the beating heart of the electronics revolution. Smaller, faster, less power-consuming amplifiers are very useful, but they are not transforming the world. Smaller, faster, less power-consuming switches are. The workstations on which this text was composed, and the variety of devices by which billions of people communicate and access the collected information and misinformation of humanity, are *boxes of switches*.

K   Electronic logic

Like other circuit components, switches can be connected in series or parallel combinations, as illustrated in Fig. VIII.17. Two switches in series function as a closed switch only if both S_1 *and* S_2 are closed. Otherwise, the combination is open. This behavior represents or realizes the logical or Boolean[48]

---

[48] George Boole, 1815–1864

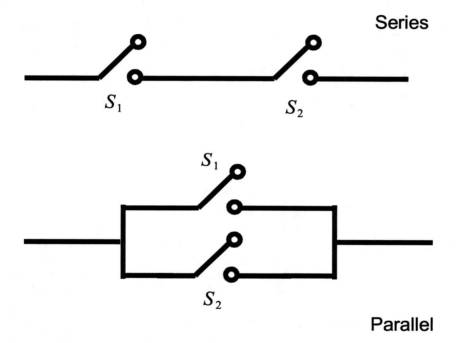

Figure VIII.17: Switches in series and in parallel.

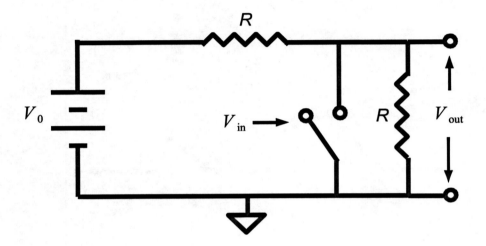

Figure VIII.18: An inverter circuit. The output is .NOT. the input.

operation AND. A circuit that employs two switches for this purpose is called an AND gate. (If the switches are transistors operated at certain standardized voltages, the circuit is a *Transistor-Transistor-Logic* or TTL AND gate.) Two switches in parallel function as a closed switch if either S_1 *or* S_2, or both, are closed; it is open if both are open. This embodies the logical operation OR, and a circuit of this design is an OR gate. Finally, a circuit such as that sketched in Fig. VIII.18, called an *inverter,* implements the logical operation NOT: If the input voltage is ON and closes the switch, the second resistor is shorted and the output voltage is zero, or OFF. Conversely, if V_{in} is OFF and the switch is open, a positive output voltage V_{out} is measured across the resistor, and the output is ON. But mathematicians know that with the operations AND, OR, and NOT, it is possible to carry out any arithmetical operation. This is the basis of all digital computation.

The same principles apply to switches of any design. Early electronic computers used *solenoid relays,* mechanical switches operated by electromagnets. (In addition to its other drawbacks, such a computer must have been very noisy in comparison to its successors!) Later machines used vacuum tubes, then discrete transistors, then IC chips, then VLSI (Very Large-Scale Integration) chips. The Central Processing Unit (CPU) of a modern desktop or laptop computer is a chip perhaps seven centimeters square; a modern supercomputer consists of cabinets full of them. Each stage of development reduced the power consumption, cooling requirements, and size of the computer—from that of a building, to a room, to a suitcase, to a pocket-sized device. More important, the development of solid-state electronics allowed *many more* switches to be packed into the same space and operated with much less power: The Apollo astronauts went to the Moon backed by less computing capacity than that of a modern smartphone.

This brings us back to the neuron example attempted in this chapter. Neurons and synapses can do what transistorized logic gates can do: A neuron can be coupled to its neighbors so that it requires two simultaneous inputs to transmit an action potential (AND), or one input can suppress another (Exclusive OR, or XOR), et cetera. Perhaps the human brain is a biologically based digital computer of surpassing sophistication. At one time, it was proposed that brain function was to be understood solely in terms of a digital-computer model, a notion termed *Ockham's Electric Razor*. This is a takeoff on *Ockham's*[49] *Razor: Entia non sunt multiplicanda praeter necessitatem.* (Beings are not to be multiplied without necessity.) This is the principle that the simplest explanations of phenomena are always to be sought in science. The label is clever, but it is not clear to this author that it is wise to bind oneself to a single model in this case. More particularly, the fact that neurons can perform digital logic does not necessarily imply that is how they work—it might be completely ancillary to their true function. And there is a deeper conundrum here. "To understand how the brain works" means to formulate *within human brains* the principles governing its operation. There is a self-referentiality here reminiscent of the Gödel[50] Incompleteness Theorems in mathematics: Is the brain congenitally capable—or incapable—of comprehending its own mechanism? To know *either* answer with certainty would be a profound understanding. But as already noted, brain science is one of the most active areas of current research. Perhaps answers to such questions are not so very far to seek.

References

All introductory physics texts devote one or more chapters to electric currents and DC circuits. Fishbane, Gasiorowicz, and Thornton [FGT96] and Halliday, Resnick, and Krane [HRK02], for example, also include brief descriptions of the operation of transistors.

Circuit theory, of course, is a subject with a vast literature—at every level— in its own right. Solid-state and digital electronics are the subjects of texts of their own. Malmstadt, Enke, and Crouch [MEC74], for example, is a wide-ranging, advanced-level text.

The electrical behavior of the axon is the subject of research going back to the 1940s. Reviews can be found in, e.g., Hobbie [Hob73], Burns and MacDonald [BM70], and the essay *Electrical Conduction in Nerve Cells,* by Stephen

---

[49]William of Ockham or Occam, 1300?–1349
[50]Kurt Friedrich Gödel, 1906–1978

C. Woods, in Tipler [Tip87].

Problems

1. A large particle accelerator drives a beam of 2.0×10^{14} protons per second through a total potential difference of 1.0×10^{12} volts. This is an electric current, just as is the flow of electrons in a wire. Calculate the current and power in the beam and the effective resistance of the accelerator.

2. Two resistors in series constitute a *voltage divider*. Given resistors R_1 and R_2 in series with an ideal EMF source producing voltage V_0, calculate the potential differences across each of the resistors in terms of these quantities.

3. Two resistors in parallel constitute a *current divider*. Given resistors R_1 and R_2 in parallel, the combination connected across an ideal EMF source producing voltage V_0, calculate the currents through each resistor in terms of these quantities.

4. The circuit in Fig. VIII.19 is called a *relaxation oscillator*. It is used in portable flashing lights such as those on highway barricades. Its operation is based on the properties of the neon lamp L_{Ne}. When the lamp is off, it has essentially infinite resistance. When the voltage across the lamp reaches 72.0 V, the gas inside ionizes and the lamp glows. The lamp will remain on until the voltage across it drops to 50.0 V, when the gas inside de-ionizes and again becomes nonconducting. The resistance of the lamp when on is small; the capacitor discharges quickly through the lamp and the lamp flashes. When the lamp goes off, the charging of the capacitor resumes and the cycle begins again.

   (a) Sketch the voltage across the lamp as a function of time, measured in units of RC.

   (b) Using the parameter values in the figure, calculate the value of capacitance C needed to make the lamp flash at intervals of 2.00 s.

5. For the relaxation oscillator of the previous problem, it should be possible to watch the voltage across the capacitor and lamp change with an ordinary voltmeter. But when a voltmeter with internal resistance 1.00 MΩ is connected across these components, the lamp stops flashing altogether. This is an example of the phenomenon known as *meter loading*.

   (a) What has happened? And what is the reading on the voltmeter?

   (b) If the voltmeter had infinite, i.e., large enough resistance, it would not affect the circuit. What is the minimum resistance the voltmeter must have for it not to stop the lamp flashing?

6. By inserting a very fine wire down the inside of an axon, experimenters can force the potential in the axon to be uniform along its length. Suppose

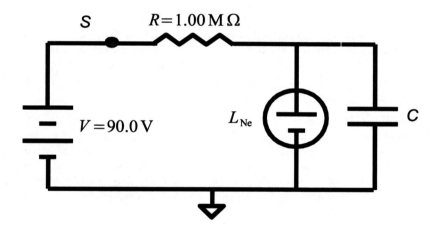

Figure VIII.19: Relaxation oscillator circuit for Problems 4 and 5.

in this situation a small potential ΔV_0 above its rest value is applied to the axon (smaller than threshold, i.e., not an action potential); the power source is then disconnected at time $t = 0$ and the potential on the axon allowed to evolve.

(a) Find the potential disturbance on the axon as a function of time, $\Delta V(t)$, for times $t \geq 0$ in terms of the internal and external resistances per unit length r_i and r_o, the membrane conductance and capacitance per unit length c_m and γ_m, and the initial disturbance ΔV_0.

(b) Using $r_o = r_i = 6.37 \times 10^9$ Ω/m, $c_m = 1.25 \times 10^{-4}$ S/m, and $\gamma_m = 0.30$ μF/m, calculate the time required for the potential disturbance to fall to half its original value.

7. Consider a small (sub-threshold) signal on a nerve fiber or a telegraph cable. The signal can be written as a sum of "harmonics":

$$V(x, t) = \sum_n V_n \cos(k_n x) f_n(t) \ ,$$

where n is some index identifying the terms, the V_n are constants, the k_n are wave numbers for the harmonics, and the f_n are time-dependence functions to be determined [with $f_n(0) = 1$ for all n]. If the harmonics all decay in time at the same rate, the signal preserves its shape as time passes. If harmonics with larger k_n values decay faster, the signal smears out with time; if those with smaller k_n values decay faster, the shape of the signal sharpens with time. Which happens?

8. Malibu[TM] lamps[51] are connected to an ideal 12.0 V power supply along a cable consisting of two parallel 18-gauge copper wires (one for outgoing current, one for incoming, separated by insulation). The lamps are connected to the cable at one-foot (30.5 cm) intervals via "vampire taps" — prongs that pierce the cable insulation on each side to make contact with each wire. Thus, the lamps form the rungs of a resistance ladder, such as used in this chapter to model the axon in the electrotonus case.

When connected individually to the power supply with wires of negligible resistance, the lamps put out 4.00 W of light power. But the resistance of 18-gauge wire is 0.110 mΩ/cm, i.e., 3.35 mΩ/ft. Assume for the purposes of this problem that the resistance of each of the lamps in operation is the same as when they are operated individually. Although actual Malibu[TM] lamps are operated with alternating current, the array can be treated here as a DC circuit.

(a) Calculate the total resistance and power consumption of an "infinite" string of Malibu[TM] lamps, e.g., lamps strung one per foot from St. Louis, Missouri, to Malibu, California. Note that in this case, in contrast to the axon model, the segments of the ladder cannot be treated as infinitesimal.

---

[51] "Malibu[TM] lamp" is a registered trademark of Intermatic Incorporated, Spring Grove, Illinois

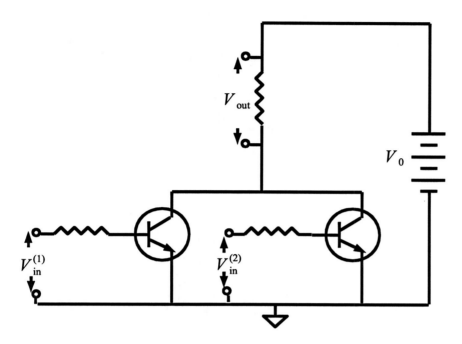

Figure VIII.20: Circuit diagram for Problem 9.

(b) How is the voltage across the lamps distributed along the length of the cable? The lamps will not "look right" if their power output is below, say, 2.00 W. How many lights of the infinite ladder will "look right"? This is the practical limit to the length of the array.

9. The input voltages $V_{in}^{(1)}$ and $V_{in}^{(2)}$ for the circuit in Fig. VIII.20 can each be either zero (OFF) or some fixed, positive value (ON) sufficient to drive the corresponding transistor to saturation.

(a) Construct a table describing the output voltage V_{out} for all possible combinations of inputs. Neglect the internal resistance of the transistors when ON.

(b) What is the significance of these results, i.e., what is this circuit?

Chapter IX

Return of the Classical Electron

- *Model the electron as a spinning, uniformly charged solid sphere.*
- *Calculate its magnetic dipole moment and gyromagnetic ratio.*
- *Compare these with their real (quantum-field-theoretic) values.*

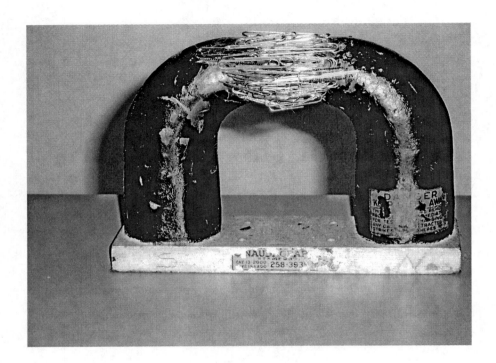

Figure IX.1: A simple observation: A magnet can pick up paper clips. But to understand *why,* one must plumb some of the deepest mysteries of physics. (Photograph by author.)

We return to the classical electron model of Ch. VII. By adding a simple feature to the model—rotation of the sphere of charge—we can hope to model the *magnetic* properties of the electron. Thus, we might understand its contribution to the well-known behavior of magnetic materials, as illustrated in Fig. IX.1. Spectacular failure awaits, but through that failure we shall be able to glimpse the very frontier of physics.

A   Magnetism: another force known in antiquity

Like the electric or amber force, the magnetic force exerted by certain metallic materials on certain other metals was known to classical thinkers. The terms *magnet* and *magnetism* derive from *Magnesia,* the name of a region in the Anatolian peninsula—now the property of modern Turkey, but historically one of the most fought-over pieces of territory on Earth. It was known to the ancients that certain minerals associated with this region, variously named *lodestone, hematite,* et cetera, could attract and pick up pieces of some metals—iron, for example, but not copper. Unlike the electric force, however, this magnetic effect was studied and put to use in the centuries between the classical period and the Enlightenment. By the thirteenth century CE, it was known that the magnetic force concentrated at points called *magnetic poles* and could be traced along field lines. Chinese seafarers were using the fact that an elongated piece of magnetized metal, suspended free to rotate, would align in a north-south direction as a navigational aid—the magnetic compass.[1] By the seventeenth century, it was understood that the Earth itself behaved as a giant magnet: William Gilbert,[2] a contemporary of Galileo, succeeded in fabricating a spherical magnet and demonstrating that its magnetic effects replicated those of the Earth on compass needles. Pivotal discoveries of the nineteenth century, to be explored in this and subsequent chapters, established the connections between magnetism and electricity. In the twentieth century, the study of electricity and magnetism led to Einsteinian relativity, quantum theory, and quantum electrodynamics.

New discoveries about the nature of *electromagnetism* and the magnetic behavior of materials continue into the twenty-first century.

The reader is cautioned that the study of magnetism is harder than that of electricity: There is no one-dimensional magnetism; magnetism is inherently three-dimensional. Nor is there any spherically symmetric magnetism. Hence, the topic places greater demands on the reader's facility with vector mathematics, as will shortly be seen.

B   Magnetic poles

It has been known since the thirteenth century CE that the magnetic force of a magnet is concentrated at two points, its *magnetic poles.* Like electric charges, these fall into exactly two categories, and all the members of each category repel each other and attract all members of the other category. These are labeled not positive and negative but "N" and "S": If a magnet is suspended free to rotate, its N pole will orient toward the north, its S pole toward the south. These designations should not be read "North pole" and "South pole," because the North Magnetic Pole of the Earth, in northern Canada, is an S pole, while the South Magnetic Pole, near the coast of Antarctica, is an N pole. The terms "North-*seeking* pole" and "South-*seeking* pole" for N and S, respectively, are clearer.

This state of affairs might suggest that magnetic poles play the role of magnetic charges, with a magnetic-force law analogous to Eq. (VII.7b): $\mathbf{F}_m = q_m \mathbf{B}$, where $\mathbf{B}$ is the traditional designation of the magnetic field. But there is one great flaw in the analogy: If a magnet is broken in two, the result is not an N piece and an S piece, but two smaller magnets, each with its own N pole

---

[1] Had later Ming emperors not mothballed their ocean-going fleet for political reasons, the Age of Exploration might have been a Chinese show, and this text might be written in Chinese.

[2] William Gilbert, 1540–1603

and S pole. Even if the breaking process were to continue right down to the atomic level, the result would be not individual N's and S's, but individual magnets with N and S poles in pairs. Isolated poles—called *magnetic monopoles*—have never been observed. It was suggested by Dirac in the 1920s that an elementary particle with the properties of a magnetic monopole might exist, and some Grand Unified Theories (GUTs) proposed in the 1970s also predicted their existence. In the 1970s and early 1980s several experimental results were reported suggesting the detection of a magnetic monopole, but this was never substantiated.[3] In the absence of such confirmation, we shall adopt here the conventional view that there are no isolated magnetic poles. The next simplest magnetic structure, then, would be a *magnetic dipole*. And the most fundamental magnetic phenomenon might not be the force on a charge, but the torque $\boldsymbol{\tau}_m = \boldsymbol{\mu} \times \mathbf{B}$ exerted on a dipole with magnetic *dipole moment* $\boldsymbol{\mu}$. Unlike its electric counterpart, this cannot be "two charges on a stick"; its meaning will become clearer as the chapter unfolds.

C   Operational definition of B: Lorentz force law

An operational definition of the magnetic field $\mathbf{B}$, i.e., one in which the field is defined by specifying how it is to be measured, cannot take the form of Eq. (VII.7a). The force law through which $\mathbf{B}$ is defined is attributed to Hendrik Lorentz, the contemporary of Einstein for whom Lorentz transformations are named. But the features of the magnetic force were deduced earlier in the nineteenth century from the careful and detailed work of experimenters such as André Ampère and Hans Örsted.[4]

Örsted's contributions are noteworthy. He was a Dutch teacher in a German *gymnasium*—essentially, a high school teacher, though his position carried more prestige than that suggests now. Apparently, he set up a classroom demonstration that electric currents produce *no* magnetic effects on a compass needle by laying a wire across the face of a compass, perpendicular to the needle, connecting the wire to a battery, and observing that the needle did not move. One of his students asked, "*Herr Studienrat,* what happens if we place the wire *parallel* to the needle?" A true scientist, Örsted did so, repeated the experiment, and observed that the needle rotated ninety degrees, proving that an electric current produces a magnetic field perpendicular to its own direction. Örsted is credited with this pivotal discovery, and the Örsted Medal, a prestigious award given in recognition of a scientific discovery made in the course of classroom teaching, is named for him. The name of the inquisitive student is lost to history.

---

[3] *Inflationary* models of the early universe, first proposed in the 1980s, offered a solution of the "monopole problem" by predicting that there should be only one magnetic monopole in the entire observable universe. These models have since been extensively revised, but if that prediction were true, it is indeed possible—however unlikely—that one of the experimenters might have found the lone monopole. But the result can never be confirmed, as the monopole is gone.

[4] Hans Christian Örsted, 1777-1851

The results of Ampère, Örsted, and others showed that the fundamental magnetic interaction is between electric *currents,* i.e., between *moving charges.* Charge q moving with velocity $\mathbf{v}$ in magnetic field $\mathbf{B}$ experiences magnetic force $\mathbf{F}_m$, with observed features:

- The force is proportional to the charge, $\mathbf{F}_m \propto q$.

- The force is proportional to the magnitude of the magnetic field and the component of the charge's velocity perpendicular to the field. Equivalently, it is proportional to the speed of the charge and the component of the magnetic field perpendicular to its velocity. These combinations can be written

$$\mathbf{F}_m \propto v_\perp \, |\mathbf{B}| = |\mathbf{v}| \, B_\perp = |\mathbf{v}| \, |\mathbf{B}| \, \sin\theta_{\mathbf{vB}} \ , \qquad \text{(IX.1)}$$

  where $\theta_{\mathbf{vB}}$ is the angle between vectors $\mathbf{v}$ and $\mathbf{B}$ placed "tail to tail."

- The force is perpendicular to both $\mathbf{v}$ and $\mathbf{B}$ in the direction selected by a *right-hand rule*—the direction indicated by the thumb of the right hand when the fingers of that hand push $\mathbf{v}$ into $\mathbf{B}$ for a positive charge, the opposite direction for a negative charge.

These features are incorporated in the cross product or vector product introduced in Ch. III Sec. F via Eqs. (III.74)–(III.76). Here, the cross product

$$\mathbf{F}_m = q\,\mathbf{v} \times \mathbf{B} \qquad \text{(IX.2)}$$

gives the magnetic force law through which the field $\mathbf{B}$ is defined.

The units for the magnetic field follow from this force law. The SI units for $\mathbf{B}$ are

$$1\,\frac{\text{N}}{\text{C\,m/s}} = 1\,\frac{\text{N}}{\text{A\,m}}$$
$$\equiv 1\,\text{T} \qquad \text{(tesla)}$$
$$= 1 \times 10^4\,\text{G} \qquad \text{(gauss)} \ . \qquad \text{(IX.3)}$$

The SI unit is named for Nikola Tesla.[5] The gauss, named for Karl Friedrich Gauss, is part of a different subset of the metric system (appropriately termed Gaussian units), but its use is very common. The size of these units can be gauged from magnetic fields encountered both in nature and in the laboratory:

- A magnetic field of magnitude $B \sim 10^{-10}$ T fills interstellar space in the Milky Way galaxy. (The existence of an inter*galactic* magnetic field is an open question.) This field has a small magntitude, but it occupies an enormous volume.

- The magnetic field of the Earth varies over its surface, but its magnitude is of the order $B \sim 50~\mu\text{T} = 0.5$ G.

- A large laboratory electromagnet might produce a field with magnitude $B \sim 1$–10 T.

---

[5] Nikola Tesla, 1857–1943

- The most powerful laboratory magnets produce fields with magnitudes $B \sim 25$ T (steady field) to $B \sim 200$ T (pulsed field).

- *Pulsars* or *neutron stars*—the cores of dead stars, collapsed to radii of the order 20 km—have magnetic fields with magnitudes near their surfaces of the order $B \sim 10^9$ T.

- Neutron stars with especially strong magnetic fields are called *magnetars*. One such specimen has a magnetic field with maximum magnitude $B \sim 4 \times 10^{10}$ T.

- The strongest known magnetic fields, apparently, are encountered in very high-energy collisions of heavy atomic nuclei. Fields of magnitude $B \sim 10^{18}$ T can be produced in small regions for brief intervals of time.

The tesla, then, is a large unit of magnetic field strength in everyday experience, but nature has no difficulty producing large numbers of them.

The complete Lorentz force law combines Eqs. (VII.7b) and (IX.2): A particle of charge q with velocity $\mathbf{v}$, in electric field $\mathbf{E}$ and magnetic field $\mathbf{B}$, is subject to force

$$\mathbf{F} = q\left(\mathbf{E} + \mathbf{v} \times \mathbf{B}\right) . \tag{IX.4}$$

This expression shows, for example, that perpendicular $\mathbf{E}$ and $\mathbf{B}$ fields will allow particles of any charge, with velocity

$$\mathbf{v}_0 = \frac{\mathbf{E} \times \mathbf{B}}{|\mathbf{B}|^2} \tag{IX.5}$$

to pass undeflected by any force. Such an arrangement is called a *velocity selector*: A beam of charged particles directed through the fields will emerge with all remaining particles traveling at velocity $\mathbf{v}_0$, all others being deflected out of the beam.

Despite its "cobbled together" appearance, force law (IX.4) is remarkable. It is the three-dimensional version of the simplest force law consistent with the *Einsteinian* version of Newton's Second Law, i.e., with Eq. (I.100).

D   Magnetic forces on moving charges

It is convenient to divide our examination of magnetic fields and forces into two parts: the *passive* aspect, i.e., the effects of magnetic forces on moving electric charges, and the *active* aspect, i.e., the production of magnetic fields by moving-charge sources. The former is treated in this section, the latter in Sec. E.

Charged particle in B field

Magnetic forces can influence the motion of otherwise-free charged particles. It is a general feature of such effects that, since the magnetic force (IX.2) is always perpendicular to the velocity of the particle, it does *no work*[6] on the charge. By the Work-Energy Theorem (II.11), then, a particle subject only to magnetic force $\mathbf{F}_m$ moves with constant kinetic energy, i.e., constant speed.

Perhaps the simplest example of motion subject to a magnetic force is a particle of mass m and charge q, say, with velocity $\mathbf{v}$ perpendicular to a uniform magnetic field $\mathbf{B}$. The magnetic force has no component in the direction of $\mathbf{B}$, so the motion of the particle remains confined to a plane perpendicular to the field. The particle moves at constant speed, so the magnitude of the magnetic force does not change, i.e., the velocity vector $\mathbf{v}$ turns at a constant rate. That is, the particle executes *uniform circular motion* perpendicular to the $\mathbf{B}$ field. Such motion in this case is called *cyclotron motion* after a device designed to make use of it.

The frequency of this circular motion or *cyclotron frequency* is set by the requirement that the magnetic force $\mathbf{F}_m$ supply the centripetal acceleration (III.35) of the particle. Equating the force magnitudes yields

$$qvB = m\,\frac{v^2}{r}\ ,\tag{IX.6a}$$

where r is the radius of the circular trajectory, and the magnitude of the cross product in Eq. (IX.2) is simplified by the fact that $\mathbf{v}$ is perpendicular to $\mathbf{B}$. Hence, the angular frequency of the motion is given by

$$\begin{aligned}\omega &= \frac{v}{r}\\ &= \frac{qB}{m}\ ,\end{aligned}\tag{IX.6b}$$

and the frequency by

$$\begin{aligned}\nu &= \frac{\omega}{2\pi}\\ &= \frac{qB}{2\pi m}\ .\end{aligned}\tag{IX.6c}$$

For example, for protons in a field of magnitude 1.000 T, the cyclotron frequency is

$$\begin{aligned}\nu &= \frac{qB}{2\pi m}\\ &= \frac{(1.602\times 10^{-19}\ \text{C})(1.000\ \text{T})}{2\pi(1.673\times 10^{-27}\ \text{kg})}\\ &= 15.24\ \text{MHz}\ ,\end{aligned}\tag{IX.6d}$$

a not quite arbitrary example. This is a frequency in the radio range between the AM (550 kHz–1700 kHz) and FM (88 MHz–108 MHz) bands. It is a key feature of cyclotron motion that ω and ν are independent of the speed of the particle and the radius of the circle: Fast particles travel on large circles, slower particles

---

[6]This opens the question of how electric motors, turned by magnetic forces on charge carriers in wires, do the work they do. We are not yet positioned to answer this question.

on smaller, but all particles with a given mass and charge (more precisely, with a given charge-to-mass ratio) cover angle and complete cycles at the same rate.

Cyclotron motion is utilized in a variety of devices. The eponymous *cyclotron* was invented by E. O. Lawrence[7] in 1931 as a source of high-energy particle beams, supplanting natural radioactive sources. It is illustrated in Fig. IX.2. Charged particles, e.g., from a source of ionized gas at the center, move in the field between the poles of an electromagnet within two hollow copper electrodes called "dees" (because of their shape). Inside the dees, the particles encounter no electric fields, but they receive a "kick" each time they cross the gap between the dees. Because the angular frequency of the particles is independent of their speed and orbit radius, it is necessary only to flip the potential difference between the dees at fixed frequency (IX.6c) to ensure that the particles receive a forward kick at each crossing. As their speed increases, the particles spiral outward from the center; at the outer edge of the dees, the beam is deflected by an electric field down an evacuated tube to its target. Originally designed for nuclear-physics experiments, cyclotrons are now also found in some hospitals to provide precision beams of radiation for cancer treatment. The dees of Lawrence's original device were five inches across; the pole faces of a modern research instrument such as that at the National Superconducting Cyclotron Laboratory at Michigan State University might be sixty inches across. Such a cyclotron can accelerate protons to kinetic energies of the order 100 MeV, and heavier ions to even higher energies.[8] The cyclotron is a *resonant* device—energy is transferred from the voltage source for the dees to the particle beam by matching frequencies. As the frequency in example (IX.6d) is a radio frequency, the dee voltage in such a machine is provided by an oscillator similar to that of a radio transmitter.

The electron beam in a traditional television picture tube is steered by cyclotron motion, rather than by electrostatic deflection as in an oscilloscope.[9] Two pairs of coils, above and below and on each side of the electron gun, provide magnetic fields that steer the beam across the screen and from top to bottom to produce the image. The *mass spectrometer,* invented by Dempster[10] in 1918 and Aston[11] in 1919, also relies on cyclotron motion: A beam of ions passes through a velocity selector (described in the preceding section) into a magnetic field, in which the particles travel on semicircular paths to strike a photographic plate or electronic detector. With q, B, and v fixed, the diame-

---

[7] Ernest Orlando Lawrence, 1901–1958

[8] The constancy of the angular frequency on which the cyclotron depends breaks down as the particles approach the speed of light, and Einsteinian mechanics comes into play; this limits the energies that can be reached. Higher energies can be attained by decreasing the frequency of the dee voltage appropriately as the beam approaches Einsteinian energies in a device called a *synchrocyclotron.* Yet higher energies are reached in a *synchrotron,* in which the particles circulate at fixed radius, periodically passing through electric fields that accelerate them while the magnetic field strength is ramped up to maintain their trajectory. The highest-energy particle accelerators in the world, e.g., the Tevatron at Fermilab and the Large Hadron Collider at CERN in Geneva, are synchrotrons.

[9] Here, a *real* oscilloscope is meant. Modern digital oscilloscopes convert voltage signals to digital data displayed on a TV screen.

[10] Arthur Jeffrey Dempster, 1886–1950

[11] Francis William Aston, 1877–1945

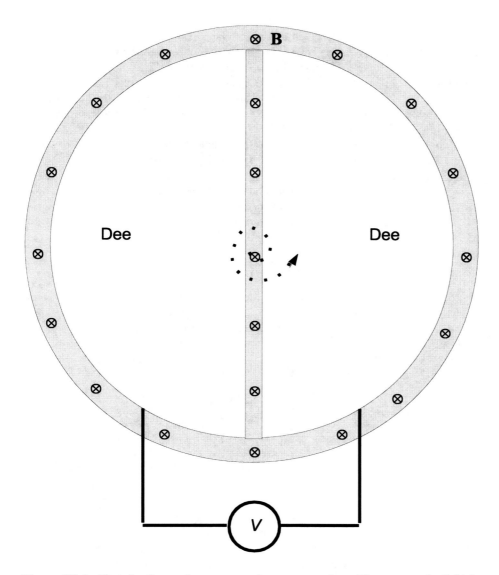

Figure IX.2: Sketch of a cyclotron, seen in cross section. The magnetic field is directed into the page. The circulating particles are accelerated by oscillating voltage V each time they pass between the "dees," spiralling outward until they are extracted.

ter of each semicircle is proportional to the mass of each particle, so the mass of each particle can be determined from its position on the film or detector. The *e/m tube* shown in Fig. VII.1 is another application of cyclotron motion: Electrons from a heated filament are accelerated through a known voltage and travel in a circular orbit in a known magnetic field. The ionization of mercury vapor in the tube makes the electron beam visible. By measuring the diameter of the circular trajectory, the charge-to-mass ratio e/m of the electrons can be determined, as Thomson did in 1897.

If charged particles in a uniform magnetic field possess a velocity component along the magnetic field, that component will be constant, as the particles' acceleration has no component in that direction. Combined with cyclotron motion, this yields uniform *helical* motion of the particles. Such motions are displayed, for example, in solar *prominences*—outpourings of ionized gas in the magnetic field of the Sun—and in the *aurorae* produced by charged particles in the magnetic fields of the Earth and the outer planets.

Forces on current-carrying wires

Magnetic forces on moving electric charges are more commonly encountered in the forces exerted on current-carrying wires, as the forces that turn electric motors are of this form. A wire carrying current I can be approximated as a line of charge of linear density λ moving at drift velocity $\mathbf{v}_d$. If the wire is subdivided into (vector) segments $d\boldsymbol{\ell}$, then force (IX.2) on charge segment $\lambda\,|d\boldsymbol{\ell}|$, moving at velocity $\mathbf{v}_d$ in magnetic field $\mathbf{B}$, can be written

$$
\begin{aligned}
\mathbf{dF}_m &= \lambda\,|d\boldsymbol{\ell}|\,\mathbf{v}_d \times \mathbf{B} \\
&= \lambda\,|\mathbf{v}_d|\,d\boldsymbol{\ell} \times \mathbf{B} \\
&= I\,d\boldsymbol{\ell} \times \mathbf{B}\;,
\end{aligned}
\tag{IX.7a}
$$

since vectors $\mathbf{v}_d$ and $d\boldsymbol{\ell}$ are both directed along the wire, and $\lambda\,|\mathbf{v}_d|$ is the current I. The total force on a current-carrying circuit, then, is given by the integral

$$
\mathbf{F}_m = \oint_{\text{Circuit}} I\,d\boldsymbol{\ell} \times \mathbf{B}\;.
\tag{IX.7b}
$$

The order of the cross product is preserved in this expression, and I is inside the integral, as it may vary over different portions of the circuit.

A simple example is provided by a current-carrying loop in a *uniform* $\mathbf{B}$ field. The net force on the loop is given by

$$
\begin{aligned}
\mathbf{F}_m &= \oint_{\text{Loop}} I\,d\boldsymbol{\ell} \times \mathbf{B} \\
&= I\left(\oint_{\text{Loop}} d\boldsymbol{\ell}\right) \times \mathbf{B} \\
&= \mathbf{0}\;,
\end{aligned}
\tag{IX.8}
$$

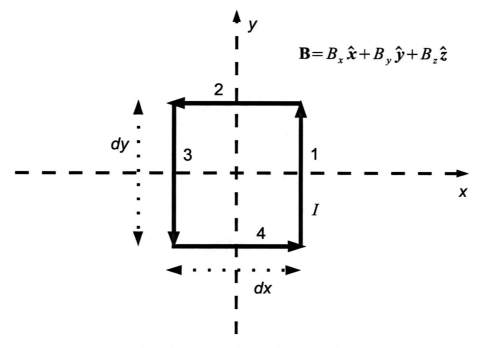

Figure IX.3: Infinitesimal rectangular current-carrying loop in a uniform magnetic field.

since the current and magnetic field are constant over the loop, and the loop is closed—net displacement from starting to finishing point on the loop is zero. However, opposing forces on different parts of the loop, acting along parallel lines, can exert torques on the loop. For example, the net torque on the rectangular loop of dimensions dx and dy, oriented in the x-y plane as illustrated in Fig. IX.3, in a uniform magnetic field of arbitrary direction, can be calculated straightforwardly. The torque about the origin on segment 1 is

$$
\begin{aligned}
d\boldsymbol{\tau}_1 &= \mathbf{r}_1 \times (I\, d\boldsymbol{\ell}_1 \times \mathbf{B}) \\
&= \tfrac{1}{2}\, dx\, \hat{\boldsymbol{x}} \times [I\, dy\, \hat{\boldsymbol{y}} \times (B_x\, \hat{\boldsymbol{x}} + B_y\, \hat{\boldsymbol{y}} + B_z\, \hat{\boldsymbol{z}})] \\
&= \tfrac{1}{2} I\, dx\, dy\, \hat{\boldsymbol{x}} \times (B_z\, \hat{\boldsymbol{x}} - B_x\, \hat{\boldsymbol{z}}) \\
&= \tfrac{1}{2} I\, dx\, dy\, B_x\, \hat{\boldsymbol{y}} \\
&= d\boldsymbol{\tau}_3 \, ,
\end{aligned}
\qquad (\text{IX.9a})
$$

where the sequence of cross products is preserved, as cross products are not associative. The torque on segment 3 is the same as that on segment 1: The current and the force are in the opposite direction, but so is the position vector.

Similarly, the torque on segment 2 is

$$
\begin{aligned}
d\boldsymbol{\tau}_2 &= \mathbf{r}_2 \times (I \, \boldsymbol{d\ell}_2 \times \mathbf{B}) \\
&= \tfrac{1}{2} \, dy \, \hat{\boldsymbol{y}} \times [-I \, dx \, \hat{\boldsymbol{x}} \times (B_x \, \hat{\boldsymbol{x}} + B_y \, \hat{\boldsymbol{y}} + B_z \, \hat{\boldsymbol{z}})] \\
&= -\tfrac{1}{2} I \, dx \, dy \, \hat{\boldsymbol{y}} \times (B_y \, \hat{\boldsymbol{z}} - B_z \, \hat{\boldsymbol{y}}) \\
&= -\tfrac{1}{2} I \, dx \, dy \, B_y \, \hat{\boldsymbol{x}} \\
&= \boldsymbol{d\tau}_4 \; .
\end{aligned}
\tag{IX.9b}
$$

The total torque about the origin on the little loop is the sum of all four contributions. It is given by

$$
\begin{aligned}
d\boldsymbol{\tau}_m &= I \, dx \, dy \, (B_x \, \hat{\boldsymbol{y}} - B_y \, \hat{\boldsymbol{x}}) \\
&= I \, dx \, dy \, \hat{\boldsymbol{z}} \times (B_x \, \hat{\boldsymbol{x}} + B_y \, \hat{\boldsymbol{y}} + B_z \, \hat{\boldsymbol{z}}) \\
&= I \, \boldsymbol{d\!A} \times \mathbf{B} \; ,
\end{aligned}
\tag{IX.9c}
$$

where the vector area $\boldsymbol{d\!A} = dx \, dy \, \hat{\boldsymbol{z}}$ of the loop is recognized, and the last term in the second line is added to complete $\mathbf{B}$, since it contributes zero to the torque $d\boldsymbol{\tau}_m$. (The choice of normal direction for $\boldsymbol{d\!A}$ is connected to the positive direction for current around the loop via a right-hand rule: With the fingers of the right hand curled in the positive direction around the loop, the thumb of that hand indicates the direction of $\boldsymbol{d\!A}$.) The torque on a general loop in a uniform field can be calculated from this by subdividing the surface bounded by the loop into "tiles" like that enclosed by the loop of Fig. IX.3. The sum of the currents around all the tiles is just that around the original loop, and the sum of the areas $\boldsymbol{d\!A}$ of the tiles is the vector area $\boldsymbol{A}$ of the surface enclosed by the loop, since areas add as vectors. The net torque on the general loop is

$$
\begin{aligned}
\boldsymbol{\tau}_m &= NI\boldsymbol{A} \times \mathbf{B} \\
&\equiv \boldsymbol{\mu} \times \mathbf{B} \; ,
\end{aligned}
\tag{IX.9d}
$$

where the factor N is included because the current-carrying wire might go around the loop N times, i.e., the loop might be a coil, and the *magnetic dipole moment* $\boldsymbol{\mu}$ of the loop is defined as $\boldsymbol{\mu} \equiv NI\boldsymbol{A}$. This has units of $\mathrm{A\,m}^2$.

Results (IX.8) and (IX.9d) are reminiscent of the force (VII.10a) and torque (VII.10b) on an electric dipole in a uniform electric field. This suggests that *a current loop is a magnetic dipole* and, by extension, that *magnetic dipoles are current loops,* though it falls short of a proof. It remains to be shown that a current loop *produces* a magnetic field in the "butterfly" pattern sketched in Fig. VII.5, corresponding to the same magnetic dipole moment $\boldsymbol{\mu}$.

For permanent magnets, the torque law $\boldsymbol{\tau}_m = \boldsymbol{\mu} \times \mathbf{B}$ *defines* the magnetic moment $\boldsymbol{\mu}$. Caution is called for here. In some older texts, this moment is divided by the length of the magnet to derive a *pole strength,* as if the magnet were two charges on a stick. But such a calculation is not meaningful, because the magnet *isn't* two charges on a stick. In other texts, the dipole moment is divided by the cross-sectional area of the magnet to define an *atomic current* responsible for the object's magnetism. This too is meaningless, because the origin of magnetism in magnetic materials at the atomic level is more subtle. It requires a deeper exploration of fundamental physics, as will be seen in subsequent sections.

The direction of torque $\boldsymbol{\tau}_m$ corresponds to a tendency to align the dipole moment $\boldsymbol{\mu}$ with the magnetic field $\mathbf{B}$, akin to the effect of an electric field on an electric dipole. This accounts for the alignment of magnetized compass needles with the Earth's magnetic field. The torque on a current-carrying coil—connected to a spring and a pointer—is used in the *d'Arsonval*[12] *galvanometer*[13] to detect the current through the coil.[14]

Like torque (VII.10b), torque $\boldsymbol{\tau}_m$ corresponds to a potential energy $U_m^{(\mathrm{dip})}$ associated with the orientation of the dipole. This is given by

$$U_m^{(\mathrm{dip})} = -\boldsymbol{\mu} \cdot \mathbf{B} \ , \tag{IX.10}$$

the magnetic analog of energy (VII.10d). But this opens up a mystery: The electric potential energy U_{dip} is simply the potential energy of the electric dipole's charges in the potential corresponding to the electric field $\mathbf{E}$. Since the magnetic field $\mathbf{B}$ does no work on moving charges, what is the source of the work represented by energy $U_m^{(\mathrm{dip})}$?

The Hall effect

A third effect of magnetic fields on moving charges was discovered by Edwin Hall[15] in 1879 and is named for him. The Hall effect arises from the displacement of charge carriers in a current-carrying conductor immersed in a magnetic field. A simple geometry in which the Hall effect appears is illustrated in Fig. IX.4. *Either* positive charge carriers moving in the direction of the current *or* negative charge carriers moving in the opposite direction would be displaced to the left side of the conductor in the figure by the magnetic force $\mathbf{F}_m$. This displacement of charge gives rise to a compensating Hall electric field $\mathbf{E}_H$. When equilibrium is attained, the Hall field has value

$$\mathbf{E}_H = -\mathbf{v}_d \times \mathbf{B} \ , \tag{IX.11a}$$

where $\mathbf{v}_d$ is the drift velocity of the charge carriers of whichever sign. This prevents further displacement of charge carriers, i.e., the fields $\mathbf{E}_H$ and $\mathbf{B}$ constitute a velocity selector for velocity $\mathbf{v}_d$. The Hall field $\mathbf{E}_H$ engenders a *Hall potential difference*

$$\begin{aligned} V_H &= |\mathbf{E}_H|\, d \\ &= |\mathbf{v}_d|\, |\mathbf{B}|\, d \end{aligned} \tag{IX.11b}$$

---

[12] Jacques Arsène d'Arsonval, 1851–1940

[13] Luigi (Aloisio) Galvani, 1737–1798

[14] Such a device was commonly connected in series or parallel with suitable resistors to construct analog voltmeters or ammeters, respectively. Modern digital meters are easier to use but much more complex: One design counts the time required for a voltage that increases at a set rate to reach equality with the input voltage, i.e., the meter is a digital clock.

[15] Edwin Herbert Hall, 1855–1938

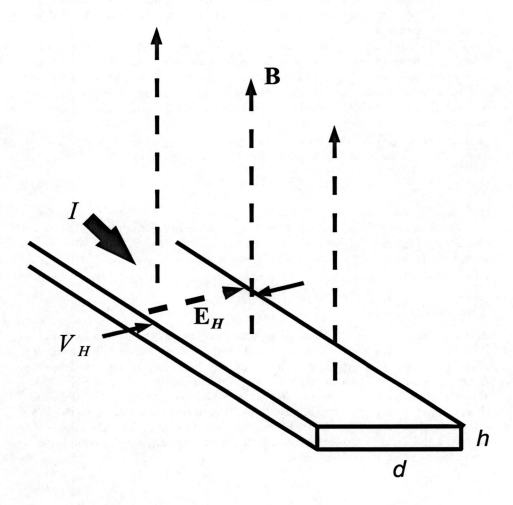

Figure IX.4: The Hall effect in a conductor carrying current I in uniform magnetic field $\mathbf{B}$. Current and field are perpendicular to simplify the analysis.

across the conductor, i.e., transverse to the current (and to the ordinary potential drop driving the current). The *sign* of V_H, however, is sensitive to the sign of the charge carriers: If the left side of the conductor in Fig. IX.4 is the higher-potential side, the charge carriers are positive; the right side is the high side for negative charge carriers. More than a century after Franklin established the standard convention for positive and negative charges, Hall's experiment established that the charge carriers in ordinary wires are negatively charged, moving in the direction opposite the current—to the chagrin of physics students ever since.

The current I in Fig. IX.4 is related to the number density n of charge carriers, their charge q, the drift velocity component v_d, and the indicated dimensions of the conductor via $I = nqdhv_d$. The Hall potential can be expressed as

$$V_H = \frac{IB}{nqh} \, . \tag{IX.11c}$$

Measurements of V_H can be used to determine the charge-carrier density n, if the magnetic field is known, or to measure B if n has already been determined. A laboratory version of the situation illustrated might involve current $I = 20.0$ A in a copper ribbon of thickness $h = 1.00$ mm, in a magnetic field of magnitude $B = 2.00$ T. The Hall potential in such case would be

$$
\begin{aligned}
V_H &= \frac{IB}{nqh} \\
&= \frac{(20.0 \text{ A})(2.00 \text{ T})}{\left(\dfrac{-96500 \text{ C}}{63.5 \text{ g}}\right)\left(\dfrac{8.93 \text{ g}}{1.00 \times 10^{-6} \text{ m}^3}\right)(1.00 \times 10^{-3} \text{ m})} \\
&= -2.95 \ \mu\text{V} \, , \tag{IX.11d}
\end{aligned}
$$

where nq is calculated here using the fact that a mole (63.5 g) of copper yields -1.00 faraday of conduction electrons and has mass density 8.93 g/(cm)^3. Such a potential difference is measurable with care. Hall-effect probes for measuring magnetic fields in the laboratory use semiconductor wafers with thickness less than 1.00 mm, and charge-carrier densities orders of magnitude less than that of copper, to produce much larger Hall potentials—of the order millivolts per millitesla.

Though discovered in the nineteenth century, the Hall effect has continued to reveal aspects of the electric and magnetic properties of matter. Charge carriers in conductors are only approximately a classical fluid. More accurately, they are quantum particles—excitations of quantum fields. Their behavior is described by quantum states in the atomic lattice of the conductor akin but not identical to the quantum states of electrons in atoms. These give rise to the *quantized Hall effect* exhibited, for example, by the *Hall resistance*. This is a quantity with units of resistance (not an ordinary resistance), given by

$$R_H \equiv \frac{V_H}{I}$$
$$= \frac{B}{nqh}$$
$$= \frac{\Phi_\mathbf{B}}{Q} \, , \tag{IX.12}$$

where $\Phi_\mathbf{B}$ is the magnetic flux through the top surface of, and Q the total mobile charge within, the conductor shown in Fig. IX.4, say. Classically, R_H is proportional to B. But for small enough changes in B, it is observed to change in discrete steps, not continuously, corresponding to the transition of a charge carrier from one quantum state to another. The last expression shows that this corresponds to "flux quantization," i.e., the magnetic flux $\Phi_\mathbf{B}$ through the surface of the conductor changes in discrete units. (The step size in R_H, known as the von Klitzig[16] constant, has been proposed as a fundamental definition of the ohm. But as the volt and the ampere already derive from fundamental SI definitions, such an additional definition might invalidate the Ohm law!) Klaus von Klitzig was awarded the Nobel Prize in Physics in 1985 for his discovery of the quantized Hall effect. A second Prize was awarded in 1995 to Robert B. Laughlin,[17] Horst L. Stormer,[18] and Daniel C. Tsui[19] for their discovery of the *fractional quantized Hall effect,* in which half-step transitions of the quantum charge-carrier fluid are observed.

E   Sources of magnetic fields: moving charges

The *active* aspect of magnetic fields relates the fields to the current or moving-charge sources that produce them.

Law of Biot and Savart

The *Law of Biot[20] and Savart[21]* or *Biot-Savart Law,* induced by those investigators from careful experiments, is the magnetic counterpart of the Coulomb law for electric fields. In modern notation, the law states that a charge q at position $\mathbf{r}_s$ produces magnetic field

$$\mathbf{B}(\mathbf{r}) = \frac{\mu_0}{4\pi} \frac{q\,\mathbf{v}_s \times (\mathbf{r} - \mathbf{r}_s)}{|\mathbf{r} - \mathbf{r}_s|^3} \, , \tag{IX.13a}$$

with *vacuum permeability* $\mu_0 \equiv 4\pi \times 10^{-7}$ N/A^2. As in the Coulomb law, the coupling constant is written in a somewhat complicated form, so that other

---

[16] Klaus von Klitzig, 1943–
[17] Robert B. Laughlin, 1950–
[18] Horst L. Stormer, 1949–
[19] Daniel C. Tsui, 1939–
[20] Jean Baptiste Biot, 1774–1862
[21] Félix Savart, 1791–1841

relations appear simpler. The value of μ_0—not to be confused with a magnetic moment μ—is exact, not within some experimental tolerance. As will be seen subsequently, the ampere of current is defined to give this value. As indicated by this relation, the magnetic field has no component along the line between field point $\mathbf{r}$ and the source charge at $\mathbf{r}_s$, and it has no component in the direction of the source velocity $\mathbf{v}_s$. This is an inverse-square field, the three powers of $|\mathbf{r} - \mathbf{r}_s|$ in the denominator because of the vector $\mathbf{r} - \mathbf{r}_s$ in the numerator. In the more commonly encountered case in which the source is moving charge carriers in a current-carrying wire, the wire can be divided into segments $d\boldsymbol{\ell}$ and $q\,\mathbf{v}_s$ replaced by $I\,d\boldsymbol{\ell}$ as in Eqs. (IX.7). The magnetic field contribution $d\mathbf{B}(\mathbf{r})$ produced by segment $d\boldsymbol{\ell}'$ at location $\mathbf{r}'$ is

$$d\mathbf{B}(\mathbf{r}) = \frac{\mu_0}{4\pi} \frac{I\,d\boldsymbol{\ell}' \times (\mathbf{r} - \mathbf{r}')}{|\mathbf{r} - \mathbf{r}'|^3} \ . \tag{IX.13b}$$

Hence, the line integral

$$\mathbf{B}(\mathbf{r}) = \frac{\mu_0}{4\pi} \oint_{\text{Source}} \frac{I\,d\boldsymbol{\ell}' \times (\mathbf{r} - \mathbf{r}')}{|\mathbf{r} - \mathbf{r}'|^3} \tag{IX.13c}$$

gives the total magnetic field produced by the source circuit.

Like the Coulomb law (VII.7c), this integral can be evaluated in closed form only for a few simple geometries. The magnetic field produced by a steady current in a long, straight wire can be calculated in this way, extending the integral to infinity in both directions along the wire. But this field will be obtained shortly via a more powerful method. A useful application of the Biot-Savart law is the calculation of the field produced by a current-carrying circular loop or coil of wire, such as that illustrated in Fig. IX.5. Calculating the *off-axis* field of the coil would require elliptic integrals beyond the scope of this text, but evaluating the field on the symmetry axis is straightforward. For the coil as illustrated, a segment $d\boldsymbol{\ell}'$ is given by

$$d\boldsymbol{\ell}' = a\,(-\sin\phi\,\hat{\boldsymbol{x}} + \cos\phi\,\hat{\boldsymbol{y}})\,d\phi \tag{IX.14a}$$

at location

$$\mathbf{r}' = a\,(\cos\phi\,\hat{\boldsymbol{x}} + \sin\phi\,\hat{\boldsymbol{y}}) \ . \tag{IX.14b}$$

The distance from this segment to field point $\mathbf{r} = z\,\hat{\boldsymbol{z}}$ is

$$|\mathbf{r} - \mathbf{r}'| = (a^2 + z^2)^{1/2} \tag{IX.14c}$$

for all ϕ. The angle ϕ runs from zero to $2\pi N$, allowing for N turns of wire around the coil. The on-axis magnetic field is given by

$$\begin{aligned}
\mathbf{B}(z\,\hat{\boldsymbol{z}}) &= \frac{\mu_0}{4\pi} \int_0^{2\pi N} \frac{I a\,(-\sin\phi\,\hat{\boldsymbol{x}} + \cos\phi\,\hat{\boldsymbol{y}}) \times [z\,\hat{\boldsymbol{z}} - a\,(\cos\phi\,\hat{\boldsymbol{x}} + \sin\phi\,\hat{\boldsymbol{y}})]}{(a^2 + z^2)^{3/2}}\,d\phi \\
&= \frac{\mu_0 I a}{4\pi(a^2 + z^2)^{3/2}} \int_0^{2\pi N} (z\,\sin\phi\,\hat{\boldsymbol{y}} + z\,\cos\phi\,\hat{\boldsymbol{x}} + a\,\hat{\boldsymbol{z}})\,d\phi \\
&= \frac{\mu_0}{4\pi} \frac{2NI(\pi a^2)\,\hat{\boldsymbol{z}}}{(a^2 + z^2)^{3/2}} \ .
\end{aligned}$$

$$\tag{IX.14d}$$

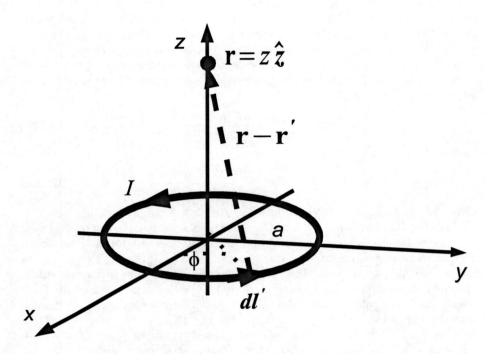

Figure IX.5: Circular loop or coil of wire of radius a, carrying current I in each turn of wire. The current can flow into and out of the coil via a pair of twisted leads (not shown), the magnetic effects of which cancel.

A magnetic dipole field would take the form

$$\mathbf{B}_{\text{dip}}(\mathbf{r}) = \frac{\mu_0}{4\pi} \frac{3(\boldsymbol{\mu} \cdot \hat{r})\,\hat{r} - \boldsymbol{\mu}}{r^3} \;,\tag{IX.15a}$$

analogous to Eq. (VII.8). If the dipole moment $\boldsymbol{\mu}$ were aligned along the $+z$ axis, then the field on that axis would be

$$\mathbf{B}_{\text{dip}}(z\,\hat{\boldsymbol{z}}) = \frac{\mu_0}{4\pi} \frac{2\boldsymbol{\mu}}{|z|^3} \;.\tag{IX.15b}$$

Hence, the circular coil or loop produces the on-axis field of a dipole, with the same dipole moment $\boldsymbol{\mu} = NI\mathcal{A}$ as in Eq. (IX.9d), at distances $|z| \gg a$ far from the loop compared to its size. It also produces the off-axis "butterfly" field of a dipole at such distances, although that has not been shown here. This completes the picture: Current loops act as magnetic dipoles—magnetic dipoles are current loops—both as receivers and as sources of magnetic fields and forces.

Gauss Law for Magnetism

Comparing the difficulty of evaluating integral (IX.13c) with that of evaluating integral (VII.7c), the reader might wish for a Gauss law for magnetism like that for electric fields. The wish is granted: There is a Gauss law for magnetism. It is called the *Gauss Law for Magnetism*. Akin to its electric counterpart, it relates the flux of magnetic field $\mathbf{B}$ through any orientable closed surface S_c to the total magnetic charge inside that surface. The catch is that absent magnetic monopoles, the total magnetic charge inside any closed surface is zero. The Gauss law for magnetism takes the form

$$\Phi_{\text{B}}(S_c) = 0$$

$$\text{i.e.,} \qquad \oint_{S_c} \mathbf{B} \cdot d\mathcal{A} = 0 \;.\tag{IX.16}$$

Magnetic field lines never end—they either form closed loops or extend to infinity. Hence, any field line that enters surface S_c must leave it, and vice versa. This is still a valid physical law; it is one of the Maxwell equations. It constrains the geometry of magnetic fields. But it does not connect magnetic fields with their sources.

The Ampère Law

An integral law that does connect magnetic fields with their sources is credited to Ampère. This law utilizes another feature of vector fields appropriated from fluid mechanics: the *circulation* of a field. This is the integral over a closed curve of the component of the field *tangential* to the curve times distance around the curve.[22] The Ampère law states that the circulation of the magnetic field $\mathbf{B}$

---

[22] Applied to the velocity field of a fluid, this describes the *rotation* inherent in its motion.

around any closed curve[23] C is proportional to the net current I_S through the surface S of which C is the boundary. This takes the form

$$\oint_C \mathbf{B} \cdot d\boldsymbol{\ell} = \mu_0 \, I_S$$

$$= \mu_0 \int_S \mathbf{j} \cdot d\boldsymbol{\mathcal{A}} \,, \qquad\qquad \text{(IX.17)}$$

where the current can be written as the flux of current density $\mathbf{j}$. The positive direction around C and the positive direction for flux through S are not independent, but are connected by a right-hand rule: If the fingers of the right hand point in the positive direction around C, the thumb of that hand indicates the positive direction through S. [This is the same rule encountered in Eqs. (IX.9c) and (IX.9d) in assigning vector areas.] The constant of proportionality is the vacuum permeability μ_0 that appears in the Biot-Savart law (IX.13), where the inclusion of the factor $1/(4\pi)$ simplifies the form of the constant here. The curve C need not correspond to a physical object. Like the closed surface S_c in the two Gauss laws, it can be freely chosen. Strictly, the Ampère law must be qualified: It applies to *steady-state* situations, as will be explained subsequently.

The Ampère law facilitates the calculation of the magnetic field produced by a current in a long, straight wire, as illustrated in Fig. IX.6. The cylindrical symmetry of this situation restricts the geometry of the magnetic field. A "bottle-brush" field—radially away from or toward the wire, the same all the way around—would be consistent with the symmetry. But such a field component is forbidden by the Gauss law for magnetism, as it would yield nonzero flux through a cylinder coaxial with the wire. A "barber-pole" field is impossible, as the charge carriers in the wire produce no field component parallel to their drift velocity, i.e., to the current. Hence, the magnetic field can only be tangential to circles centered on the wire, lying in planes perpendicular to the wire, and the magnitude of the field can depend only on radial distance from the wire. The circulation of the field around such a circle is given by

$$\oint_C \mathbf{B} \cdot d\boldsymbol{\ell} = 2\pi r B_\phi$$

$$= \mu_0 I \qquad \text{by Ampère,} \qquad\qquad \text{(IX.18a)}$$

with B_ϕ the azimuthal component of $\mathbf{B}$, which is its only component. The magnetic field, then, is

$$\mathbf{B} = \frac{\mu_0 I}{2\pi r} \, \hat{\boldsymbol{\phi}} \,, \qquad\qquad \text{(IX.18b)}$$

with $\hat{\boldsymbol{\phi}}$ the unit vector in the azimuthal direction. The direction of B is connected to that of I via the right-hand rule described previously. A similar calculation can be used to determine the magnetic field *inside* a cylindrical,

---

[23]The curve must be—at least piecewise—smooth enough to have a well-defined tangent vector, C^1 in the parlance of mathematicians.

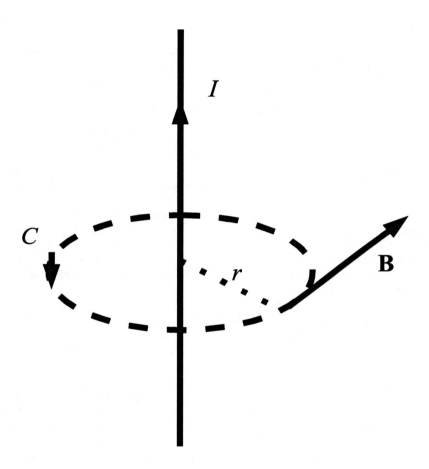

Figure IX.6: Magnetic field produced by a steady current in a long, straight wire.

current-carrying wire if the current density distribution $\mathbf{j}(r)$ inside the wire is known.

This implies the important result that currents in parallel wires exert magnetic forces on each other. Two long, straight, parallel, current-carrying wires are illustrated in Fig. IX.7. The magnetic field produced by wire 1 at wire 2 is

$$\mathbf{B}_{\text{at }2} = \frac{\mu_0 I_1}{2\pi d}\,\hat{\otimes}\ , \tag{IX.19a}$$

with $\hat{\otimes}$ a unit vector into the page. The resulting magnetic force on the "infinite" wire is infinite, but the magnetic force *per unit length,* from Eqs. (IX.7), is

$$\frac{\mathbf{F}_{\text{on }2}}{\ell_2} = \frac{\mu_0 I_1 I_2}{2\pi d}\,\hat{\leftarrow}\ . \tag{IX.19b}$$

Here $\hat{\leftarrow}$ denotes a unit vector to the left, i.e., toward wire 1. Parallel currents—with I_1 and I_2 either both positive or both negative—attract, while antiparallel currents repel. Hence, for example, the current streams within a current-carrying conductor exert attractive forces on each other that compress the conductor, a phenomenon known as *magnetostriction.*

The magnetic force between parallel currents provides the fundamental definition of the ampere of current and, consequently, the coulomb of charge first used in Ch. VII. The ampere is defined as *that steady current that, in long parallel conductors one meter apart in vacuum, gives rise to a force of exactly* 2×10^{-7} *newtons per meter on each.* This definition establishes the value of μ_0 as exactly $4\pi \times 10^{-7}$ N/A^2. This standard is implemented by connecting an ammeter in series to a *current balance,* e.g., at the laboratories of the National Institute of Standards and Technology (NIST) in Boulder, Colorado. The current is run through parallel conductors and the force between them is measured, thus providing a precision calibration of the ammeter in terms of the fundamental definition of the ampere.

Another application of the Ampère law that will prove useful later is the calculation of the magnetic field inside a long, straight *solenoid*: a close-wound cylindrical coil of wire of length L much greater than its radius R. This is sketched in Fig. IX.8. If the solenoid is long enough that the changes in the magnetic field at its ends ("end effects") can be neglected, then the field inside is parallel to the solenoid axis. Outside, the opposite sides of the solenoid act as "infinite" sheets of current in opposite directions, and the resulting magnetic field is zero. That is, near the solenoid and well away from its ends, the field is essentially confined within the cylinder. The Ampère law, applied to the rectangular *Amperian loop* shown in the figure, implies

$$B_{\text{in}}x = \mu_0 nxI\ , \tag{IX.20a}$$

where n is the number of turns of wire per unit length on the solenoid, each turn carrying current I. This yields interior magnetic field

$$B_{\text{in}} = \mu_0 nI\ , \tag{IX.20b}$$

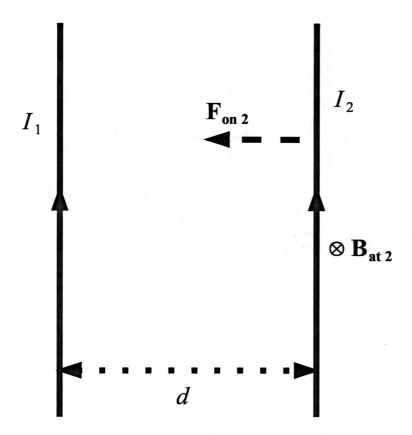

Figure IX.7: Parallel current-carrying wires exert magnetic forces on each other.

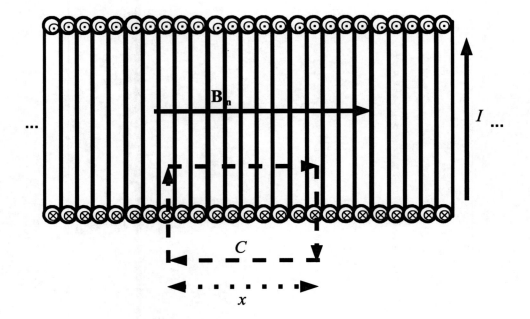

Figure IX.8: A long, straight, current-carrying solenoid, seen in cross section. The Amperian loop used to calculate the interior magnetic field is shown.

which, of course, does not depend on the arbitrary loop width x. It also does not depend on the depth of the loop into the solenoid. That is, to this approximation, the solenoid produces a uniform magnetic field inside, just as the parallel-plate capacitor produces an approximately uniform electric field. The direction of **B** within the solenoid is given by another application of the right-hand rule: If the fingers of the right hand curl about the solenoid in the direction of the current, the thumb of that hand indicates the direction of the interior field. This is consistent with previous applications of the right-hand rule.

The *on-axis* field of the solenoid can be explored in more detail using the Biot-Savart law, i.e., by integrating result (IX.14d) over the length of the solenoid. Features of the field include:

- At each end of a long solenoid, with $L \gg R$, the on-axis field drops to half its interior value, i.e., $B_{\text{end}} \cong \frac{1}{2} B_{\text{in}}$.

- Outside the solenoid, far from either end compared with its length, the field behaves as the dipole field (IX.15a) of any current-carrying coil, i.e., $\mathbf{B} \cong \mathbf{B}_{\text{dip}}$.

- An exact result for the on-axis field can be obtained for arbitrary values of L and R, incorporating the end effects. This result will not be needed here.

The Ampère law can also be used to calculate the interior field of a *toroidal solenoid,* i.e., a solenoid curved into a closed torus or doughnut shape. While an important practical application, this result also will not be needed here.

The Ampère-*Maxwell* Law

Despite its utility, the Ampère law has one drawback: It is, or can be, internally inconsistent. This can be illustrated in the situation sketched in Fig. IX.9. The chosen Amperian loop is the curve C, and the surface through which the current is to be calculated is the portion S of the ellipsoid to the right of this curve. The Ampère law requires

$$\oint_C \mathbf{B} \cdot d\boldsymbol{\ell} = \int_S \mathbf{j} \cdot d\boldsymbol{\mathcal{A}} \,. \tag{IX.21a}$$

In a sequence of such calculations, in which C is gradually shifted to the left (like sliding a rubber band toward the left end of the ellipsoid), the circulation of **B** must approach the limit

$$\oint_C \mathbf{B} \cdot d\boldsymbol{\ell} \to 0 \,, \tag{IX.21b}$$

as **B** remains finite while C becomes vanishingly small. The flux of **j** approaches the limit

$$\int_S \mathbf{j} \cdot d\boldsymbol{\mathcal{A}} \to \oint_{S_c} \mathbf{j} \cdot d\boldsymbol{\mathcal{A}} \,, \tag{IX.21c}$$

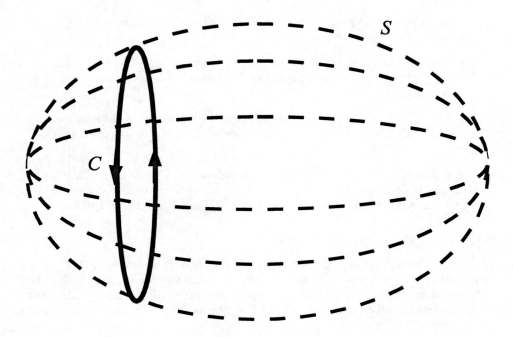

Figure IX.9: Loop C and surface S for a sequence of Ampère-law calculations. The entire ellipsoid is the closed surface S_c.

where S_c is the closed surface represented by the entire ellipsoid. If the two quantities are always equal, they must approach the same limit. But the flux of **j** through the closed surface S_c is guaranteed to be zero *only in steady-state situations*. It will not be zero if the charge inside S_c is increasing or decreasing with time, a contradiction.

The reader will have noticed that in the original statement (IX.17) of the Ampère law, it was not specified *which* of the infinity of surfaces S with boundary C was to be used to calculate the current. This is the same conundrum, because any two such surfaces, taken together, constitute a closed surface. It is not guaranteed that the current through the two surfaces will be the same—the net current through the combined surface will be zero—except in the steady state.

James Clerk Maxwell discovered that the Ampère law could be "patched," in modern computer-science parlance, to remove the inconsistency in all cases. His patch was to supplement the current source in the Ampère law with an additional source term, such that the flux of the combination through any closed surface was guaranteed to be zero. His reasoning is straightforward: The continuity equation (VIII.6) and the Gauss law (VII.13) imply that the flux of **j** through any closed surface S_c is given by

$$
\begin{aligned}
\oint_{S_c} \mathbf{j} \cdot d\boldsymbol{\mathcal{A}} &= -\frac{dQ_{\text{in}}}{dt} \\
&= -\frac{d}{dt}\left(\epsilon_0 \oint_{S_c} \mathbf{E} \cdot d\boldsymbol{\mathcal{A}}\right) \\
&= -\epsilon_0 \oint_{S_c} \frac{\partial \mathbf{E}}{\partial t} \cdot d\boldsymbol{\mathcal{A}}\,.
\end{aligned}
\tag{IX.22a}
$$

Hence, the combined flux

$$
\oint_{S_c}\left(\mathbf{j} + \epsilon_0 \frac{\partial \mathbf{E}}{\partial t}\right) \cdot d\boldsymbol{\mathcal{A}} = 0
\tag{IX.22b}
$$

through any closed surface is zero in any case. The additional term, which can variously be written

$$
\begin{aligned}
\int_S \epsilon_0 \frac{\partial \mathbf{E}}{\partial t} \cdot d\boldsymbol{\mathcal{A}} &= \epsilon_0 \frac{d\Phi_{\mathbf{E}}(S)}{dt} \\
&= \int_S \frac{\partial \mathbf{D}}{\partial t} \cdot d\boldsymbol{\mathcal{A}} \\
&= \frac{d\Phi_{\mathbf{D}}(S)}{dt}
\end{aligned}
\tag{IX.22c}
$$

for any surface S, is called the *displacement current*. Here **D** is the same *electric displacement* field introduced in Eq. (VII.42) for dealing with dielectrics. The patched *Ampère-Maxwell* law includes both true-current and displacement-

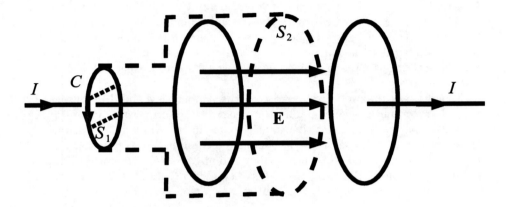

Figure IX.10: A charging parallel-plate capacitor, with loop and two choices of surface for the Ampère-Maxwell law.

current sources in the relation

$$\oint_C \mathbf{B} \cdot d\boldsymbol{\ell} = \mu_0 \left(I_S + \epsilon_0 \, \frac{\Phi_{\mathbf{E}}(S)}{dt} \right)$$
$$= \mu_0 \int_S \left(\mathbf{j} + \epsilon_0 \, \frac{\partial \mathbf{E}}{\partial t} \right) \cdot d\boldsymbol{\mathcal{A}} \, . \qquad \text{(IX.23)}$$

This reduces to the Ampère law (IX.17) in the steady state, in which the time derivatives are zero. But it is consistent in all cases for any choice of surface S with boundary C.

Some texts suggest that Maxwell pulled the displacement current "out of a hat," or in order to make the Ampère law more symmetric with the Faraday law (the subject of the next chapter). This is misleading. As the preceding results show, both the presence of the displacement current and its precise form are mathematically necessary.

An example of displacement current in action can be found in a charging capacitor, such as the parallel-plate capacitor sketched in Fig. IX.10. If the capacitor is being charged with steady current I, then the Ampère-Maxwell law applied to surface S_1 yields

$$\oint_C \mathbf{B} \cdot d\boldsymbol{\ell} = \mu_0 \int_{S_1} \mathbf{j} \cdot d\mathbf{A}$$
$$= \mu_0 I \ . \tag{IX.24a}$$

On surface S_2, there is no true current density, but the changing electric field between the plates gives displacement current

$$\oint_C \mathbf{B} \cdot d\boldsymbol{\ell} = \mu_0 \epsilon_0 \frac{d}{dt} \int_{S_2} \mathbf{E} \cdot d\mathbf{A}$$
$$= \mu_0 \epsilon_0 \frac{d}{dt} \left(\frac{Q}{\epsilon_0 \mathcal{A}} \mathcal{A} \right)$$
$$= \mu_0 I \ . \tag{IX.24b}$$

Here, the electric field $\mathbf{E}$ and the plate charge Q are related as in Eq. (VII.37a), and the current I is dQ/dt. Hence, the two results are completely consistent.

The displacement current between the capacitor plates gives rise to a magnetic field there, with concentric circular field lines parallel to the plates. There is indeed such a magnetic field between the plates of a charging capacitor, although because of the factor ϵ_0 in the displacement current, the field is very small in ordinary laboratory situations. This field was not measured directly until 1928, more than sixty years after Maxwell's work predicted it.[24]

Of significance is the further expansion of the field concept implicit in the Ampère-Maxwell law. As pointed out at the end of Ch. VII Sec. G, the electric field cannot be regarded as a mere bookkeeping device. Not only does it possess energy in its own right, but a changing $\mathbf{E}$ field can act as the source of a $\mathbf{B}$ field. That the reverse is also true will be seen in Ch. X.

F   Magnetic moment of the electron

The ideas of the preceding sections can be used to enhance the classical model of Ch. VII in an attempt to incorporate the *magnetic* properties of the electron.

Classical magnetic moment for solid-sphere model

If the electron is modeled as a *rotating,* uniform-density sphere of charge, then each bit of charge constitutes a current loop about its rotation axis and contributes a magnetic dipole moment along that axis. The simplest way to calculate the total magnetic moment of the electron in this model is to envision the sphere sliced into disks perpendicular to the rotation axis, and each disk divided into concentric rings, as illustrated in Fig. IX.11. The magnetic moment

---

[24]Of course, Maxwell's ideas were not in doubt all that time. Their most important consequences were confirmed by Hertz in 1886. By 1895, they had led to the invention of radio, and by 1920, commercial radio was on the air.

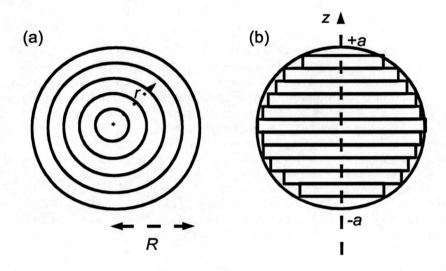

Figure IX.11: (a) A disk of charge subdivided into concentric rings, and (b) a sphere subdivided into stacked disks.

of a uniform disk of radius R, charge q, and angular speed ω is the sum of ring contributions

$$
\begin{aligned}
\mu_z^{\text{(disk)}} &= \int_0^R \pi r^2 \, \frac{1}{2\pi/\omega} \, \frac{q(2\pi r \, dr)}{\pi R^2} \\
&= \frac{q\omega}{R^2} \int_0^R r^3 \, dr \\
&= \tfrac{1}{4} q\omega R^2 \, ,
\end{aligned}
\tag{IX.25a}
$$

where the last factor in the first line is the fraction of the disk's charge in the ring of radius r and thickness dr, and the *component* of the magnetic moment along the rotation axis is calculated for later convenience. The total magnetic moment of a (rigidly) rotating, uniform sphere of total charge Q and radius a is the sum of disk contributions

$$
\begin{aligned}
\mu_z^{\text{(sph)}} &= \int_{-a}^a \tfrac{1}{4}\omega R^2 \, \frac{Q(\pi R^2) \, dz}{\frac{4}{3}\pi a^3} \\
&= \frac{3}{16} \frac{Q\omega}{a^3} \int_{-a}^a (a^2 - z^2)^2 \, dz \\
&= \tfrac{1}{5} Q a^2 \omega \\
&= \frac{Q}{2m} S_z^{\text{(sph)}} \, ,
\end{aligned}
\tag{IX.25b}
$$

where the last factor in the first line is the fraction of the sphere's charge in the disk of radius R at height z, where these are connected as illustrated in Fig. IX.11 (b), and $S_z^{\text{(sph)}} = \tfrac{2}{5} m a^2 \omega$ is the component of the *spin angular momentum* of a uniform sphere of mass m, radius a, and angular speed ω. For an electron of charge $-e$ and mass m_e, this yields

$$
\mu_z^{(e)} = \frac{-e}{2m_e} S_z^{(e)} \, ,
\tag{IX.25c}
$$

where the negative sign means that the vectors $\boldsymbol{\mu}^{(e)}$ and $\mathbf{S}^{(e)}$ point in opposite directions. The classical radius a of the electron calculated in Ch. VII Sec. F does not appear explicitly in this form of the result.

Gyromagnetic ratio and reduced gyromagnetic ratio

At this point, we could borrow a value for $S_z^{(e)}$ from quantum mechanics and compare the resulting value of $\mu_z^{(e)}$ with measured values. It is customary, however, to express this result in terms of the *gyromagnetic ratio* $\mu_z^{(e)}/S_z^{(e)}$ of the electron because of the precision to which this quantity can be measured. The ratio can be written

$$
\frac{\mu_z^{(e)}}{S_z^{(e)}} = g_e \, \frac{-e}{2m_e} \, ,
\tag{IX.26}
$$

where the dimensional factors have been collected, and g_e is the dimensionless *reduced gyromagnetic ratio* or *g-factor*. This serves as a "fudge factor," reflecting the extent to which the gyromagnetic ratio differs from a classical value.

Result (IX.25c) shows that our classical model yields $g_e = 1$. The measured value is slightly greater than 2. Hence, the model's result seems reasonable until one realizes that this quantity can be measured and calculated to some sixteen decimal places. Our classical result is therefore in error by the order of a *quadrillion* standard deviations, sufficient to fail even the most lenient physics laboratory course.

So spectacular a failure illuminates the need for new physics. The electron is not a classical ball of charge. It is governed by Einsteinian, not Newtonian, mechanics, and it is properly described by quantum, not classical, mechanics. Bringing both descriptions to bear is not easy: The two conceptual frameworks are *almost* inconsistent.[25] In 1927, P. A. M. Dirac formulated a quantum wave equation for electrons consistent with Lorentzian kinematics and Einsteinian dynamics. Called the *Dirac equation,* this is the chemists' go-to equation when they want to describe electrons in atoms correctly. Electron spin is built into the Dirac description, not added on as in a Schrödinger quantum-mechanical treatment. For electrons interacting with magnetic fields, the Dirac equation predicts $g_e = 2$. This is a considerable improvement—only a trillion standard deviations in error.

The physics still missing is in the treatment of the electric and magnetic fields with which the electron interacts. These are automatically Einsteinian, not Newtonian. But a quantum-mechanical treatment of the fields is needed. This is known as *Quantum Electrodynamics,* or QED. It is an example of a *quantum field theory,* mentioned briefly in Ch. V Sec. G. The fundamental entities in such a theory are fields governed by quantum mechanics. Particles such as the electron are quantum excitations of the normal modes of these fields. The intrinsic spin of particles appears in Einsteinian quantum field theories *as a bookkeeping device,* as the answer to the question, "How many states—i.e., what is the dimension of the vector space of quantum states—to describe a particular particle at rest?" Since these states can be transformed into one another by rotations of the coordinate system, they are labeled by quantum numbers with the characteristics of angular momentum—hence, particle spin. One state (dimension one) means a scalar particle of spin zero. Dimension two means a spinor particle, spin one-half, like the electron. Dimension three corresponds to a vector particle with spin one, and so on. The question "How many states...?" must have an answer for each species of particle, ergo, particles must have spin.[26]

An apparent drawback to quantum field theories is that, for example, fields have an infinite number of normal modes. As a consequence, such theories tend to yield infinite results in the calculation of any physical quantity. In 1948,

---

[25] The bad news is that it is difficult to construct Einsteinian quantum mechanics. The good news is that if one finds a way, it is probably correct—there cannot be many competitors.

[26] The more comprehensive and penetrating physical theory becomes, the more aspects of nature appear *necessary,* the fewer accidental or incidental. Many thinkers have speculated that in an ultimate description of nature, *everything* would be necessary, nothing incidental. So far, that remains speculative.

Richard Feynman,[27] Julian Schwinger,[28] and Tomonaga Shin-ichiro,[29] working independently, discovered methods through which quantum electrodynamics could be made to yield finite results for physical quantities. The three men shared the 1965 Nobel Prize in Physics for their achievement. Their methods, collectively termed *renormalization,* became a centerpiece of physical theory throughout the second half of the twentieth century. Too arcane to be illustrated in this text, renormalization can be described in a variety of ways. For example: A theory with finite values for its fundamental parameters yields infinite predictions. By allowing the fundamental parameter values to become infinite in the right way, the infinities can be made to cancel, making the physical predictions finite. This may seem rather haphazard, but it can yield predictions of unrivaled precision and accuracy.[30]

The calculation of the gyromagnetic ratio of the electron in quantum electrodynamics is a perturbation calculation: a sum of correction terms with higher and higher powers of the *fine-structure constant*

$$\alpha \equiv \frac{e^2}{4\pi\epsilon_0 \hbar c}$$
$$= (7.2973531 \pm 0.0000003) \times 10^{-3} , \qquad \text{(IX.27)}$$

where $\hbar$ is Planck's reduced constant and c is the vacuum speed of light. The higher the power—the greater precision desired—the more terms involved. The present best value involves the calculation of hundreds of millions of terms. (In the 1970s and 1980s this led to the extraordinary situation that it was more expensive to calculate the next decimal place in g_e than to measure it!) An example value—not the current record for precision—is

$$g_e = 2.00231930439 \pm (2. \times 10^{-11}) . \qquad \text{(IX.28)}$$

Is this a calculated or measured value? Both! Nowhere in science is there an agreement between prediction and measurement of greater precision and accuracy than in quantum electrodynamics. The precision of the present result is now so high that it is necessary to include the contributions of *nuclear interactions* to push the calculations further.

Particle magnetic moments

The magnetic moments of fundamental particles are of importance in a wide variety of applications. The component of the electron's magnetic moment

---

[27]Richard P. Feynman, 1918–1988

[28]Julian Schwinger, 1918–1994

[29]Tomonaga Shin-ichiro, 1906–1979

[30]A theory for which this works is said to be *renormalizable.* For many years, this feature was considered a *sine qua non* of any fundamental physical theory. At present, however, its necessity in an ultimate theory—if one exists—is not certain.

along its spin axis, say, is given by

$$\mu_z^{(e)} = -\frac{g_e}{2}\frac{e\hbar}{2m_e}$$
$$= -9.2847701 \times 10^{-24} \text{ J/T} .\qquad\text{(IX.29a)}$$

Components are used here because the treatment of vector quantities in quantum mechanics introduces additional complications. Planck's reduced constant appears here because the spin angular momentum component for the electron is $S_z^{(e)} = \hbar/2$. The customary units J/T are shown; these are equivalent to $A\,m^2$. The collection of dimensional factors in the second fraction of the first line here is labeled the *Bohr magneton*, given by

$$\mu_B \equiv \frac{e\hbar}{2m_e}$$
$$= 9.2740154 \times 10^{-24} \text{ J/T} .\qquad\text{(IX.29b)}$$

This is the natural magnetic-moment unit at the atomic level, e.g., in calculating the magnetic dipole moments associated with the *orbital* motions of electrons in atoms.

The proton and neutron, or *nucleons,* also possess intrinsic magnetic moments. The proton has the magnetic-moment component along its spin axis

$$\mu_z^{(p)} = 1.41060761 \times 10^{-26} \text{ J/T}$$
$$= 2.792847386\,\mu_N .\qquad\text{(IX.29c)}$$

The neutron, although electrically neutral, has the magnetic-moment component

$$\mu_z^{(n)} = -0.96623707 \times 10^{-26} \text{ J/T}$$
$$= -1.91304275\,\mu_N .\qquad\text{(IX.29d)}$$

The negative sign indicates that the vector magnetic moment of the neutron, like that of the electron, is opposite to its spin angular momentum. The dimensional factor μ_N appearing in both of these expressions is the *nuclear magneton,* given by

$$\mu_N \equiv \frac{e\hbar}{2m_p}$$
$$= 1.41060761 \times 10^{-26} \text{ J/T} .\qquad\text{(IX.29e)}$$

All of these are some three orders of magnitude smaller than their electronic counterparts because of the much larger proton mass. The numerical factors appearing in the second lines of Eqs. (IX.29c) and (IX.29d) are the values of $g/2$ for the proton and neutron, respectively. The reduced gyromagnetic ratios for these particles are far from the Dirac value of 2. This is strong evidence that the proton and (especially) the neutron are not elementary particles, but possess internal electrical structure—now understood in terms of *quarks* and *gluons.*[31]

---

[31] Richard Feynman once told this author and others, over lunch, that he would retire once he could calculate the magnetic moment of the proton from first principles—from *quantum*

The numerical values in Eqs. (IX.29) are not necessarily the best available measurements. But they serve to indicate just how well these quantities are known.

EPR, NMR, and MRI

These acronyms stand for *Electron Paramagnetic Resonance, Nuclear Magnetic Resonance,* and *Magnetic Resonance Imaging.* Both phenomena—the third term refers to the application of nuclear magnetic resonance to medical imaging; it was coined because patients were afraid of the word "nuclear"—are based on the fact that electrons or nucleons in a magnetic field are subject to magnetic torques of form (IX.9d), tending to align their magnetic moments with the field. But because these particles possess angular momentum, the torques instead induce *gyroscopic precession,* like that described in Ch. III Sec. F, about the field lines. If energy is supplied to the precessing particles at their precession frequency, e.g., via radio-frequency electric fields, resonant absorption of this energy occurs that can be detected. (In quantum-mechanical terms, the particles undergo *spin flips* to higher-energy states.) NMR was discovered in 1946 by Felix Bloch[32] and Edward Purcell,[33] who received the Nobel Prize for it in 1952. It and its counterpart EPR now underlie some of the widest-ranging measurement techniques in science. By measuring the resonant frequency or tracking the absorption, we can use these phenomena to measure particle gyromagnetic ratios, magnetic fields, nuclear structure, molecular structure, crystalline structure, proton concentrations, et cetera. This last is important because it is used (as MRI) to map the structure of soft tissues in the body, a now-common imaging technique complementary to diagnostic x-rays.

G Magnetic effects in bulk matter

Our journey to the frontiers of physics has brought us full circle to the point where we can address the oldest problems in the study of magnetism: the magnetic properties of materials. These remain areas of active research; this section will only introduce some of the language with which these problems can be discussed.

Describing magnetic response

A varied terminology has been developed to describe the response of bulk matter to magnetic fields. The description is reminiscent of that of the *polarization* of dielectric materials by electric fields: In magnetic materials, applied magnetic

---

chromodynamics (QCD), the quantum field theory that governs quarks and gluons—to match these measured values. Professor Feynman died in 1988 before he could accomplish this feat. Approximate calculations of $\mu_z^{(p)}$ still fall far short of the precision of the measurements.

[32] Felix Bloch, 1905–1983
[33] Edward Mills Purcell, 1912–1997

fields either *induce* magnetic dipoles in the atoms of the material or *align* atomic dipoles already present, resulting in the *magnetization* of the material.

Suppose a material is exposed to an external magnetic field $\mathbf{B}_0$. Its response will alter the magnetic field in the material in some way, resulting in a new net field $\mathbf{B}$. The relationship between the two fields can be expressed in various ways, e.g.,

$$
\begin{aligned}
\mathbf{B} &= K_m\,\mathbf{B}_0 && \text{(relative permeability } K_m\text{)}\,, \\
&= (1 + \chi_m)\,\mathbf{B}_0 && \text{(magnetic susceptibility } \chi_m \equiv K_m - 1\text{)}\,, \\
&= (1 + \chi_m)\,\mu_0\,\mathbf{H} && \text{(magnetic intensity } \mathbf{H} \equiv \mathbf{B}_0/\mu_0\text{)}\,, \\
&= \mu\,\mathbf{H} && \text{(permeability } \mu \equiv K_m\mu_0\text{)}\,, \\
&= \mu_0\,(\mathbf{H} + \mathbf{M}) && \text{(magnetization } \mathbf{M} \equiv \chi_m\mathbf{H}\text{)}\,.
\end{aligned}
\tag{IX.30}
$$

Here, relative permeability K_m, susceptibility χ_m, and permeability μ (not to be confused with a dipole moment) all characterize the material. The magnetic intensity $\mathbf{H}$ can be considered an independent field, akin to the electric displacement $\mathbf{D}$ in Eqs. (VII.42) and (VII.43). The magnetization $\mathbf{M}$, akin to the electric polarization $\mathbf{P}$, is the net magnetic dipole moment per unit volume in the material. It, like H, has SI units of A/m, equivalent to $\mathrm{A\,m^2}$ (or J/T) per $\mathrm{m^3}$.

Caution is needed here: Relations (IX.30) can make magnetization appear simpler than it is. The relationships shown can be nonlinear, in some cases not even a function. The coefficients K_m, χ_m, and μ can depend on $\mathbf{H}$, and in some cases not just on its instantaneous value but on its entire history $\mathbf{H}(t)$. Some of these complications are illustrated in this section.

Magnetic materials can be grouped into four categories. These are termed *paramagnetic, diamagnetic, ferromagnetic,* and *antiferromagnetic.*

Paramagnetic materials

Paramagnetic materials are characterized by small, positive magnetic susceptibilities: $\chi_m \sim +1 \times 10^{-5}$ is a typical value. Oxygen, for example, is paramagnetic. Such materials are attracted to the poles of a magnet.

The atoms or molecules of paramagnetic materials possess intrinsic magnetic dipole moments, arising from the *spin* magnetic moments of unpaired electrons. The magnetization of such materials consists of the alignment of these intrinsic moments by an applied magnetic field in competition with the thermal agitation of the atoms or molecules. A *magnetization curve* typical of a paramagnetic material is shown in Fig. IX.12. The graph shows $B - B_0 = \mu_0 M$ as a function of $B_0 = \mu_0 H$, in the notation of Eqs. (IX.30). The slope of the curve near the origin is the susceptibility χ_m. Because of the competition between magnetic alignment and thermal motion, this is inversely proportional to the absolute temperature of the material, a relationship known as the Curie[34,35] Law. (The

---

[34]Marja Sklodowska (Marie) Curie, 1867–1934
[35]Pierre Curie, 1859–1906

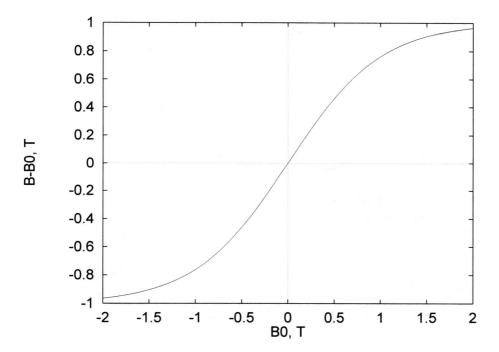

Figure IX.12: Magnetization curve for a paramagnetic material.

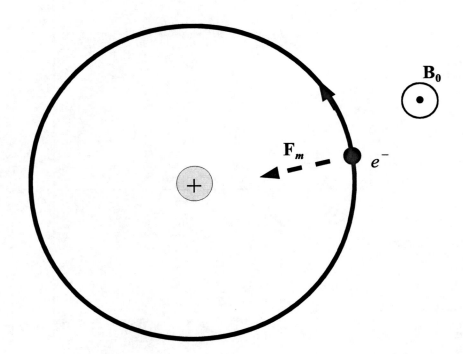

Figure IX.13: Classical rendition of an electron in an atom subject to magnetic forces from an external field $\mathbf{B}_0$.

slope of the curve in the figure is exaggerated for clarity, i.e., it corresponds to a material at very low temperature.) The curve asymptotically approaches a maximum magnetization value in both directions. This *saturation* condition corresponds to the nearly complete alignment of the atomic or molecular dipoles, beyond which further magnetization is not possible.

Diamagnetic materials

Diamagnetic materials are characterized by negative susceptibilities of small magnitude: $\chi_m \sim -1 \times 10^{-5}$ is a typical value. Bismuth, the most diamagnetic of the elements at room temperature, has susceptibility $\chi_m^{(\text{Bi})} = -2.8 \times 10^{-4}$. Diamagnetic materials are repelled from magnetic poles. Diamagnetism most closely resembles the behavior of dielectric materials.

Diamagnetic behavior is evident in materials with atoms or molecules possessing no intrinisic magnetic moments, because their electron spins are all paired. Magnetic effects on the orbital motion of the electrons *induce* magnetic moments that oppose the applied field. This is crudely sketched in Fig. IX.13. An applied magnetic field out of the page increases the inward force on the electron orbiting as shown. To maintain its orbit, the electron must speed up, giving rise to an increased magnetic moment into the page. An electron orbiting in the opposite direction would have to slow down, decreasing the magnetic moment directed out of the page. Even if the two electrons were paired, giving zero net moment, both changes would engender an induced moment into the page, opposing the applied field. This yields a negative susceptibility and diamagnetic character.

This diamagnetic induction effect occurs in all atoms. But in those with unpaired electron spins, it is overwhelmed by the effect of lining up the intrinsic magnetic moments, yielding paramagnetic character.

There is an exception to the rule that diamagnetic susceptibilities are small. Superconducting materials behave as perfect diamagnets: They completely exclude magnetic fields from their interiors, which corresponds to susceptibility $\chi_m = -1$. This is termed the *Meissner*[36] *effect* or *Meissner-Ochsenfeld*[37] *effect*, discovered by Meissner and Ochsenfeld in 1933. A sufficiently strong field—the critical value is a characteristic of the material—can force itself into a superconductor. In such instance, the material ceases to be superconducting, just as if it had been heated above its critical temperature.

Ferromagnetic materials

The longest-known and most familiar magnetic materials, termed *ferromagnetic* from the Latin *ferrum,* iron, are also the most difficult to understand—because in these materials, magnetic effects are not small. Ferromagnetic materials have large positive susceptibilities; values in excess of 100,000 are possible. Moreover, the susceptibility depends on the temperature of the material and the applied magnetic field—not the instantaneous value, but the entire previous magnetic history of the material.

Like those in paramagnetic materials, the atoms in ferromagnets have unpaired electron spins and intrinsic magnetic moments. But in ferromagnets, strong interactions between the atoms—not magnetic forces, but much stronger, quantum-mechanical *exchange interactions*—cause the atoms to align their magnetic moments with those of neighboring atoms, forming *domains* trillions of atoms strong. Under the proper circumstances, the domains can be large enough to be seen under a microscope. In unmagnetized material, the different domains have random magnetic orientations. They are separated by *domain walls* a few hundred atoms in thickness, within which the orientation of the atomic moments shifts from that in one domain to that in another. This is sketched in Fig. IX.14. An applied magnetic field rotates nearly aligned atoms into closer alignment with the field. This rotates favorably oriented domains and shifts the domain walls so that favorably aligned domains gain atoms at the expense of those far from alignment with the field. The result is a large net alignment or magnetization effect. Sufficiently large applied fields can also "flip" unfavorably aligned domains.

---

[36] Fritz Walther Meissner, 1882–1974
[37] Robert Ochsenfeld, 1901–1993

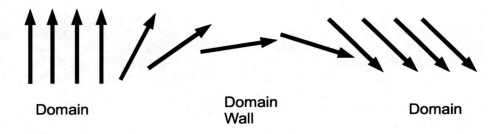

Figure IX.14: Schematic representation of ferromagnetic domains with aligned atomic magnetic moments. The orientation shifts from that of one domain to that of another across the thin domain wall.

The alignment of the atoms in this case is not in competition with their thermal motion, as the interatomic forces have already aligned the atoms in their domains. Of course, at high enough temperatures thermal agitation must win: Each ferromagnetic material has a characteristic *Curie temperature,* above which the interatomic forces are overcome and the material is paramagnetic, below which domains form and ferromagnetic character appears. The change is a first-order phase transition, like the melting or freezing of a solid.

The magnetization curve for a ferromagnetic material is sketched in Fig. IX.15. If an external magnetic field (or magnetic intensity) is applied to the unmagnetized material, the magnetization increases from the origin toward the saturation value. If the applied field is reduced, the magnetization does not retrace the middle portion of the curve. Rather, it decreases more slowly along the upper portion of the curve, as indicated. Even when the applied field returns to zero, the magnetization does not: It retains *remnant magnetization B_r,* as shown. The applied field must be driven to negative, i.e., opposite values to force the magnetization back to zero, then toward saturation in the opposite direction. If the field is then increased toward zero and positive values, the magnetization traces the lower portion of the curve, retaining remnant value $-B_r$ at zero field, until forced back to zero by positive values of B_0 or H. If the applied field is then cycled back and forth, the magnetization continues to trace the upper and lower portions of the curve.

Such a curve is called an *hysteresis loop,* from the Greek word for "lag": Changes in magnetization lag behind the changes in the applied field. The magnetization curve is not the graph of a function. The value of $B - B_0$ or M does not depend only on the value of B_0 or H, but also on the previous magnetic history of the material. The material "remembers" its previous magnetization. The metaphor is apt, as this hysteresis phenomenon is the basis for all magnetic information-storage devices or "memory"—wires, audio and video tape, disks floppy and hard, at one time arrays of tiny iron rings interlaced by wires to magnetize the rings and to read the stored magnetization. It is also, of course, the basis of permanent magnetization, the original magnetic phenomenon.

Ferromagnetic materials fall into two classes, *hard* and *soft.* Hard ferromagnets have "fat" hysteresis curves, with remnant magnetization a large fraction of the saturation value. Such materials are used for permanent magnets. Soft ferromagnets have much narrower hysteresis curves, with much smaller values of B_r. Such materials are useful in devices such as electric transformers, in which the material is cycled repeatedly through its hysteresis loop. It takes work to do that, work that is lost as heat dissipated in the material. It is shown in more advanced treatments that this work is proportional to the *area* of the loop. Hence, soft ferromagnets—the limiting cases are the *superparamagnets,* with magnetization curves almost as narrow as Fig. IX.12—are used to minimize such losses.

The scales in Figs. IX.12 and IX.15 are roughly correct. The saturation magnetization values of paramagnetic and ferromagnetic materials are similar: Their intrinsic atomic moments are comparable, and full alignment is the same for both. It is the strong interatomic exchange forces that give rise to the

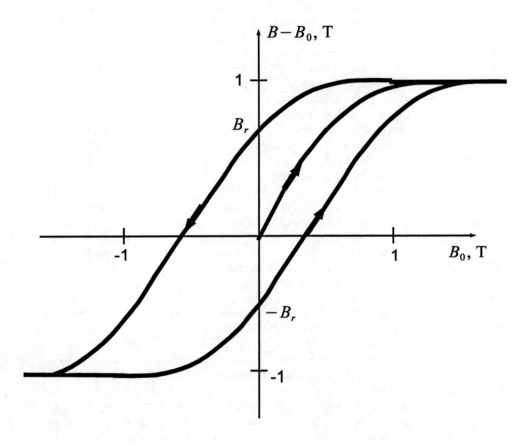

Figure IX.15: Magnetization curve for a ferromagnetic material.

distinctive behavior of ferromagnets.

There are materials in which interatomic forces align intrinsic atomic *electric* dipole moments in domains. Such materials are called *ferroelectric*; they exhibit hysteresis, memory, and permanent polarization. Pieces of such material, when polarized, are called *electrets,* the electric analog of magnets.

Antiferromagnetic materials

A fourth class of materials is termed *antiferromagnetic*. In such materials, exchange forces align the atoms in domains, with their intrinsic moments not aligned but antiparallel to their neighbors. That is, the orientations of adjacent moments alternate through the domain.

References

All introductory physics texts devote one or more chapters to magnetism, to the effects and sources of magnetic fields, though details vary. The magnetic properties of matter are described more thoroughly in intermediate and advanced texts on electromagnetism and solid-state physics. Reitz, Milford, and Christy [RMC08] and Kittel [Kit04] provide intermediate-level treatments. Goodstein [Goo75] and Landau, Pitaevskii, and Lifshitz [LPL84] describe these properties at advanced levels. These texts also discuss nuclear magnetic resonance, which has its own extensive literature.

The gyromagnetic ratio of the electron is a topic reserved for advanced-level texts in Einsteinian quantum mechanics and quantum field theory. Bjorken and Drell [BD64, BD65] are classic references; the magnetic moment of the electron is discussed in the first volume. Landau, Lifshitz, and Pitaevski [LLP90], Volume 4 of the *Course of Theoretical Physics,* is another such classic text. The anthology *Selected Papers on Quantum Electrodynamics,* edited by Julian Schwinger [Sch58], contains some of the seminal papers on the subject, including Schwinger's original 1948 paper on the electron magnetic moment.

Problems

1. Calculate the ratio of the magnitudes of the magnetic and electric forces that one particle exerts on another when both—with identical charges q and masses m—are traveling side by side on parallel straight lines, at speed v much less than that of light, separated by distance d.

   (a) The ratio $|\mathbf{F}_m|/|\mathbf{F}_e|$ is proportional to v^2. The constant of proportionality must be the inverse of a squared speed, since the ratio is dimensionless. What speed?

   (b) What are the implications of result (a)?

2. A cyclotron with dees of diameter 1.500 m accelerates protons to kinetic energy 100.0 MeV before they are extracted at the outer edge of the dees. Assume Newtonian mechanics is in play throughout.

   (a) What magnetic field is required?

   (b) What frequency AC voltage must be applied to the dees?

3. Although our classical model of the electron did not fare so well in this chapter, consider a classical model of the hydrogen atom, in which an electron orbits a proton in a circular orbit of radius $a_0 = 5.29 \times 10^{-11}$ m. Take the plane of the orbit to be perpendicular to the spin angular momentum of the proton. The spin angular momentum of the electron is allowed to be either parallel or antiparallel to that of the proton. Both particles have spin angular-momentum components along their spin axes equal to $\hbar/2$, where $\hbar$ is Planck's reduced constant. Calculate the energy difference between these two configurations, owing to the magnetic interaction between the spin magnetic moment of the electron and the magnetic field produced by the spin moment of the proton. Disregard the magnetic effects of the electron's orbital motion around the proton. What is the significance of this *hyperfine splitting* of the ground state of the electron in the hydrogen atom?

4. The Earth's magnetic field is approximately a dipole field, with dipole moment $\mu_\oplus = 8.0 \times 10^{22}$ J/T.

   (a) The source of this field is not fully understood. It almost certainly is not permanent magnetism, as temperatures in the Earth's interior are higher than the Curie temperatures of ferromagnetic materials. The circulation of ions in the Earth's molten outer core is a possible source. Treating this as a circular current loop and looking up the necessary data, estimate the current required.

   (b) Assume that the Earth's dipole moment is (anti-)parallel to its spin axis, i.e., pointing from the North Pole to the South Pole, and assume that the field is centered on the center of the Earth. That is, neglect the *declination* of the Earth's field. Calculate the *magnitude* and *dip angle* (angle below the horizontal) of the magnetic field at the Earth's surface at: (1) the equator; (2) the North Pole; and (3) St. Louis, Missouri, latitude 38°38′ North. Treat the Earth as a perfect sphere of radius 6371 km.

5. The two coils enclosing the e/m tube in Fig. VII.1 are called *Helmholtz coils*. These are identical, coaxial circular coils in parallel planes *separated by a distance equal to their radius*. They carry the same current in the same direction. Helmholtz coils are designed to produce a very uniform magnetic field in the space between the coils. Calculate the on-axis field of such a pair of coils in the vicinity of the point midway between the coils in terms of the coil radius R, the number of turns N in each coil, and the current I in each turn. Express your result as a Taylor series in

the displacement x along the axis from the midpoint carried out to *fourth order* in x.

6. Why did our classical electron model fail so spectacularly? Using the classical radius of the electron calculated in Ch. VII, calculate the angular speed ω required to give a uniform-density, solid-sphere classical electron spin angular momentum $S_z^{(e)} = \hbar/2$, as indicated by quantum mechanics. Then calculate the speed of a point on the "equator" of the spherical electron, spinning at this rate. What difficulty does this present?

7. Nuclear Magnetic Resonance (NMR) and Electron Paramagnetic Resonance (EPR) are based on the fact that a particle with spin angular momentum and a magnetic moment, immersed in a magnetic field, will precess about the magnetic field.

    (a) Calculate the angular frequency of this precession in terms of these quantities.

    (b) A proton has spin angular-momentum component $S_z^{(p)} = \hbar/2$. The NMR spectrometer in a modern physics lab detects protons precessing at frequency (not angular frequency) $\nu = 15.35850$ MHz. This is the radio frequency at which resonance occurs, and it is known to this precision. Calculate the magnitude of the magnetic field in which the protons are immersed.

    (c) Calculate the precession frequency for *electrons* in the same magnetic field, i.e., in an EPR experiment using the same magnetic field.

8. Molecular hydrogen gas has magnetic susceptibility $\chi_m^{(H_2)} = -3.98 \times 10^{-6}$.

    (a) What is the spin configuration of the two electrons in the H_2 molecule?

    (b) What is the magnetic character of *atomic* hydrogen gas (H)?

9. A certain material consists of atoms with intrinsic magnetic moments of magnitude μ and number density per unit volume N. The atoms have two allowed states in an external magnetic field $\mathbf{B}_0$: with magnetic moment aligned with the field or antialigned. In accord with classical thermodynamics, the probability for an atom to be in a state of energy E at absolute temperature T is proportional to $\exp[-E/(k_B T)]$, where k_B is the Boltzmann constant, and the constant of proportionality is determined by the requirement that the probabilities of all allowed states sum to unity.

    (a) Calculate the magnetization (magnetic moment per unit volume) of this material in field $\mathbf{B}_0$ at temperature T.

    (b) Calculate the low-field limit ($|\mathbf{B}_0| \rightarrow 0$) of the magnetic susceptibility χ_m of this material as a function of temperature.

    (c) Lithium (Li) atoms contain a single unpaired electron (along with two electrons with paired, i.e., opposite spins). Assuming a lithium atom thus has the intrinsic dipole moment of a single electron, and looking up any other necessary data, estimate the magnetic susceptibility of lithium at temperature $T = 293.$ K.

10. Consider a classical model of a helium (He) atom: a nucleus consisting of two protons and two neutrons with two electrons orbiting in opposite directions in a circular orbit of radius $a_1 = 2.65 \times 10^{-11}$ m. The spins of the electrons are paired, i.e., the atom has no intrinsic magnetic dipole moment. Estimate the magnetic susceptibility of helium gas at 1.000 atm pressure and temperature $T = 293.$ K using this model. For simplicity, assume that the applied magnetic field is perpendicular to the plane of the electrons' orbit, and that the radius of the orbit does not change.

Chapter X

The Solenoid

- *Consider a long, cylindrical solenoid with a close-fitting ferromagnetic core, of relative permeability $K_m \gg 1$, which can be slid in and out. Assuming the solenoid is operated at constant current I, calculate the force exerted on the core by the solenoid.*

375

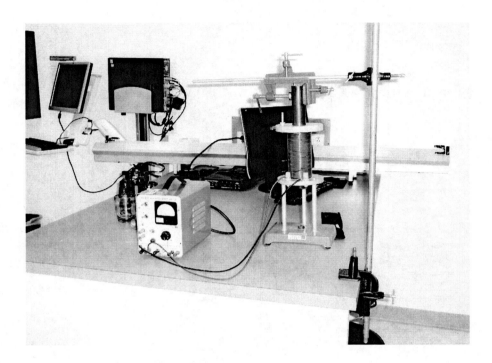

Figure X.1: This solenoid can exert a magnetic force on its iron core sufficient to lift its own weight. How can this force be calculated? (Photograph by author.)

From the fundamental properties of the electron, we turn to a more prosaic application: the solenoid as an electromechanical device. Introduced in Sec. E of the previous chapter, the straight solenoid can be used to produce an approximately uniform magnetic field. But it is far more commonly used to exert a force on a ferromagnetic core free to slide in and out of the solenoid, thus performing mechanical work. Switches operated in this way are called *solenoid relays*. These are used in the horn and headlight circuits of automobiles and other high-current circuits and formed the basis of some early digital computers. A solenoid pulls the gear that connects the starter motor of a car to the flywheel of its internal-combustion engine.[1]

The magnetic force exerted by a solenoid is easily demonstrated, as shown in Fig. X.1. Our goal in this chapter is to calculate this force from the specifications of the solenoid. This will introduce an important and general technique for treating complicated forces. The calculation requires the application of another

---

[1] Few drivers could safely start a modern car with a hand crank, as was done a century ago.

relationship between electric and magnetic fields, the final piece of the classical electromagnetic puzzle.

A   The Faraday Law of Induction

The great British experimentalist Michael Faraday and his American counterpart Joseph Henry[2] discovered this law independently.[3] It describes the observation that *a changing magnetic field can produce an electric field.* This parallels the effect of the displacement current (IX.22c)—a changing electric field producing a magnetic field. But Maxwell introduced the displacement current two decades after the discovery of Faraday *induction,* or *electromagnetic induction,* as the process described here is known.

Induced EMF

Faraday and Henry did not originally describe the law of induction in terms of magnetic and electric *fields.* In fact, much of our modern vector-field mathematics and notation was invented later in the nineteenth century to treat electromagnetic theory. Instead, the law was originally couched in terms of *electromotive force* $\mathcal{E}$, e.g., the EMF driving a current in a loop, and *magnetic flux* $\Phi_{\mathbf{B}}$. The Faraday law states that the *induced* EMF around a loop C is equal to the negative of the *time rate of change* of the magnetic flux through a surface S bounded by the loop. This can be written

$$\mathcal{E}_C = -(N)\,\frac{d\Phi_{\mathbf{B}}(S)}{dt}\;, \qquad\qquad (\text{X.1a})$$

or in modern vector notation,

$$\oint_C \mathbf{E} \cdot d\boldsymbol{\ell} = -(N)\,\frac{d}{dt}\int_S \mathbf{B} \cdot d\boldsymbol{\mathcal{A}}\;. \qquad\qquad (\text{X.1b})$$

In the latter form, the parallel with the Ampère-Maxwell law (IX.23) is clear: The EMF $\mathcal{E}_C$ is the circulation of the electric field $\mathbf{E}$, while the rate of change of magnetic flux is the counterpart of the displacement current. The factor N, the number of times around loop C used to calculate $\mathcal{E}_C$, appears here because the Faraday law is often applied to a coil of wire, say, that circles the loop multiple times. (Some older texts incorporate the factor N into the definition of magnetic flux, but it is clearer to define the flux as the straightforward surface integral and display N explicitly.) As with the Ampère-Maxwell law, however, it is not *necessary* that C correspond to a material object. It can be any smooth— piecewise C^1, in mathematical parlance—closed curve in space. The negative

---

[2] Joseph Henry, 1797–1878

[3] Nowadays the law might be called the "Faraday-Henry law," but at the time, Faraday enjoyed far greater prestige. In the nineteenth century, the United States was a scientific backwater compared with Great Britain. Still, Henry did all right: He has an SI unit named for him, as described in Sec. B following.

sign, which contrasts with the Ampère-Maxwell law, will be seen to be of great importance.

This law reveals a novel feature. The circulation of $\mathbf{E}$ around any closed curve should vanish for a conservative force field, as described in Ch. II Sec. E. Apparently, this induced electric field is *non*conservative! The Law of Conservation of Energy is not superseded; apparently, some other energy contribution is in play in this case.

The magnetic flux $\Phi_{\mathbf{B}}$ is encountered often enough, and is of sufficient importance, that it has its own units. The SI unit of magnetic flux is the *weber* (Wb), named for Wilhelm Weber[4] and defined as

$$1 \text{ Wb} \equiv 1 \text{ T m}^2$$
$$= 1 \times 10^8 \text{ G cm}^2$$
$$\equiv 1 \times 10^8 \text{ maxwell} . \qquad \text{(X.2)}$$

The last two lines define the maxwell, the unit of magnetic flux in the Gaussian system. In fact, in some texts, the magnetic field is termed the *flux density,* measured in units of Wb/m^2 or maxwells per square centimeter. The flux represents a count of field lines; these are regarded in older treatments almost as physical objects, although from a modern viewpoint they are seen simply as a graphical device to depict the vector field.

Motional EMF

An important class of phenomena featuring Faraday induction is that of conductors in motion in magnetic fields. The EMF produced in such a situation is called *motional EMF.* Much of the electricity we encounter on an everyday basis is driven by motional EMF produced by electric generators.

A somewhat artificial but simple example of motional EMF is illustrated in Fig. X.2: A metal bar sliding frictionlessly on metal rails, moving at constant velocity $\mathbf{v}$, the entire apparatus immersed in a constant, uniform magnetic field $\mathbf{B}$. For maximum simplicity, the magnetic field, the bar, and its velocity are taken to be mutually perpendicular. The charge carriers in the bar are subject to magnetic force

$$\mathbf{F}_m = q\,\mathbf{v} \times \mathbf{B} , \qquad \text{(X.3)}$$

which would drive positive charge carriers upward or negative ones downward in the figure. If switch S is open, then the separation of charge in the bar will simply give rise to a compensating electric field, with value

$$\mathbf{E}_{\text{comp}} = -\mathbf{v} \times \mathbf{B} \qquad \text{(X.4)}$$

at equilibrium. This field corresponds to a potential difference (EMF) between the ends of the bar, given by

$$\mathcal{E} = |\mathbf{E}_{\text{comp}}|\,\ell$$
$$= \ell\,|\mathbf{v}|\,|\mathbf{B}| \qquad \text{(X.5)}$$

---

[4]Wilhelm Eduard Weber, 1804–1891

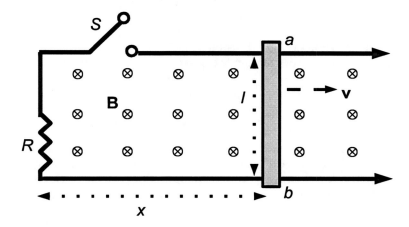

Figure X.2: Motional EMF: A conducting bar slides on conducting rails at constant velocity **v**, immersed in uniform magnetic field **B** directed into the page. The bar has length ℓ and resistance r. The rails completing the loop, and any associated load, have resistance R. Points a and b denote the ends of the bar where they make contact with the rails.

in the geometry shown. If the switch is closed, current I will flow counterclockwise in the loop formed by the bar and rails. The potential difference between the ends of the bar will be reduced by the voltage drop along its length to value

$$V_{ab} = \mathcal{E} - Ir \ . \tag{X.6a}$$

This potential difference drives the current through the rails and load, satisfying

$$V_{ab} = IR \ . \tag{X.6b}$$

These imply

$$I = \frac{\mathcal{E}}{R + r} \ , \tag{X.6c}$$

suggesting that the induced EMF $\mathcal{E}$ can be treated as an ordinary EMF, applied to the entire circuit at a (any) point.

What actually does the work on the circulating charge carriers in the closed circuit? The magnetic field $\mathbf{B}$ cannot. To see what does, a more detailed description of the charges' motion is needed. Their (average) velocity is actually not $\mathbf{v}$, but $\mathbf{v} + \mathbf{v}_d$, where $\mathbf{v}_d$ is the drift velocity associated with the current. Hence, the charges are subject to magnetic force

$$\mathbf{F}_m = q\,(\mathbf{v} + \mathbf{v}_d) \times \mathbf{B} \ . \tag{X.7}$$

The smaller $q\mathbf{v}_d \times \mathbf{B}$ term would, if unopposed, drive the charge carriers horizontally out of the bar. But charge carriers are bound in conducting materials by strong electrostatic forces: An electric field $\mathbf{E}_{\text{hold}} = -\mathbf{v}_d \times \mathbf{B}$ would arise to hold the carriers in the bar. This field does work on a charge carrier at rate

$$\begin{aligned}
P_{\text{hold}} &= q\,\mathbf{E}_{\text{hold}} \cdot (\mathbf{v} + \mathbf{v}_d) \\
&= -q\,(\mathbf{v}_d \times \mathbf{B}) \cdot \mathbf{v} \\
&= \mathbf{v}_d \cdot (q\mathbf{v} \times \mathbf{B}) \ ,
\end{aligned} \tag{X.8}$$

where the "box product" of three vectors in the second line has been written as a determinant of components and its rows suitably permuted. This is exactly the rate corresponding to EMF $\mathbf{E}$ and current I, i.e., the rate at which force (X.3) would do work on the flowing charge carriers if it could. In fact, it is the electric field $\mathbf{E}_{\text{hold}}$, not the magnetic field $\mathbf{B}$, that does all the actual work of driving the current!

Where does the energy come from? When the current is flowing, the magnetic field exerts force

$$\mathbf{F}_{\text{react}} = I\,\boldsymbol{\ell} \times \mathbf{B} \tag{X.9a}$$

on the bar, where $\boldsymbol{\ell}$ is the vector length of the bar, extending from point b to point a in the direction of the current. This force is directed to the left, opposing the motion of the bar. To maintain its constant velocity, external force $\mathbf{F}_{\text{ext}} = -\mathbf{F}_{\text{react}}$ must be applied to the bar. This does work at rate

$$
\begin{aligned}
P_{\text{ext}} &= -I \left(\boldsymbol{\ell} \times \mathbf{B} \right) \cdot \mathbf{v} \\
&= I \boldsymbol{\ell} \cdot (\mathbf{v} \times \mathbf{B}) \\
&= I \mathcal{E} \\
&= I^2 (R + r) \,,
\end{aligned}
\qquad \text{(X.9b)}
$$

where again the box product in the first line has been suitably permuted. Hence, all the electrical energy dissipated in the circuit is supplied by the source of the external force $\mathbf{F}_{\text{ext}}$.

How does the Faraday law apply to this situation? The loop in question is the rectangle formed by the rails and the moving bar. If the counterclockwise direction of the current is taken to be positive, then by the right-hand rule the positive direction for flux through the rectangle is out of the page. Hence, the negative of the rate of change of the magnetic flux is given by

$$
\begin{aligned}
-\frac{d\Phi_{\mathbf{B}}}{dt} &= -\frac{d}{dt}(-\ell x |\mathbf{B}|) \\
&= \ell \frac{dx}{dt} |\mathbf{B}| \\
&= \ell |\mathbf{v}| |\mathbf{B}| \\
&= \mathcal{E} \,,
\end{aligned}
\qquad \text{(X.10)}
$$

exactly in accord with the Faraday law.

Induction without motion

In situations in which the loop C, whether it corresponds to a physical object or not, is not in motion, the Faraday law takes the form

$$
\oint_C \mathbf{E} \cdot d\boldsymbol{\ell} = -(N) \int_S \frac{\partial \mathbf{B}}{\partial t} \cdot d\boldsymbol{\mathcal{A}} \,.
\qquad \text{(X.11)}
$$

A time-dependent magnetic field by itself can serve as a source of an electric field.

Generators and motors

Perhaps the most widespread applications in which electromagnetic induction is encountered are electric generators and motors. Remarkably, these are the same devices operated "in the opposite order." The design of generators and motors is a complex and highly developed field of engineering. Only the basics will be described here. A very simple version is sketched in Fig. X.3: A rectangular coil of N turns of (insulated) wire turns on an axle between the poles of a magnet, which can be either a permanent magnet or an electromagnet. The coil is connected into an electric circuit via, say, carbon blocks called *brushes* attached to the leads of the coil, sliding against conducting rings as shown.

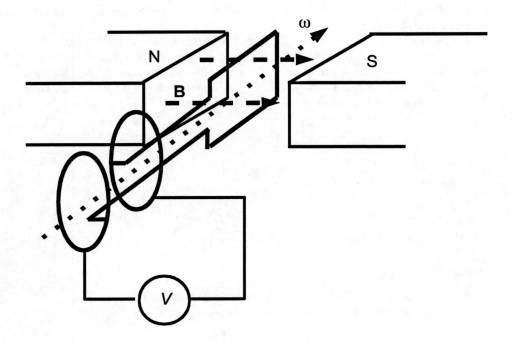

Figure X.3: A simple version of an electric generator or motor. The rectangular coil turns on its axle in the field of the magnet. Electrical contact with the coil is made through carbon brushes sliding along the conducting rings.

For a *generator,* the input to the device is external torque to spin the coil on its axle, and the output is the induced EMF. The torque can be provided by a hand crank, a spinning bicycle tire, an automobile engine, a windmill, a waterfall, a steam turbine,[5] et cetera. If the coil has area $\mathcal{A}$ and rotates with angular speed ω—so that the angle between the (positive) normal to the coil and the magnetic field is ωt—then the magnetic flux through the coil is

$$\Phi_{\mathbf{B}} = \mathcal{A}B \cos(\omega t) \ , \tag{X.12a}$$

where B is the magnitude of the magnetic field, assumed uniform here. The Faraday law indicates that the EMF induced in the coil and supplied to the circuit is

$$\mathcal{E} = N\mathcal{A}B\omega \sin(\omega t) \ . \tag{X.12b}$$

This is an alternating-current (AC) voltage.[6] The generator is an *alternator,* like those that supply electrical power for automobile engines.

If the magnetic field of the generator is supplied by electromagnets, these *stator coils* can be operated off the output of the generator itself, in series or in parallel with the rotating (rotor) coil, once the generator is started from an external source. In contrast, a generator with magnetic field provided by permanent magnets is called a *magneto.* These are used, for example, in airplane engines, since they can operate in the event of complete battery failure.[7]

In operation, the generator rotor is a current-carrying coil in a magnetic field, hence, subject to a magnetic torque. This torque never assists the rotation of the coil. If it did, the generator could be made to turn itself, supplying energy to the circuit with no source—in violation of the Law of Conservation of Energy. Instead, the magnetic torque always opposes the coil's rotation. The external source of torque for the generator must do work against the magnetic torque on the rotor. This work appears as the energy supplied to the electrical circuit by the generator, plus any work lost to friction in the generator bearings or elsewhere.

For a *motor,* the input to the device is an external EMF, and the output is mechanical torque on the magnetic dipole moment of the rotor coil. It is necessary to reverse the direction of the current through the coil each time its magnetic moment turns into alignment with the magnetic field (or to flip the field) to keep the motor turning. This might be done with AC input to the coil or with a device called a *commutator,* which reverses the connections to the coil every half turn. For example, if the connector rings in the figure were split into semicircles, left and right, with the left-hand semicircles connected to one side of the circuit and the right-hand semicircles to the other, this *split-ring commutator* would reverse the coil current as required. The repeated breaking

---

[5]Coal, oil, natural gas, and nuclear power plants the world over differ, essentially, only in the heat source that boils the water for the turbines that turn their generators.

[6]Generators supplying a network or grid must be operated so that their voltages are in the proper phase relationships, i.e., their cycles are synchronized appropriately. That is no small engineering task.

[7]This is a safety feature: If your car engine stops, you stop moving. If your airplane engine stops, you fall out of the sky.

and making of the connections gives rise to the sparks often seen at the brushes of an electric motor in operation.

As the rotor turns, the motor simultaneously acts as a generator. The EMF induced in the rotor coil, called *back EMF*, never assists the flow of current through the rotor, always opposes it—else the motor could be made to turn itself without external input. The external EMF source must do work against this back EMF to maintain the rotor current, work that appears as the mechanical work done by the motor. The faster the motor turns, the greater the back EMF and the less current driven by a given input EMF. The equilibrium speed of the motor is that at which the difference between the input EMF and the back EMF is just sufficient to drive the current through the ordinary resistance of the rotor coil. This is why household lights are seen to dim momentarily when the compressor motor of a refrigerator or air conditioner starts and its current draw is greatest. A motor is at greatest risk of burning out not when it is turning fastest, but if its bearings seize and it is stopped in operation.

As for a generator, the magnetic field for a motor can be provided by permanent magnets or by electromagnets—stator coils, which can be in the same circuit as the rotor coil. If these are in series with the rotor, the motor is a *series motor*; if in parallel, the motor is a *shunt motor*. Different designs are used to produce different operating characteristics, e.g., maximum torque at low or at high speed.[8]

The Lenz Law

The Lenz law is a generalization about the *direction* of electromagnetic induction effects. It is illustrated in all the examples described in this section: The magnetic force on the sliding bar opposes its motion; the current driven through the bar and rails produces a magnetic field through the rectangle opposing the external field (for the direction of motion shown); the torque on a generator rotor opposes its rotation; the EMF induced in a motor coil opposes the current through the coil. The Lenz law states that *induction effects always oppose the changes that produce them*. Reminiscent of the Le Châtelier Principle of Spite in chemistry, the Lenz law is in fact a facet of the Law of Conservation of Energy: If induction effects instead *reinforced* the changes that produced them, they could drive themselves endlessly once started, and limitless work and energy could be obtained with no input.

The Lenz law is not an independent law. The negative sign in the Faraday law, along with the convention relating the positive directions for circulation and flux, will assign the same directions to induction effects as the Lenz law in every case. That negative sign is more significant than it appears. Though not apparent at present, it is intimately connected with the structure of spacetime— with our experiences of past and future, cause and effect. As that negative sign

---

[8]Such requirements account for the use of *diesel-electric,* rather than diesel, engines in modern railroad locomotives. A diesel engine turns an electric generator, which powers electric motors attached to the axles of the locomotive in order to produce the torque necessary to set a massive train into motion without stalling.

incorporates the Lenz law, a consequence of energy conservation, the connection between energy conservation and the fundamental nature of existence begins to emerge.

B   Inductance

Inductance describes the magnetic effects of elements in electric circuits. The Lenz law indicates that these effects constitute a "magnetic inertia," i.e., resistance to change.

General definition

The *inductance* of circuit elements has an elegant general definition. Consider a general circuit with various loops, coils, and components—including, perhaps, a loop around the entire circuit—around which EMF can be calculated and through which magnetic flux can be determined. Let these elements be numbered by an index. Current in any element can create a magnetic flux through all the elements, including itself. Each of these flux contributions is proportional to the current that produces it. Hence, the total flux through element i is given by

$$(N_i)\,\Phi_{\mathbf{B}}^{(i)} = \sum_j L_{ji}\,I_j \; , \tag{X.13a}$$

where N_i, the number of times around loop i, is included for later convenience. The I_j are the currents through all the elements, including element i, and the coefficients L_{ji} (the order of indices is traditional) are constants dependent only on the geometry and nature of the elements. Coefficients with $j \neq i$ are called *mutual inductances* and—defined as here—are indeed mutual, satisfying $L_{ji} = L_{ij}$. The coefficients with $j = i$ are the *self inductances*, or (where mutual inductance is not in play) simply *inductances* of the loops or elements. In simple cases where only a single element is affected by significant magnetic flux, the sum can be reduced to a single term

$$N\,\Phi_{\mathbf{B}} = LI \tag{X.13b}$$

involving only self-inductance L.

Inductance has its own units. Relations (X.13) imply that the SI units of inductance are

$$\begin{aligned}
1\,\frac{\mathrm{T\,m^2}}{\mathrm{A}} &= 1\,\frac{\mathrm{Wb}}{\mathrm{A}} \\
&= 1\,\frac{\mathrm{N\,m}}{\mathrm{A^2}} \\
&= 1\,\frac{\mathrm{J}}{\mathrm{A^2}} \\
&\equiv 1\,\mathrm{H} \quad \text{(henry)} \; ,
\end{aligned} \tag{X.14}$$

named for Joseph Henry.

Inductance of long solenoid

A useful example illustrating the magnitude of the henry is the (self) inductance of a long, straight solenoid. Neglecting end effects—because most of the turns of the solenoid are far from the ends—the magnetic field in the solenoid is given by result (IX.20b): $B = \mu n I$, allowing for the presence of a magnetic core with nonvacuum permeability μ, as in Eq. (IX.30), and with n the number of windings per unit length of the solenoid. The magnetic flux through the entire solenoid is given by

$$
\begin{aligned}
N\,\Phi_{\mathbf{B}} &= (n\ell)(\mu n I)\,\mathcal{A} \\
&= \mu n^2 \ell \mathcal{A}\, I \ ,
\end{aligned}
\tag{X.15a}
$$

with ℓ the length and $\mathcal{A}$ the cross-sectional area of the solenoid. Compared with Eq. (X.13b), this implies inductance

$$
\begin{aligned}
L &= \mu n^2\, \ell \mathcal{A} \\
&= \mu n^2\, \mathcal{V} \ ,
\end{aligned}
\tag{X.15b}
$$

where $\mathcal{V} = \ell\mathcal{A}$ is the *volume* of the solenoid. This result illustrates why *inductors*—circuit components designed specifically for their inductance—unlike resistors, capacitors, diodes, and transistors, are not etched into integrated-circuit (IC) chips, but remain discrete components: Inductance requires physical volume.

For example, consider an empty solenoid with length $\ell = 1.00$ m and radius $r = 10.0$ cm (a stove pipe), closely wound with 1.00 mm-diameter wire, i.e, with $n = 1.00 \times 10^3$ m^{-1}. This device—which actually requires a great deal of wire—has inductance

$$
\begin{aligned}
L &= \mu_0 n^2 \ell (\pi r^2) \\
&= (4\pi \times 10^{-7}\ \text{N/A}^2)(1.00 \times 10^3\ \text{m}^{-1})^2(1.00\ \text{m})\pi(0.100\ \text{m})^2 \\
&= 39.4\ \text{mH} \ .
\end{aligned}
\tag{X.16}
$$

Inductances much larger than this can be obtained in much smaller devices: A one-henry inductor can be made from a close-wound coil (with multiple layers of wire) with an iron core, which can be held in one hand. Likewise, a thirty-henry inductor can be made from a close-wound coil with many layers and an iron core the size of a large cinnamon bun but rather heavier.

A straightforward calculation along the lines of that for Eqs. (X.15) can give the mutual inductance, say, of two coaxial solenoids, one inside the other, and show that their mutual inductances are equal. Such a calculation will not be explored further here.

Inductors in circuits

The importance of inductors in circuits arises from the EMFs induced when *time-dependent* currents flow through the circuit elements. The Faraday law and definition (X.13a) imply that in general, the induced EMF in element i is

$$\mathcal{E}^{(i)} = -\sum_j L_{ji} \frac{dI_j}{dt} \ . \tag{X.17a}$$

For a single element, this takes the form

$$\mathcal{E}_L = -L \frac{dI}{dt} \ , \tag{X.17b}$$

with only self inductance in play.

Inductor voltage (X.17b) can be compared with the voltages across a resistor and a capacitor:

$$\mathcal{E}_R = IR \tag{X.18a}$$

and

$$\begin{aligned} \mathcal{E}_C &= \frac{Q}{C} \\ &= \frac{1}{C} \int_0^t I(t') \, dt' \ , \end{aligned} \tag{X.18b}$$

say. These voltages are proportional to the derivative of the current $I(t)$, the current itself, and its time integral, respectively. Hence, by setting up an appropriate circuit and measuring voltages across certain components, it is possible to perform calculus operations on an input signal. These are examples of *analog computation,* a form of electronic computing distinct from the *digital computation* of switches and logic gates. As late as the 1970s analog and digital techniques competed for the future of computing. Subsequent advances in digital electronics have made the digital approach ubiquitous, although analog computers are still used for specialized applications.

A circuit in which resistance and inductance both contribute is called an *LR circuit,* in comparison to the RC circuits of Ch. VIII Sec. G.[9] A simple LR circuit is sketched in Fig. X.4. The symbol for inductor L is a stylized representation of a solenoid, although the actual component may be different. Inductor L and resistor R_1 are represented here as discrete *lumped components,* although inductance and resistance would be combined attributes of any real device. If switch S is closed at time $t = 0$, say, current I_L will begin to flow clockwise through the inductor branch of the circuit; a separate current will flow through resistor R_2. The Kirchhoff loop rule, applied to the outer loop of the circuit, implies

$$V_0 - I_L R_1 - L \frac{dI_L}{dt} = 0 \ , \tag{X.19a}$$

where the sign of the derivative term accords with the Lenz law—the induced EMF opposes the increasing current I_L. Applying the loop rule calls for some

---

[9]Circuits in which inductance and capacitance both contribute are called *LC circuits*; if all three attributes are in play, the circuit is an *LRC circuit.* These circuits exhibit oscillatory behavior, involving *alternating-current* (AC) currents and voltages. This important branch of circuit theory must be left to another text.

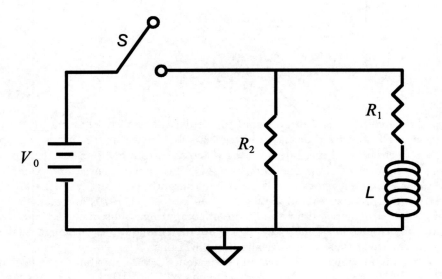

Figure X.4: A simple LR circuit.

consideration, because the loop rule is the assertion that electric potential is well-defined in the circuit. But since induced electric fields are nonconservative, the potential is *not* well-defined everywhere in this circuit. However, as was seen in the bar-and-rails example, induced EMF can be treated as applied to the circuit at a point. Hence, if we regard the inductor as a point and refrain from asking exactly how the potential might be distributed within it, loop-rule result (X.19a) can still be used. This is akin to using the junction rule in RC circuits—valid as long as we treat capacitors as point components and do not consider what happens at the capacitor plates, where charges can pile up or drain away.[10] This relation can be rearranged into the differential equation

$$\frac{dI_L}{dt} + \frac{R_1}{L} I_L = \frac{V_0}{L} \ , \tag{X.19b}$$

or

$$\frac{d}{dt} \left(\frac{V_0}{R_1} - I_L \right) = -\frac{R_1}{L} \left(\frac{V_0}{R_1} - I_L \right) \ . \tag{X.19c}$$

As with the capacitor-charging circuit of Fig. VIII.9 and Eqs. (VIII.28) and (VIII.29), this implies that the quantity in parentheses varies exponentially in time. A little more rearrangement yields the solution

$$I_L(t) = \frac{V_0}{R} \left[1 - \exp\left(-\frac{R_1}{L} t \right) \right] \ , \tag{X.19d}$$

with $I_L(0) = 0$. The current through the inductor does not switch on instantaneously—that would give rise to an infinite back EMF. Rather, it builds up asymptotically to final value V_0/R_1. The difference between I_L and this value decreases by factor $e^{-1} \doteq 36.8\%$ every *LR time constant* $\tau_{\mathrm{LR}} \equiv L/R_1$. This solution suggests a useful trick for treating LR circuits and networks: At the instant the circuit is switched on, the inductors draw no current and can be treated as open switches. After a long time (compared to τ_{LR}), the inductors can be treated as closed switches—the currents stop changing and the induced EMFs vanish.

If switch S is opened again after a long time $t_O \gg L/R_1$, current will continue to flow through the right-hand loop of Fig. X.4. The Kirchhoff loop rule applied to this loop yields

$$-L \frac{dI_L}{dt} = I_L (R_1 + R_2) \ , \tag{X.20a}$$

where the induced EMF now *drives* I_L, opposing the decrease in the current. This has solution

$$I_L(t) = \frac{V_0}{R_1} \exp\left(-\frac{R_1 + R_2}{L} (t - t_O) \right) \ , \tag{X.20b}$$

---

[10]In fact, current flow is continuous even within capacitors if displacement current is included in combination with ordinary current. A comparable modification can restore rigor to the loop rule even when induced fields are involved, but such a refinement will not be needed here.

with $I_L(t_O)$ approximated by V_0/R_1. The current in the inductor does not switch off immediately, with an infinite rate of decrease—it decays exponentially to zero, with time constant appropriate to the new configuration of the circuit.

Resistor R_2 is in the circuit to allow this exponential decay to occur. It is called a *bleeder resistor*. Without it, opening the switch would cause a large induced "spike" in voltage and a spark. Sometimes this is desired: Automobile engines used to use a *distributor,* in which a rotor, turned by the engine, would repeatedly make and break contact with metal *points*. This would close and open a circuit with the car battery, a coil, and the spark plug in each cylinder in the appropriate sequence. The resulting spark at each spark plug fired the fuel/air mixture in each cylinder's power stroke. The same function is performed by electronic ignition in modern engines, eliminating the need to replace the points regularly.[11] In other applications involving large inductances, however, one is cautioned to *do nothing quickly*: Circuits are turned on and off using variable resistors adjusted slowly to avoid creating dangerously large induced voltages.

A spectacular demonstration of this took place in 1978 at Michigan State University.[12] A large electromagnet for a new cyclotron was being tested, drawing some hundreds of amperes of current. The magnet had its own power supply controlled by what was then called a microcomputer. The magnet could not simply be switched off, of course. It was shut down by allowing the current to decay away through a bleeder resistor that had to be water-cooled. Through some mishap, the cooling water boiled away, and the bleeder resistor melted, opening the circuit. The resulting voltage spike created an arc to ground something like a lightning bolt. It blasted a hole the size of two fists in a steel cabinet door—one could see the steel that had been melted around the edges of the hole afterward—and blew out the power-supply computer, literally: The computer's IC chips had craters a centimeter wide blown out of them.[13] Fortunately, no one was hurt in the incident.

Energy in magnetic fields

The behavior of inductors shows quite clearly that there is energy stored in a current-carrying inductor. This energy can be attributed to the magnetic field in the inductor, just as the energy in a charged capacitor can be attributed to its electric field. The work required to build up the current in an inductor to some value I, performed by an external power source against the induced EMF, is given by

$$\Delta W = \int_{I'=0}^{I} \left(+L\,\frac{dI'}{dt} \right) I'\,dt'$$
$$= \tfrac{1}{2} L I^2 , \tag{X.21a}$$

---

[11] The older system had the advantage that the car could be rendered virtually unstealable by the simple expedient of removing the rotor arm, a device the size of a modern flash drive.

[12] This author did not witness the actual event but did see some of the evidence afterward.

[13] One might wonder what that microcomputer "thought" in those last CPU cycles after the resistor melted, the voltage started to rise, and it knew it was going to die.

i.e., the external *power* required, integrated over time. This is the energy stored in the inductor. The work done against the ordinary resistance of the inductor is dissipated and is not counted. For example, the solenoid dangling in Fig. X.1 has inductance $L = 100.$ mH with its core fully inserted. With current $I = 10.0$ A running through it, the solenoid stores energy

$$
\begin{aligned}
\Delta W &= \tfrac{1}{2}LI^2 \\
&= \tfrac{1}{2}(0.100 \text{ H})(10.0 \text{ A})^2 \\
&= 5.00 \text{ J} .
\end{aligned} \tag{X.21b}
$$

Capacitors are usually used in preference to inductors as energy-storage devices, because a current-carrying inductor—unless it were a superconducting device— loses energy to Ohmic heating.

The stored energy can be taken to reside in the magnetic field in the inductor rather than in the current. For example, if the inductor is a long, straight solenoid, the stored energy takes the form

$$
\begin{aligned}
\Delta W &= \tfrac{1}{2}(\mu n^2 \mathcal{V})\, I^2 \\
&= \frac{(\mu n I)^2}{2\mu}\, \mathcal{V} \\
&= \frac{B_{\text{in}}^2}{2\mu}\, \mathcal{V} ,
\end{aligned} \tag{X.22a}
$$

using results (X.15b), (IX.20b), and (IX.30). This suggests a field energy density

$$
u_B = \frac{B^2}{2\mu} \tag{X.22b}
$$

for a field with magnitude B in a medium with permeability μ, a result that is generally applicable. For example, a field with magnitude $B = 1.0$ T in vacuum has energy density

$$
\begin{aligned}
u_B &= \frac{B^2}{2\mu_0} \\
&= \frac{(1.0 \text{ T})^2}{2(4\pi \times 10^{-7} \text{ N/A}^2)} \\
&= 4.0 \times 10^5 \text{ J/m}^2 .
\end{aligned} \tag{X.22c}
$$

This is four orders of magnitude larger than the energy density associated with the breakdown electric field of air, displayed in Eq. (VII.52). In this sense, this magnetic field is one hundred times larger than that electric field, although the comparison may seem odd because the two fields are measured in different units. In any case, the magnetic field, like the electric field, has taken on the attributes of a physical entity in its own right.

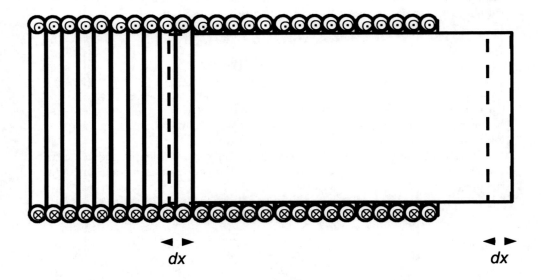

Figure X.5: A solenoid carrying constant current pulls its ferromagnetic core in a distance dx, exerting force F on the core.

C   Force exerted by a solenoid

These results provide the ideas necessary to calculate the force exerted by a long, straight solenoid on a close-fitting, ferromagnetic core. The problem with a direct force calculation is that the force arises from the non-uniform, complicated magnetic fields at the end of the core inside the solenoid and the end of the solenoid into which the core is inserted. This problem can be circumvented by a method that, in general, is very useful for treating complicated forces: We envision the solenoid pulling the core in an infinitesimal distance dx, as illustrated in Fig. X.5, and calculate increment dW of work done by determining the change in energy of the system. The force is deduced from the work as dW/dx.

An advantage of this approach is that the complicated field configurations at the ends of the core and the solenoid are unchanged. The effect of the displacement is simply to replace $n\,dx$ turns of empty solenoid with $n\,dx$ turns filled with core material of magnetic susceptibility χ_m. This increases the magnetic-field energy stored in the system, which appears paradoxical: Positive work is done pulling the core, and the energy of the system *increases*. Something for nothing? No, because the increased magnetic flux through the turns of the solenoid will induce an EMF that, by Lenz, would tend to reduce the current through the solenoid. If the solenoid is operated at constant current, the external power source must do work against this back EMF. The energy balance for the system takes the form

$$\mathcal{E}_{\text{ext}} I \, dt = dU + F \, dx \ , \tag{X.23a}$$

with dU the increment of magnetic field energy in the solenoid. This is given by the difference in energy density (X.22b) between filled and empty turns times the affected volume $\mathcal{A} \, dx$, with $\mathcal{A}$ the cross-sectional area of the solenoid and the core. The external EMF $\mathcal{E}_{\text{ext}}$ must oppose that provided by Faraday induction, so the energy balance can be recast as

$$+\frac{N\Phi_{\mathbf{B}}}{dt} I \, dt = \left(\frac{B^2}{2\mu} - \frac{B_0^2}{2\mu_0} \right) \mathcal{A} \, dx + F \, dx \ , \tag{X.23b}$$

where B is the field magnitude in filled turns of the solenoid, B_0 that in empty turns, and μ and μ_0 the corresponding permeabilities (*not* dipole moments). The number of affected turns is $N = n \, dx$, and the change in flux through each is $\Delta\Phi_{\mathbf{B}} = (B - B_0)\mathcal{A}$. Hence, the balance takes the form

$$n \, dx \, (B - B_0) \, \mathcal{A}I = \left(\frac{B^2}{2\mu} - \frac{B_0^2}{2\mu_0} \right) \mathcal{A} \, dx + F \, dx \ , \tag{X.23c}$$

with each term proportional to dx. Results (IX.20b) and (IX.30) yield $B = \mu nI$, $B_0 = \mu_0 nI$, and $\mu - \mu_0 = \chi_m \mu_0$. With these, the balance can be simplified to

$$\mu_0 \chi_m n^2 \mathcal{A} I^2 = \tfrac{1}{2}\mu_0 \chi_m n^2 \mathcal{A} I^2 + F \ . \tag{X.23d}$$

This yields the desired result,

$$F = \tfrac{1}{2}\mu_0 \chi_m n^2 \mathcal{A} I^2 \ , \tag{X.23e}$$

for the force applied to the core. Results (X.23d) and (X.23e) imply that the work supplied by the external power source is divided evenly between the magnetic energy of the solenoid and the mechanical work done on the core. For actual devices, end effects, et cetera, may contribute correction terms. But with this result, the approach and the calculation must be reckoned successful.

The solenoid in Fig. X.1 has filled inductance $L = 100.$ mH and empty inductance $L_0 = 7.0$ mH, hence, core susceptibility $\chi_m = 13$. It has length $\ell = 10.$ cm. It is wound in multiple layers, making n and $\mathcal{A}$ difficult to determine precisely, but the product $\mu_0 n^2 \mathcal{A}$ is given by

$$\begin{aligned}
\mu_0 n^2 \mathcal{A} &= \frac{L_0}{\ell} \\
&= \frac{7.0 \times 10^{-3} \text{ H}}{0.18 \text{ m}} \\
&= 3.9 \times 10^{-2} \text{ N/A}^2 \ .
\end{aligned} \tag{X.24}$$

Operated with a current $I = 10.$ A, the solenoid should provide force

$$F = \tfrac{1}{2}\mu_0 \chi_m n^2 \mathcal{A} I^2$$
$$= \tfrac{1}{2}(3.9 \times 10^{-2} \text{ N/A}^2)(10. \text{ A})^2$$
$$= 25. \text{ N} .$$

(X.25)

This is the weight of a brick or, by rough estimation, the weight of the solenoid[14] seen suspended in Fig. X.1. Victory!

References

Electromagnetic induction and inductance are linchpins of classical electromagnetism. They are treated in physics texts, or texts on electricity and magnetism, at every level. The energy calculation of the force exerted by a solenoid is featured in Reitz, Milford, and Christy [RMC08]. The method is treated more generally in Jackson [Jac98].

Problems

1. Prove that the Faraday law does *not* suffer from the inconsistency—or equivalently, the ambiguity associated with the choice of surface S bounded by loop C—encountered with the Ampère law.

2. A coil of N turns, and area A, carrying a constant current I, flips in an external magnetic field $\mathbf{B}_{\text{ext}}$ from having its dipole moment opposite $\mathbf{B}_{\text{ext}}$ to alignment with that field. Calculate the work done by the coil's power supply to maintain the constant current during this process. What does this say about the energy of a magnetic dipole in a magnetic field?

3. An ideal battery (with no internal resistance) with EMF $\mathcal{E}$ is connected to a superconducting (resistanceless) coil of inductance L at time $t = 0$. Find the current in the coil as a function of time, $I(t)$.

4. Consider the LR circuit of Fig. X.4. Calculate the power output of the voltage source V_0 as a function of time after the switch is closed at $t = 0$, assuming the source is ideal and maintains a constant voltage. Calculate the *rates* at which energy is (a) dissipated in resistor R_1, (b) dissipated in resistor R_2, and (c) stored in the inductor as functions of time.

---

[14]In 1996 this author assigned this problem to four students—Mr. Jason Pattie, Mr. Patrick Farrell, Mr. Dennis Howie and Mr. Chris Jones—as a term project. The students connected the solenoid to a power supply, attached the core to a spring scale, and *measured* the force. They were able to observe the I^2 dependence of force (X.23e) very clearly. Their measured force at current $I = 10.0$ A was approximately $(40. \pm 2.)$ N under conditions somewhat different than those of this idealized calculation. But the calculated and measured results certainly differ by much less than a billion standard deviations of the measurement. Hence, in comparison to the preceding chapter, the analysis carried out here must be judged a staggering success.

5. Consider again the LR circuit of Fig. X.4. Calculate by integration the total energy dissipated in resistors R_1 and R_2 after switch S is opened, having first been closed for a time long compared to L/R_1. Compare this with the energy stored in the inductor before the switch is opened.

6. In Ch. VII, Problem 1, the magnitude of the electric field near the surface of the Earth was given as $E_\oplus = 150.$ N/C. The magnitude of the Earth's magnetic field varies over its surface, but $B_\oplus = 50.0\ \mu$T is a typical value. Calculate and compare the energy densities associated with these two fields. Take the electric permittivity and magnetic permeability of air to be those of the vacuum.

7. Electric and magnetic fields in vacuum have configurations

$$\mathbf{E}(\mathbf{r}, t) = \mathbf{E}_0 \cos(\mathbf{k} \cdot \mathbf{r} - \omega t)$$
$$\mathbf{B}(\mathbf{r}, t) = \mathbf{B}_0 \cos(\mathbf{k} \cdot \mathbf{r} - \omega t) \ ,$$

with $\mathbf{k}$ a constant vector perpendicular to the constant $\mathbf{E}_0$, constant $\mathbf{B}_0$ given by $\mathbf{B}_0 = \mathbf{k} \times \mathbf{E}_0$, and ω a constant satisfying $\omega/|\mathbf{k}| = (\mu_0 \epsilon_0)^{-1/2}$. Calculate and compare the energy densities associated with these fields. The significance of these field configurations will be made clear in the next chapter.

8. How could a solenoid like that considered in this chapter be configured to *push*, rather than pull, its core? Why is such an arrangement not used?

Chapter XI

Submarine Radio Communication

- *Consider an electromagnetic plane wave impinging perpendicularly on the planar surface of a region filled with an ohmic medium of permittivity ϵ, permeability μ, and conductivity σ. Part of the wave is reflected; part is transmitted into the medium. Analyze the propagation of the transmitted wave.*

- *Apply these results to the problem of radio communication with submerged submarines.*

Figure XI.1: Is this submarine in radio communication with its base? What physical phenomena might make that difficult? (Illustration by Andreus.)

The quest approaches the culmination of classical electromagnetism, indeed, of classical physics: James Clerk Maxwell's synthesis of electricity and magnetism—which set the future course of physics—and his most spectacular prediction that electric and magnetic forces propagate through space as waves.

Maxwellian *electrodynamics* undergirds all our broadcast communications technologies—radio, television, radar, et cetera—and encompasses the entirety of classical optics, described in the next chapter. The submarine depicted in Fig. XI.1 faces a physical challenge in trying to maintain communications with the Earth's surface while submerged. We shall examine this challenge to explore the sweep and power of Maxwell's discoveries.

A   The Maxwell Equations

By 1873, Maxwell[1] had indentified four relationships as central to the description of all the electric and magnetic phenomena known to him. The reader has already encountered all of them. They are: the Gauss Law,

$$\oint_{S_c} \epsilon \mathbf{E} \cdot d\mathbf{A} = Q_{\text{in}} \; ; \qquad\qquad (\text{XI.1a})$$

the Gauss Law for Magnetism,

$$\oint_{S_c} \epsilon \mathbf{B} \cdot d\mathbf{A} = 0 \; ; \qquad\qquad (\text{XI.1b})$$

the Faraday Law of Induction,

$$\oint_{C} \mathbf{E} \cdot d\boldsymbol{\ell} = -(N) \frac{d}{dt} \int_{S} \mathbf{B} \cdot d\mathbf{A} \; ; \qquad\qquad (\text{XI.1c})$$

and the Ampère-Maxwell Law,

$$\oint_{C} \frac{1}{\mu} \mathbf{B} \cdot d\boldsymbol{\ell} = \int_{S} \mathbf{j} \cdot d\mathbf{A} + \frac{d}{dt} \int_{S} \epsilon \mathbf{E} \cdot d\mathbf{A} \; . \qquad\qquad (\text{XI.1d})$$

Here, some of the constants have been moved from previous versions, and allowance is made for the presence of dielectric or magnetic materials. These four equations, plus the Lorentz force law that defines the fields $\mathbf{E}$ and $\mathbf{B}$, govern all of classical electricity and magnetism. So simple and yet so all-encompassing is this synthesis that Ludwig Boltzmann was moved to ask—quoting Goethe's[2] *Faust—War es ein Gott, der diese schrieb?* ("Was it a god who wrote these?") A bit of *Sturm und Drang*, perhaps, but the equations do have an "engraved on stone tablets" aspect.

As is detailed in Sec. C, these equations imply a remarkable prediction: Electric and magnetic fields, *combined,* can propagate through space or matter as waves. The equations specify that the waves travel in vacuum at speed

$$\frac{1}{(\mu_0 \epsilon_0)^{1/2}} = \{(4\pi \times 10^{-7} \text{ N/A}^2)[8.854187817\ldots \times 10^{-12} \text{ C}^2/(\text{N m}^2)]\}^{-1/2}$$

$$= 299,792,458 \text{ m/s} \qquad (\text{exact}),$$

$$= c \; , \qquad\qquad (\text{XI.2})$$

the speed of light *in vacuo,* leading to the conclusion that light waves are themselves electromagnetic and governed by the Maxwell equations. Heinrich Hertz confirmed the existence of electromagnetic waves in the laboratory by 1886, although an explicit demonstration of the presence of electric and magnetic fields

---

[1] Maxwell also made important contributions to thermodynamics and solid mechanics. He is said to be the last person able to master all of the physics known in his time.

[2] Johann Wolfgang von Goethe, 1749–1832

in waves of visible light was not accomplished until 1960. By then, of course, the soundness of Maxwell's predictions was long established.

Which brought physics to a crisis. One might say a Greek tragedy was played out. The Greek tragic hero is never brought low by bad luck or bad timing, but by the inexorable march of fate brought about by a tragic flaw in the hero's nature—an aspect of the very virtue that made him a hero in the first place. In the nineteenth century, classical physics, built upon the twin pillars of Newtonian mechanics and Maxwellian electrodynamics, appeared to offer a complete, all-encompassing understanding of nature, the goal of the effort from the beginning. But it contained within itself a fatal flaw, a crack in the foundation of the mighty edifice: The two pillars are logically inconsistent with one another. Galilean invariance demands that Newtonian mechanics can never predict the (absolute) speed of anything. The speed of one thing relative to another, yes; the unqualified speed of anything, no. Yet the Maxwell equations predict, unambiguously, the speed of his waves.

Maxwell and his contemporaries recognized the problem immediately and spent decades trying to reconcile the two sets of ideas. Attempts to interpret the speed of electromagnetic waves as a speed relative to source or observer—*ballistic theories*—foundered. It was reasoned that just as the Newtonian equations for sound waves predicted the speed of the waves relative to their medium of propagation, that speed (XI.2) was speed relative to a medium, the *luminiferous ether*. This notion had problems—what medium could be stiff enough to support waves at such a speed, yet offer no resistance to the motion of heavenly bodies?— and ran afoul of some famous experiments, which will not be detailed here. As seen in Ch. I Sec. H, Einstein resolved the crisis in 1905 by revising *mechanics,* demonstrating the necessity of replacing Galilean with Lorentzian kinematics and modifying dynamics accordingly. Einstein did not pull his new mechanics "out of a hat": It follows by rigorous logic from the Maxwell equations and their consequences.

B   Electromagnetic-wave solutions

Before we explore the wave dynamics that follows from the Maxwell equations, we can examine some of the features of electromagnetic waves. These take a great variety of forms, but much can be understood by examining the simplest wave solutions.

A plane-polarized, monochromatic plane wave

Each of those adjectives is important. A *plane-polarized* wave is one in which the electric field (and separately, the magnetic field) oscillates at each point along a single line, the same direction throughout the wave. A *monochromatic* wave is one in which the functional dependence of the fields is sinusoidal, a sine or cosine function. As was seen in Ch. VI Sec. B, any other functional form is a combination of waves of different wavelengths and frequencies—*colors* in an optical context. A *plane* wave is one in which the surfaces of constant phase, e.g., the wave crests and troughs, are flat planes in three-dimensional space. (Electromagnetic waves exist without any of these properties.) One plane-polarized, monochromatic plane-wave solution of the Maxwell equations consists of the fields

$$\mathbf{E}(\mathbf{r}, t) = \mathbf{E}_0 \cos(\mathbf{k} \cdot \mathbf{r} - \omega t)$$
$$\mathbf{B}(\mathbf{r}, t) = \mathbf{B}_0 \cos(\mathbf{k} \cdot \mathbf{r} - \omega t) \ , \tag{XI.3a}$$

where $\mathbf{E}_0$, $\mathbf{B}_0$, and $\mathbf{k}$ are constant vectors and ω a constant scalar. The Maxwell equations demand that the wave be *transverse,* satisfying

$$\mathbf{k} \cdot \mathbf{E}_0 = \mathbf{k} \cdot \mathbf{B_0}$$
$$= 0 \ , \tag{XI.3b}$$

hence, similarly for $\mathbf{E}$ and $\mathbf{B}$, wave vector $\mathbf{k}$ specifying the direction of wave propagation. The vector amplitudes $\mathbf{E}_0$ and $\mathbf{B}_0$ are not independent; they must satisfy the relation

$$\mathbf{B}_0 = \frac{\mathbf{k} \times \mathbf{E}_0}{\omega} \ . \tag{XI.3c}$$

The wave fields $\mathbf{E}$ and $\mathbf{B}$ satisfy a similar relation, i.e., they are *crossed fields.* They oscillate in space and time, changing direction simultaneously, so that $\mathbf{E}$, $\mathbf{B}$, and $\mathbf{k}$ always form a mutually perpendicular triad. Wave vector $\mathbf{k}$ and angular frequency ω are not independent either; they are required to satisfy the *dispersion relation*

$$\frac{\omega}{|\mathbf{k}|} = \frac{1}{(\mu\epsilon)^{1/2}} \ , \tag{XI.3d}$$

where μ and ϵ are the permeability and permittivity, respectively, of the medium through which the wave is traveling.

Virtually every introductory physics text displays a diagram of a plane-polarized, monochromatic plane electromagnetic wave featuring sinusoidal graphs of $\mathbf{E}$ and $\mathbf{B}$. Such a figure does not appear here because it is misleading: It gives the impression that the wave has that shape in three-dimensional space. (If the figures were actual graphs of $\mathbf{E}$ and $\mathbf{B}$, that would be all right, but the axes perpendicular to the direction of propagation are invariably labeled with the other two spatial coordinates.) Nothing has that shape in space, nor does the wave move on such a path. Perhaps the best way to envision this wave is as a stack of flat slabs, perpendicular to wave vector $\mathbf{k}$, in each of which the fields have constant, perpendicular values. The fields vary from slab to slab according to wave functions (XI.3a), and the stack moves past any fixed observer, in the direction of $\mathbf{k}$, at phase speed (XI.3d).

The speed of light

The speed of light—and other electromagnetic waves—is a quantity with a remarkable history. It is much greater than any speed perceptible by human senses. It was not even known to be finite until late in the 17th century. Galileo attempted to measure it by flashing lantern light to students on nearby hilltops, but the time intervals to be measured were shorter than the precision of his instruments. He was forced to conclude that "...if not instantaneous, it is exceedingly rapid." Dutch astronomer Ole Römer[3] obtained the first known value in 1675 by timing the moons of Jupiter and recognizing that his measurements included the light-travel time across the Earth's orbit around the Sun. He deduced a value about two-thirds of the modern value, rather impressive since he did not have a precise value of the size of the orbit. Hippolyte Fizeau[4] made the first laboratory measurement of the speed of light in 1849 using spinning gears and mirrors; he was also able to measure the speed in various materials and moving media. Into the twentiethth century the speed was determined more and more precisely, e.g., by measuring simultaneously the frequencies and wavelengths of microwaves, but by 1986, the measurement of time intervals had become so much more precise than that of distances that a new definition of the meter was adopted, making the speed of light a *defined* value. That value is shown in Eq. (XI.2). Along with the defined value of μ_0, it also defines the value of ϵ_0, which is why that quantity, first shown in Ch. VII, has no experimental uncertainty.

The Maxwell equations imply that the speed of electromagnetic waves in any medium depends on the electric and magnetic properties of the material. That speed, $\tilde{c}$ (to distinguish it from the vacuum value c), is given by dispersion relation (XI.3d). It is

$$\begin{aligned}
\tilde{c} &= \frac{1}{(\mu\epsilon)^{1/2}} \\
&= \frac{c}{(K_m\kappa)^{1/2}} \\
&= \frac{c}{n(\omega)} \, .
\end{aligned} \qquad \text{(XI.4)}$$

Here, K_m and κ are the relative permeability and dielectric constants, respectively, of the material. The square root of their product is recognized as the *index of refraction* of the material. This was introduced in 1621 by Willebrord Snell[5] to describe the bending of light rays at interfaces between media. He could have had no knowledge of its connection with the speed of light; Christian Huygens made the correct connection more than fifty years later. And only two centuries later did its relation to the electric and magnetic properties of materials become clear. The last line indicates that the index, hence speed $\tilde{c}$, can depend on the angular frequency of the wave[6]—in *dispersive* media, in the language of Ch. VI Sec. B.

---

[3]Ole (Olaus) Römer, 1644–1710

[4]Armand Hippolyte Louis Fizeau, 1819–1896

[5]Willebrord Snellius or Snell van Royen, 1591–1626

[6]For example, the dielectric constant κ of water for *static* fields is about 80., but its optical index of refraction is 1.33. (Its relative permeablity K_m is very close to unity.) The dielectric constant of water *at optical frequencies* is approximately 1.78.

The electromagnetic spectrum

The frequencies and wavelengths of electromagnetic waves vary over many orders of magnitude; specific upper and lower bounds are not known. This range or *electromagnetic spectrum* is poetically termed "Maxwell's Rainbow," as the real rainbow displays the range of frequencies or wavelengths of visible light. It is divided into human-delineated categories according to the source of the radiation and the uses to which it is put. Waves with wavelengths of roughly a meter or greater [frequencies of the order 300 MHz or lower; the two quantities are related via Eq. (VI.2b), viz., $\nu\lambda = c$ for waves in vacuum] are labeled *radio waves* and are produced by macroscopic electric currents. Wavelengths of the order centimeters or millimeters characterize *microwaves,* also produced by macroscopic currents. Between the micron and the millimeter are *infrared* wavelengths, associated with molecular vibrations. The wavelength range between 400. nm and 700. nm, approximately, is *visible light,* produced by electron transitions in the outer shells of atoms. The range 100. nm to 400. nm, roughly, is *ultraviolet light,* and 1 nm to 100 nm, again roughly, are *x-ray* wavelengths. Both of these are produced by electronic transitions in the inner shells of atoms. Shorter-wavelength waves are produced by nuclear transitions and are termed *gamma radiation.*

It is difficult to contemplate this variety of electromagnetic waves without wondering what human experience would be like if human vision, like human hearing, ranged over many octaves instead of the less than a single octave represented by visible light. (An *octave* interval is a factor of two in frequency.) How many more words for colors would we need—not the millions of *mixtures* of colors, which, say, a modern television or computer monitor can produce, but many more *different* colors? But it cannot be, and physics explains why. Our vision is a *resonant* phenomenon, as described in Ch. V Sec. E. It depends on a matching of frequencies between light and chemical transitions in our retinas. The height of the resonance peak necessary to allow us to see at all demands the narrow bandwidth to which our eyes are sensitive.

Sources of electromagnetic radiation

Electric charges at rest produce electrostatic fields. Steady currents produce steady magnetic fields. Electromagnetic radiation is produced by *accelerating* charges. For example, the charged particles circulating in an accelerator such as a cyclotron or synchrotron emit radiation. Such *synchrotron radiation* is becoming an important source of x-rays. A beam of electrons striking a metal target can give up their kinetic energy as *bremsstrahlung* ("braking radiation"), also a standard source of x-rays.

It is much more common, though, to encounter accelerating charges in oscillating charge or current distributions, from nuclear to astronomical scales. A dipole distribution, for example, produces a "butterfly" field pattern like that sketched in Fig. VII.5. As the dipole moment oscillates through zero and changes direction, the field lines form closed loops and propagate away, the electric and magnetic fields sustaining each other through Faraday induction and displacement currents. One feature of such radiation is that an oscillating dipole emits no radiation along the direction of its dipole moment.

Intensity of radiation

Electromagnetic waves transport energy, an assertion that accords with our sensory experience of visible light and infrared radiation. The rate of this energy transport can be described in terms of the field variables of Eqs. (XI.3). Such a wave has field energy densities

$$u_E = \tfrac{1}{2}\epsilon |\mathbf{E}|^2$$
$$\text{and} \qquad u_B = \tfrac{1}{2}\frac{|\mathbf{B}|^2}{\mu} \qquad\qquad\qquad \text{(XI.5a)}$$
$$= u_E\ ,$$

this last by virtue of Eqs. (XI.3b)–(XI.3d). The energy of the wave is equally distributed between its electric and magnetic fields. The propagation of this energy-density distribution yields (instantaneous) energy flux

$$\tilde{c}\,(u_E + u_B) = \tilde{c}\epsilon|\mathbf{E}|^2$$
$$= \frac{\tilde{c}|\mathbf{B}|^2}{\mu}$$
$$= \frac{|\mathbf{E}||\mathbf{B}|}{\mu}\ , \qquad\qquad \text{(XI.5b)}$$

as the two terms are equal and can be expressed in terms of either field.

This flux can be expressed in vector form in terms of the *Poynting*[7] *vector*

$$\mathbf{S} \equiv \frac{1}{\mu}\,\mathbf{E} \times \mathbf{B}\ . \qquad\qquad \text{(XI.5c)}$$

This has magnitude equal to the instantaneous flux, and its direction is that of the wave propagation, i.e., of wave vector $\mathbf{k}$. Some care must be exercised in interpreting this vector, however. It is true in general that the flux of $\mathbf{S}$ through any closed surface is the rate of energy transport out of the surface. But the local or pointwise interpretation of $\mathbf{S}$ as the energy flux can lead to strange results. For example, in the simple case of an ideal battery driving current I through a cylindrical resistor R, the total flux of S through the surface of the resistor is $-I^2R$, an influx equal to the rate of Ohmic heating. But the vector $\mathbf{S}$ is actually radially inward around the surface of the resistor, suggesting—if interpreted "literally"—that the energy flows into the resistor from the surrounding space and not through the circuit! Nonetheless, for the case of electromagnetic wave fields, the Poynting vector gives a useful description of energy transport.

---

[7] John Henry Poynting, 1852–1914

Except for astronomical *micropulsations,* which can have periods of the order seconds, electromagnetic waves oscillate too fast for the instantaneous energy flux to be perceptible. The *intensity I* of a wave is the average of this flux over many periods. This can be written in various forms, such as

$$I = \langle |\mathbf{S}| \rangle$$

$$= \frac{|\mathbf{E}_0||\mathbf{B}_0|}{2\mu}$$

$$= \tfrac{1}{2} \left(\frac{\epsilon}{\mu} \right)^{1/2} |\mathbf{E}_0|^2$$

$$= \frac{\tilde{c}|\mathbf{B}_0|^2}{2\mu} \, , \qquad (\text{XI.5d})$$

et cetera. Here, the value $\langle \cos^2(\mathbf{k}\cdot\mathbf{r} - \omega t) \rangle = \tfrac{1}{2}$, noted in Ch. VI Sec.C, has been used. For example, the intensity of sunlight impinging on the top of the Earth's atmosphere is known as the *Solar Constant,* with value $K = 1.37$ kW/m^2. A plane-polarized plane wave, traveling in vacuum, of this intensity has field amplitudes

$$|\mathbf{E}_0| = (2K)^{1/2} \left(\frac{\mu_0}{\epsilon_0} \right)^{1/4}$$

$$= [2(1.37 \times 10^3 \text{ W/m}^2)]^{1/2} \left(\frac{4\pi \times 10^{-7} \text{ N/A}^2}{8.854 \times 10^{-12} \text{ C}^2/(\text{N m}^2)} \right)^{1/4}$$

$$= 1.02 \times 10^3 \text{ N/C} \qquad (\text{XI.6a})$$

and

$$|\mathbf{B}_0| = \left(\frac{2\mu_0 K}{c} \right)^{1/2}$$

$$= \left(\frac{2(4\pi \times 10^{-7} \text{ N/A}^2)(1.37 \times 10^3 \text{ W/m}^2)}{2.998 \times 10^8 \text{ m/s}} \right)^{1/2}$$

$$= 3.39 \ \mu\text{T} \, . \qquad (\text{XI.6b})$$

As they correspond to equal energy densities, these fields are "the same size" despite their different units.[8] In this sense, the tesla is a much larger unit than the newton per coulomb.

Electromagnetic waves also transport momentum. A wave impinging on some material will exert both electric and magnetic forces on the charges within the material. Because of the relationships (XI.3), the rate at which the magnetic field imparts momentum to a charge q with velocity $\mathbf{v}$, i.e., the force $F_m = q\mathbf{v} \times \mathbf{B}$, and the rate $q\mathbf{E} \cdot \mathbf{v}$ at which the electric field does work on the charge

---

[8]In some systems, the Lorentz force law is written $\mathbf{F} = q[\mathbf{E} + (\mathbf{v}/c) \times \mathbf{B}]$, so that $\mathbf{E}$ and $\mathbf{B}$ are expressed in the same units. In such systems, electromagnetic-wave fields have equal amplitudes.

are always in the ratio $|\mathbf{B_0}|/|\mathbf{E_0}| = 1/\tilde{c}$. Hence, for example, a finite wave packet with energy ΔU also carries linear momentum

$$\mathbf{\Delta P} = \frac{\Delta U}{\omega}\, \mathbf{k}\, , \qquad\qquad\text{(XI.7a)}$$

with magnitude $|\mathbf{\Delta P}| = \Delta U/\tilde{c}$. A wave with intensity I also transports force per unit area, i.e., exerts *radiation pressure*

$$P_{\mathrm{rad}} = \begin{cases} I/\tilde{c} & \text{(total absorption);} \\ 2I/\tilde{c} & \text{(total reflection).} \end{cases} \qquad\text{(XI.7b)}$$

The pressure is doubled for reflected radiation because the momentum of the wave is reversed; hence, by momentum conservation, twice as much momentum is imparted to the reflecting material. This can be demonstrated with a device called a *Crookes*[9] *radiometer,* a pinwheel in an evacuated glass bulb with vanes black on one side and white or mirrored on the other. The higher radiation pressure on the reflective sides spins the pinwheel with the reflective sides trailing.[10] The radiation pressure of sunlight at the top of the Earth's atmosphere is

$$\begin{aligned} P_{\mathrm{rad}} &= \frac{K}{c} \\ &= \frac{1.37 \times 10^3 \text{ W/m}^2}{2.998 \times 10^8 \text{ m/s}} \\ &= 4.57\ \mu\text{Pa} \qquad\qquad\qquad\text{(XI.7c)} \end{aligned}$$

for total absorption. This is comparable to the pressure amplitude of sound waves at the threshold of hearing, imperceptible to our sense of touch. But radiation pressure shapes parts of the tails of comets as they approach and recede from the Sun, and it plays an important role in the dynamics of massive stars.

The *inverse-square law* is a familiar feature of electromagnetic waves of every type. For a *small, isotropic* source—i.e., a source small compared to its distance, radiating equally in all directions—both the intensity I and the radiation pressure P_{rad} vary with distance r from the source as $1/r^2$, the inverse square. This is a straightforward consequence of energy and momentum conservation and the three-dimensionality of space: The areas of the spheres centered on the source through which the same energy and momentum must pass vary with distance as r^2.

Polarization of electromagnetic radiation

Electromagnetic waves, being transverse, possess a *polarization* feature or "degree of freedom." This is usually described in terms of the direction of the

---

[9]William Crookes, 1832–1919

[10]The versions found in novelty shops usually spin the other way. They contain a small amount of air; the greater heating of the dark sides of the vanes provides gas pressure to spin the pinwheel greater than the radiation pressure.

electric-field vector. For example, two plane-polarized waves propagating in the x direction, with electric fields

$$\mathbf{E}_1(x,t) = E_0\,\hat{\boldsymbol{y}}\,\cos(kx - \omega t)$$
$$\mathbf{E}_2(x,t) = E_0\,\hat{\boldsymbol{z}}\,\cos(kx - \omega t)\;, \tag{XI.8}$$

each with corresponding magnetic field, represent two independent polarization states. Any other plane polarization for waves traveling in that direction can be constructed as a linear combination of these two. This feature of the waves can be manipulated in a wide variety of applications.

One means of utilizing the polarization of light waves is by means of *birefringent materials*. These are *anisotropic* dielectrics, such as crystalline minerals, with different responses to electric fields in different directions. Hence, these materials sustain different wave speeds—have different indices of refraction— for waves of different polarizations. Perhaps the best known of these is calcite, crystalline calcium carbonate. Unpolarized light, i.e., an equal mixture of orthogonal polarizations, passing through such a crystal is split into two beams, each of a different polarization. *Polarizing materials* combine birefringence with preferential absorption, splitting unpolarized light into two polarized beams, then absorbing one of them. For example, certain plastics composed of long, parallel molecules have this property. Electric fields can drive currents along the molecules. The energy of a wave with its electric field in that direction is dissipated by these currents. Waves with perpendicular electric fields pass unhindered through the material. Sheets of such material appear gray, as the intensity of light passing through them is reduced. The intensity of unpolarized light is reduced by half. Polarized light, with electric-field direction at angle θ to the preferred (passed) direction of the material, is reduced by the factor

$$I_{\text{out}} = I_{\text{in}}\,\cos^2\theta\;, \tag{XI.9}$$

a relationship known as the *Malus*[11] *Law*. It follows from the facts that only the vector component of the electric field along the preferred direction passes through the material and that the intensity is proportional to the squared magnitude of the electric field.

This behavior can give rise to a remarkable phenomenon. Two such polarizing sheets, parallel to one another with preferred directions perpendicular, will block 100% of the light sent through them. But if a third sheet is inserted between them, with preferred direction at, say, 45° to theirs, then one-eighth of the intensity of the original (unpolarized) light incident on the first sheet will pass through the combination. Projecting the electric field of the light polarized by the first sheet first on the preferred direction of the intermediate sheet, then on that of the final sheet *regenerates* a component in the preferred direction of the last sheet.[12]

---

[11]Étienne Louis Malus, 1775–1812

[12]A similar regeneration phenomenon is observed with the quantum states corresponding to the long- and short-lived versions of the *kaon* or *K*-meson particle.

Some materials possess the property that the polarization direction of light passing through the material is rotated, the amount of rotation depending on the wavelength of the light and the thickness of the layer passed through. Such materials are termed *optically active*. These need not be exotic materials: Ordinary corn syrup has this property. Liquid crystals—materials with crystalline order in some directions, liquid disorder in others—can have optical activity that can be altered by electric fields in the material. Hence, by applying potentials to a liquid-crystal cell, light can be passed through polarizers on each side of the cell or blocked. This is the basis of liquid-crystal display technology.

How is polarized radiation produced? In a variety of ways. Radiation from a single oscillating dipole, e.g., the current in the dipole antenna of a radio transmitter, is polarized parallel to the dipole moment. Unpolarized waves can also be polarized by *scattering,* partially by scattering at any angle, completely by scattering at right angles. In a classical picture, electromagnetic waves impinging on scattering objects cause the charges within the material of the scatterers to oscillate. These oscillating charges then reradiate the waves in new directions. This is represented in Fig. XI.2. Unpolarized waves, an equal mixture of both polarizations, drive the charges in a scattering center into oscillation. But for scattering at 90°, as illustrated, one of the polarizations causes the charges to oscillate in the direction of the scattered waves. As already observed, an oscillating dipole does not radiate in the direction of its dipole moment. Hence, that polarizaton is absent from the scattered waves, which emerge entirely with the other polarization. Sky light is polarized in this way, most strongly 90° from the position of the Sun. Apparently, ancient Viking seafarers made use of this for navigation, using birefringent calcite crystals to observe the degree of polarization of the light.

Waves can also be polarized by *reflection* at a smooth interface, which is actually a particular form of scattering. This is depicted, in the traditional way, in Fig. XI.3. Unpolarized waves impinge on the interface between two materials, with different indices of refraction n_i (*i* for *incident*) and n_t (*t* for *transmitted*), along the *incident ray* shown. The waves oscillate the charges in the atoms at the interface and are reradiated. Part of the wave energy is reflected along the *reflected ray,* the rest transmitted along the *transmitted ray.* The angles θ_i, θ_r (*r* for *reflected*), and θ_t that these rays make with the normal to the interface are related by the *Law of Equiangular Reflection* and the *Snell Law of Refraction,* described in greater detail in Ch. XII. If the reflected and transmitted rays are perpendicular, the polarization in the *plane of incidence* containing the three rays causes charges to oscillate in the direction of the reflected ray. Hence, that polarization is absent from the reflected waves, which are completely polarized normal to the plane. All the energy in the waves polarized in the plane is transmitted through the interface. (At other angles, the reflected waves are partially polarized.) The incident and reflected angles at which full polarization

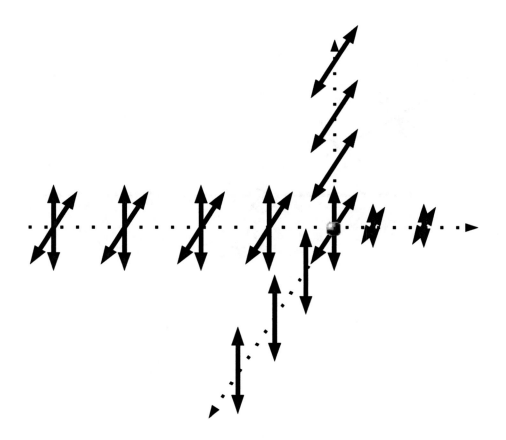

Figure XI.2: Polarization by scattering. The arrows represent the directions of the electric fields in the incoming and outgoing waves.

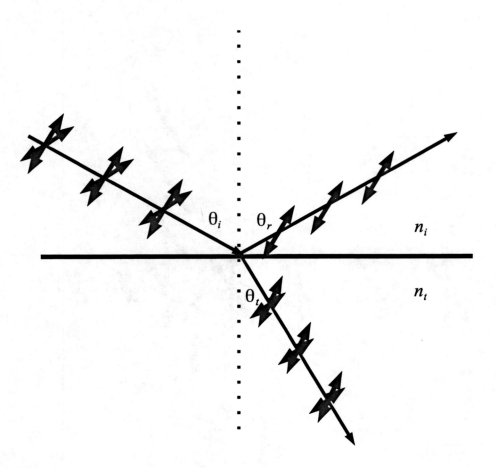

Figure XI.3: Polarization by reflection. Incident, reflected, and transmitted rays at a smooth interface are shown; the angles of incidence, reflection, and transmission are measured from the normal to the interface. When the reflected and transmitted rays are perpendicular, the reflected waves are polarized perpendicular to the plane of incidence.

occurs are specified by the *Brewster*[13] *Law,*

$$\theta_i^{(\text{pol})} = \theta_r^{(\text{pol})}$$

$$= \text{Tan}^{-1}\left(\frac{n_t}{n_i}\right) \ . \tag{XI.10}$$

This value is known as the *Brewster angle.* Polarizing sunglasses are designed to take advantage of this effect: Made of polarizing material mounted with its preferred direction vertical, they block reflected glare from horizontal surfaces. Photographers also use polarizing filters, e.g., either to suppress or to enhance images reflected from water surfaces.

In constrast to plane polarization, electromagnetic waves can also feature *circular polarization.* For example, in the combination of plane-polarized waves with electric field

$$\mathbf{E}(x,t) = E_0\left[\hat{\boldsymbol{y}}\,\sin(kx - \omega t) + \hat{\boldsymbol{z}}\,\cos(kx - \omega t)\right] , \tag{XI.11}$$

the electric field at any location rotates in the *y-z* plane. Circularly polarized light can be produced by sending unpolarized light through a *quarter-wave plate.* This is made of birefringent material of such thickness that waves of the two polarizations are separated by one-quarter cycle (phase shift $\pi/2$) when they emerge.[14] Circularly polarized waves carry angular momentum as well as energy and linear momentum. A wave packet of energy ΔU has angular momentum of magnitude $\Delta L = \Delta U/\omega$, with direction specified by a right-hand rule—i.e., the thumb of the right hand indicates the direction of $\mathbf{\Delta L}$ when the fingers are curled in the direction of rotation of the $\mathbf{E}$ field at any point. For example, the wave with electric field (XI.11) has angular momentum $\mathbf{\Delta L}$ in the $+x$ direction.

C   Electromagnetic wave dynamics

To explore the dynamics of electromagnetic waves, it is useful to cast the Maxwell equations in a different form.

The Maxwell equations in differential form

The integral forms (XI.1) of the Maxwell equations can be transformed into differential equations by applying them to infinitesimal regions. For example, the flux of $\mathbf{E}$ through the surface of an infinitesimal rectangular solid is illustrated in Fig. XI.4(a). The outward flux through the top and bottom faces, say, is the *difference* in component E_z between top and bottom times the area of one face. The combined contributions of all three pairs of faces take the form

$$\oint \mathbf{E} \cdot d\mathbf{\mathcal{A}} = \left(\frac{\partial E_x}{\partial x} + \frac{\partial E_y}{\partial y} + \frac{\partial E_z}{\partial z}\right) dx\,dy\,dz \ . \tag{XI.12a}$$

---

[13] David Brewster, 1781–1868
[14] If the two polarizations have different amplitudes, the light is *elliptically polarized.*

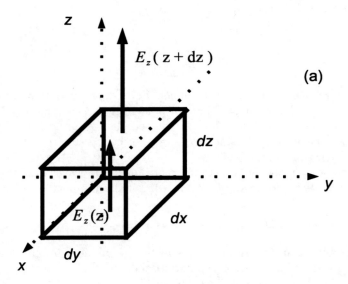

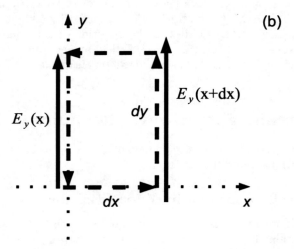

Figure XI.4: (a) Flux through the surface of an infinitesimal volume. (b) Circulation around an infinitesimal loop.

The quantity in parentheses is called the *divergence* of the vector field **E**, as it measures the "outflow" represented by the field. It is represented as $\nabla \cdot \mathbf{E}$ because of its resemblance to the "dot product" of the vector *differential operator*

$$\nabla = \hat{\boldsymbol{x}} \frac{\partial}{\partial x} + \hat{\boldsymbol{y}} \frac{\partial}{\partial y} + \hat{\boldsymbol{z}} \frac{\partial}{\partial z} \tag{XI.12b}$$

with the vector field.[15] So the flux through the surface of the infinitesimal solid is

$$\oint \mathbf{E} \cdot d\boldsymbol{\mathcal{A}} = \nabla \cdot \mathbf{E} \, dx \, dy \, dz \ . \tag{XI.12c}$$

The charge inside the solid is $\rho \, dx \, dy \, dz$, where ρ is the volume charge density. Hence, the Gauss law (XI.1a) applied to the solid implies the local, pointwise result

$$\nabla \cdot \mathbf{E} = \frac{\rho}{\epsilon} \ . \tag{XI.13a}$$

A similar application of the Gauss law for magnetism (XI.1b) yields

$$\nabla \cdot \mathbf{B} = 0 \tag{XI.13b}$$

as the local form of that law.

The circulation of **E** about an infinitesimal rectangular loop is shown in Fig. XI.4(b). The contribution of the vertical sides is the difference of E_y between right and left times the length of one side. That of the horizontal sides is the *negative* of the difference of E_x on top and bottom (because of the direction of circulation around the loop) times the length of one side. The total circulation is

$$\oint \mathbf{E} \cdot d\boldsymbol{\ell} = \left(\frac{\partial E_y}{\partial x} - \frac{\partial E_x}{\partial y} \right) dx \, dy \ . \tag{XI.14a}$$

The quantity in parentheses is the z component of the *curl* of **E**, the vector

$$\nabla \times \mathbf{E} = \left(\frac{\partial E_z}{\partial y} - \frac{\partial E_y}{\partial z} \right) \hat{\boldsymbol{x}} + \left(\frac{\partial E_x}{\partial z} - \frac{\partial E_z}{\partial x} \right) \hat{\boldsymbol{y}} + \left(\frac{\partial E_y}{\partial x} - \frac{\partial E_x}{\partial y} \right) \hat{\boldsymbol{z}}$$

$$= \begin{vmatrix} \hat{\boldsymbol{x}} & \hat{\boldsymbol{y}} & \hat{\boldsymbol{z}} \\ \dfrac{\partial}{\partial x} & \dfrac{\partial}{\partial y} & \dfrac{\partial}{\partial z} \\ E_x & E_y & E_z \end{vmatrix} \, , $$

$$\tag{XI.14b}$$

where the determinant is simply a mnemonic for remembering the component derivatives.[16] The vector area of the loop is $d\boldsymbol{\mathcal{A}} = dx \, dy \, \hat{\boldsymbol{z}}$, so the circulation

---

[15]The resemblance is exact only in Cartesian, i.e., rectangular coordinates. In other, curvilinear coordinate systems, the componentwise form of the divergence is more complicated.

[16]Here, too, the resemblance to a cross product is exact in Cartesian coordinates, with more complicated forms in other coordinate systems.

can be written in the coordinate-independent form

$$\oint \mathbf{E} \cdot \boldsymbol{d\ell} = (\nabla \times \mathbf{E}) \cdot \boldsymbol{d\mathcal{A}} \, . \qquad \text{(XI.14c)}$$

The rate of change of the flux of $\mathbf{B}$ through the infinitesimal loop is simply

$$\frac{d}{dt} \int \mathbf{B} \cdot \boldsymbol{d\mathcal{A}} = \frac{\partial \mathbf{B}}{\partial t} \cdot \boldsymbol{d\mathcal{A}} \, , \qquad \text{(XI.14d)}$$

and the circulation is calculated for one trip around the loop. Hence, since the Faraday law of induction (XI.1c) must hold for *any* infinitesimal loop, it implies

$$\nabla \times \mathbf{E} = -\frac{\partial \mathbf{B}}{\partial t} \qquad \text{(XI.15a)}$$

at every point. Similarly, applying the Ampère-Maxwell law to an infinitesimal loop yields

$$\nabla \times \mathbf{B} = \mu \mathbf{j} + \mu \epsilon \frac{\partial \mathbf{E}}{\partial t} \, , \qquad \text{(XI.15b)}$$

the local form of that law.

Readers familiar with vector analysis will recognize that the differential forms (XI.13) and (XI.15) of the Maxwell equations follow from the application of the *divergence theorem* and *Stokes theorem* to the integral forms (XI.1). These differential forms are the versions of the Maxwell equations used for most applications.[17] The integral forms can also be derived from the differential forms by treating finite volumes and surfaces as sums (integrals) of infinitesimal blocks and tiles, respectively, i.e., by applying the divergence and Stokes theorems in reverse. Hence, the two forms of the Maxwell equations are logically equivalent—they convey the same information.

An electromagnetic wave equation

In differential form, the Maxwell equations are coupled, first-order equations for the fields $\mathbf{E}$ and $\mathbf{B}$. A single second-order equation for $\mathbf{E}$ alone can be obtained by equating the curls of both sides of Eq. (XI.15a), which yields

$$\nabla \times (\nabla \times \mathbf{E}) = -\nabla \times \left(\frac{\partial \mathbf{B}}{\partial t} \right)$$
$$= -\frac{\partial}{\partial t} (\nabla \times \mathbf{B}) \, , \qquad \text{(XI.16a)}$$

since space and time partial derivatives commute, for well-behaved functions. *In vacuum or matter but away from any sources ρ or $\mathbf{j}$,* form (XI.15b) for $\nabla \times \mathbf{B}$ implies

$$\nabla \times (\nabla \times \mathbf{E}) = -\mu\epsilon \frac{\partial^2 \mathbf{E}}{\partial t^2} \, . \qquad \text{(XI.16b)}$$

---

[17]Advanced treatments may introduce the electric displacement $\mathbf{D}$ into Eqs. (XI.13a) and (XI.15b) and the magnetic intensity $\mathbf{H}$ into Eqs. (XI.13b) and (XI.15b).

The curl of the curl of $\mathbf{E}$, i.e., the curl of the vector field $\nabla \times \mathbf{E}$, requires a bit of care. The x component of this vector is given by

$$
\begin{aligned}
[\nabla \times (\nabla \times \mathbf{E})]_x &= \frac{\partial}{\partial y}(\nabla \times \mathbf{E})_z - \frac{\partial}{\partial z}(\nabla \times \mathbf{E})_y \\
&= \frac{\partial}{\partial y}\left(\frac{\partial E_y}{\partial x} - \frac{\partial E_x}{\partial y}\right) - \frac{\partial}{\partial z}\left(\frac{\partial E_x}{\partial z} - \frac{\partial E_z}{\partial x}\right) \\
&= \frac{\partial}{\partial x}\left(\frac{\partial E_y}{\partial y} + \frac{\partial E_z}{\partial z}\right) - \left(\frac{\partial^2 E_x}{\partial y^2} + \frac{\partial^2 E_x}{\partial z^2}\right) \\
&= \frac{\partial}{\partial x}(\nabla \cdot \mathbf{E}) - \left(\frac{\partial^2 E_x}{\partial x^2} + \frac{\partial^2 E_x}{\partial y^2} + \frac{\partial^2 E_x}{\partial z^2}\right) \\
&= \frac{\partial}{\partial x}(\nabla \cdot \mathbf{E}) - \nabla^2 E_x \ ,
\end{aligned}
\tag{XI.16c}
$$

where a term $\partial^2 E_x/\partial x^2$ is added to and subtracted from the third line to produce the fourth line, and the *Laplacian* operator

$$
\begin{aligned}
\nabla^2 &= \nabla \cdot \nabla \\
&= \frac{\partial^2}{\partial x^2} + \frac{\partial^2}{\partial y^2} + \frac{\partial^2}{\partial z^2}
\end{aligned}
\tag{XI.16d}
$$

is introduced[18] in the last line. Of course, there is nothing special about x; this is the x component of the vector identity

$$
\nabla \times (\nabla \times \mathbf{E}) = \nabla(\nabla \cdot \mathbf{E}) - \nabla^2 \mathbf{E} \ .
\tag{XI.16e}
$$

This suggests that the curl operator satisfies the BAC-CAB identity of vector cross products after a fashion. Away from sources, the Gauss law (XI.13a) implies $\nabla \cdot \mathbf{E} = 0$, so Eq. (XI.16b) takes the form

$$
\left(\nabla^2 - \mu\epsilon \frac{\partial^2}{\partial t^2}\right) \mathbf{E} = \mathbf{0} \ ,
\tag{XI.17a}
$$

or, explicitly,

$$
\left(\frac{\partial^2}{\partial x^2} + \frac{\partial^2}{\partial y^2} + \frac{\partial^2}{\partial z^2} - \mu\epsilon \frac{\partial^2}{\partial t^2}\right) \mathbf{E} = \mathbf{0} \ .
\tag{XI.17b}
$$

In a similar manner, $\mathbf{E}$ can be eliminated from Eqs. (XI.15), yielding an identical equation for $\mathbf{B}$ alone.

This is an hyperbolic, second-order partial-differential equation in the language of Ch. VI Sec. C. That is, it is a wave equation, and its solutions propagate as waves. That is the basis of Maxwell's remarkable discovery.[19]

---

[18]Again, the coordinate form shown here is appropriate to Cartesian coordinates.

[19]There is a logo, appearing on posters, t-shirts, cups, et cetera, which paraphrases Genesis: *And God said—$\nabla \cdot \mathbf{E} = \rho/\epsilon_0$, ...—and there was light.* The logo is clever. But it is likely that the Maxwell equations were not proclaimed from on high in these forms, but as the Lorentzian tensor identities $d\mathbf{F} = \mathbf{0}$ and $d * \mathbf{F} = \mu_0 * \mathbf{J}$, where $\mathbf{F}$ is the *Faraday field tensor*.

The negative sign of the time-derivative term in these equations, which makes them wave equations, is the product of the sign of the displacement current in the Ampère-Maxwell law and the negative sign in the Faraday law of induction. Without that negative sign, the equations would be elliptic in character, describing equilibrium configurations. But in the spacetime geometry of Einstein's mechanics, the propagation of electromagnetic waves delineates past, present, and future regions. That negative sign—innocuous as it appears—thus undergirds our experiences of past, present, and future; of cause and effect; even of such notions as consequence, responsibility, and morality. The negative sign in the Faraday law is the embodiment of the Lenz law, which is a facet of energy conservation. This chain of connections is the reason it can be asserted that the Law of Conservation of Energy is far more profound than a mere theorem in mechanics—it is fundamental to the nature of existence itself.

Solutions

The wave equation determines dynamical features of the waves. For example, $\mathbf{E}(\mathbf{r}, t) = \mathbf{f}(\mathbf{k} \cdot \mathbf{r} - \omega t)$ is a solution of Eqs. (XI.17) for any functional form $\mathbf{f}$, provided that the angular frequency ω and wave vector $\mathbf{k}$ satisfy dispersion relation (XI.3d).

The wave fields $\mathbf{E}$ and $\mathbf{B}$ satisfy separate wave equations. However, they must also satisfy the original Maxwell equations, which connect the two fields. For example, for sinusoidal fields of forms

$$
\begin{aligned}
\mathbf{E}(\mathbf{r}, t) &= \mathbf{E}_0 \, \cos(\mathbf{k} \cdot \mathbf{r} - \omega t) \\
\mathbf{B}(\mathbf{r}, t) &= \mathbf{B}_0 \, \cos(\mathbf{k} \cdot \mathbf{r} - \omega t + \phi) \,,
\end{aligned}
\qquad \text{(XI.18)}
$$

allowing for the possibility that the fields have different phases, the Gauss laws (XI.13) impose the transverse requirements (XI.3b), while the Faraday law (XI.15a) imposes condition (XI.3c), as well as the requirement $\phi = 0$, i.e., that the electric and magnetic fields oscillate in phase.

D   Submarine communication

Radio communication with submerged submarines is complicated by the fact that seawater is a conducting medium.[20] This is a different physical situation from that described by Eqs. (XI.17). It calls for a different wave equation, the solutions of which display novel features.

Wave equation in ohmic medium

For electromagnetic waves in an *ohmic* medium of conductivity σ, the Ampère-Maxwell law (XI.15b) includes the ordinary-current term $\mu\mathbf{j} = \mu\sigma\mathbf{E}$. When the Faraday and Ampère-Maxwell laws are combined as in the previous section,

---

[20]A submarine can trail a radio antenna on the surface, but then it risks detection.

Eq. (XI.16b) acquires the additional term $-\mu\sigma\left(\partial\mathbf{E}/\partial t\right)$ on the right-hand side. The resulting equation for $\mathbf{E}$ takes the form

$$\left(\nabla^2 - \mu\sigma\frac{\partial}{\partial t} - \mu\epsilon\frac{\partial^2}{\partial t^2}\right)\mathbf{E} = \mathbf{0} , \qquad (XI.19)$$

with an identical equation for $\mathbf{B}$. Though different from Eqs. (XI.17), this is still an hyperbolic second-order equation, i.e., a wave equation.

Boundary conditions

Perhaps the simplest scenario we can construct to investigate the submarine-communication problem is that of a plane-polarized plane electromagnetic wave in air, impinging vertically downward on the flat surface of the ocean. Part of the wave's energy is reflected upward; the rest is transmitted into the water, as sketched in Fig. XI.5. It is the propagation of the transmitted wave that is of interest here.

This is a *boundary-value problem*. The wave is a solution of wave equation (XI.19), subject to the *boundary conditions* that its fields must match those of the combined incident and reflected waves at the surface—the Maxwell equations demand this. If we take the surface to be the x-y plane, with coordinate z increasing downward into the ocean, we can write this boundary condition in the form

$$\mathbf{E}(z=0,t) = E_0\,\cos(\omega t)\,\hat{\boldsymbol{y}} , \qquad (XI.20)$$

and similarly for $\mathbf{B}$. Here, E_0 is a constant and ω is the angular frequency of the incoming wave. The wave fields of the solutions must satisfy these boundary conditions for all times t.

Solution

Tackling this problem calls for one of the two chief methods for solving differential equations—in this case, a judicious guess. It suffices to consider just the electric field. The field

$$\mathbf{E}(z,t) = \Re\{E_0\,\exp[i(kz - \omega t)]\}\,\hat{\boldsymbol{y}} \qquad (XI.21)$$

satisfies the boundary condition, while the use of the complex exponential function simplifies the evaluation of the derivatives.

This is a solution of wave equation (XI.19) if and only if dispersion relation

$$k^2 = \mu\epsilon\omega^2 + i\mu\sigma\omega \qquad (XI.22)$$

is satisfied. The value of k is most easily obtained by casting the complex right-hand side of this expression in polar form, as detailed in Eqs. (V.9), and using trigonometric half-angle formulae to display the square root. Solution (XI.21) ultimately takes the form

$$\mathbf{E}(z,t) = E_0\,e^{-\Im kz}\,\cos(\Re kz - \omega t)\,\hat{\boldsymbol{y}} , \qquad (XI.23a)$$

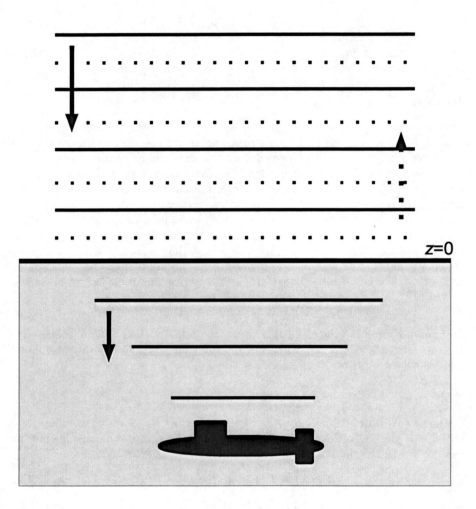

Figure XI.5: Incident, reflected, and transmitted radio wave fronts at the surface of the ocean ($z = 0$).

with

$$\Re k = \frac{\omega}{\tilde{c}} \left(\frac{\left(1 + \frac{\sigma^2}{\omega^2\epsilon^2}\right)^{1/2} + 1}{2} \right)^{1/2}$$

$$\Im k = \frac{\omega}{\tilde{c}} \left(\frac{\left(1 + \frac{\sigma^2}{\omega^2\epsilon^2}\right)^{1/2} - 1}{2} \right)^{1/2} ,$$

(XI.23b)

and $\tilde{c} = (\mu\epsilon)^{1/2}$, as in Eq. (XI.4). This shows behavior not seen in the waves of Secs. B and C. The wave decays away exponentially as it propagates into the conducting medium, its energy dissipated by the currents driven in the medium by the wave fields.

Skin depth

The distance within a conducting medium at which the wave fields are reduced to $e^{-1} \doteq 36.8\%$ of their surface amplitudes is called the *skin depth* of the medium for those waves. The skin depth δ is given by

$$\delta \equiv \frac{1}{\Im k}$$

$$\cong \left(\frac{2}{\sigma\mu\omega} \right)^{1/2} ,$$

(XI.24)

where the last (approximate) expression is appropriate to the limit $\sigma \gg \epsilon\omega$. For a given medium, the skin depth decreases with increasing wave frequency.

This has a useful consequence for microwave technologies. For example, silver has conductivity $\sigma_{\mathrm{Ag}} \approx 3. \times 10^7$ S/m and permittivity and permeability near their vacuum values at microwave frequencies such as $\nu = 1. \times 10^{10}$ Hz. These satisfy condition $\sigma \gg \epsilon_0\omega$, and the corresponding skin depth is

$$\delta_{\mathrm{Ag}} = \left(\frac{1}{\pi\sigma\mu_0\nu} \right)^{1/2}$$

$$= \left(\frac{1}{\pi(3. \times 10^7 \text{ S/m})(4\pi \times 10^{-7} \text{ N/A}^2)(1.0 \times 10^{10} \text{ Hz})} \right)^{1/2}$$

$$= 0.9 \ \mu\mathrm{m} .$$

(XI.25)

Hence, microwave components such as waveguides, resonators, and antennae made of brass or aluminum, say, with silver plating a few microns thick will yield the same performance as high-conductivity, solid silver components at a fraction of the cost.

Our submarine is surrounded by seawater, with conductivity $\sigma_{\text{sea}} \doteq 4.3$ S/m at low radio frequencies and permeability near μ_0. For a skin depth of, say, $\delta_{\text{sea}} = 10.$ m, the corresponding frequency is

$$
\begin{aligned}
\nu &= \frac{1}{\pi \sigma_{\text{sea}} \mu_0 \delta_{\text{sea}}^2} \\
&= \frac{1}{\pi (4.3 \text{ S/m})(4\pi \times 10^{-7} \text{ N/A}^2)(10. \text{ m})^2} \\
&= 590 \text{ Hz} .
\end{aligned}
\tag{XI.26}
$$

At this frequency $\sigma \gg \epsilon \omega$ is still satisfied, so expression (XI.24) for δ_{sea} remains valid. At a depth of $3.0\,\delta_{\text{sea}}$, for example, the intensity of a radio signal—proportional to $|\mathbf{E}|^2$—is reduced by a factor $e^{-6.0} \doteq 0.25\%$. Hence, such extremely low frequencies are needed to communicate with submarines running submerged. This is why the U.S. Navy, in the 1970s, intended to bury an extremely-low-frequency radio-antenna grid under hundreds of square miles of the Upper Peninsula of Michigan to communicate with its ballistic-missile submarines on patrol in the oceans of the world.[21]

The low-frequency requirement (XI.26), three orders of magnitude lower than AM broadcast frequencies, also imposes limitations on the *rate* at which information can be transmitted. The basic calculations presented here—which serve to show the Maxwell equations in use—only touch upon the engineering challenges presented by the problem of submarine communication.

E Quantum wave equations

Before we leave this topic, it behooves us to compare electromagnetic wave equations (XI.17) and (XI.19) with some other wave equations of importance. In particular, such equations are central to quantum mechanics, the overarching framework of most of current physics.

Principles of quantum mechanics

Quantum mechanics is built on two complementary principles: Matter classically regarded as particulate is properly described by waves. At the same time, these waves have quantized excitations that impart to matter its particulate nature. The links between the particle and wave descriptions are found in the *de Broglie*[22] *relations* between the dynamical variables appropriate to particles and corresponding wave properties. Momentum is linked to the wave vector via relation

$$
\mathbf{p} = \hbar \mathbf{k} .
\tag{XI.27a}
$$

---

[21]This proposal, known as Project Seafarer, was very controversial at the time.
[22]Louis-Victor de Broglie, 1892–1987

This is expressed as a differential operation on a wave function ψ in the form

$$\mathbf{p}\psi = \frac{\hbar}{i}\,\nabla\psi\;.\tag{XI.27b}$$

Applied to a wave function of form $\psi(\mathbf{r}, t) = \exp[i(\mathbf{k} - \omega t)]$, describing a state of definite momentum, this reproduces the previous relation. Energy is associated with angular frequency via

$$E = \hbar\omega\;,\tag{XI.27c}$$

or the form

$$E\psi = i\hbar\,\frac{\partial\psi}{\partial t}\tag{XI.27d}$$

as a differential operation on wave function ψ.

Schrödinger wave equation

The Schröinger equation fundamental to ordinary quantum mechanics is simply the Newtonian energy relation

$$\frac{|\mathbf{p}|^2}{2m} + U(\mathbf{r}) = E\tag{XI.28a}$$

for a single particle of mass m with potential energy $U(\mathbf{r})$ and total energy E, transformed via the de Broglie relations into the famous form

$$-\frac{\hbar^2}{2m}\,\nabla^2\psi + U\,\psi = E\,\psi\;.\tag{XI.28b}$$

As noted in Ch. VI Sec. C, this *is* a wave equation despite its parabolic form because of the imaginary factor i in the time-derivative term. Its solutions are wavelike, not diffusive, in character.

Klein-Gordon equation

Transforming the *Einsteinian* energy-momentum relation

$$E^2 - |\mathbf{p}|^2 c^2 = m^2 c^4\tag{XI.29a}$$

via the de Broglie relations yields the *Klein-Gordon equation*

$$\left(\nabla^2 - \frac{1}{c^2}\,\frac{\partial^2}{\partial t^2}\right)\varphi - \frac{m^2 c^2}{\hbar^2}\,\varphi = 0\;.\tag{XI.29b}$$

This equation governs the wave function φ describing a spinless, free particle of mass m. Potentials can be introduced into the Klein-Gordon equation, but that calls for a more sophisicated treatment than is appropriate here. The interpretation of this equation and its solutions is rather involved, the proper subject of another text.

References

The Maxwell equations and electromagnetic waves are central topics in all introductory physics texts and in texts on electricity and magnetism at every level. The example tackled in this chapter, viz., the propagation of waves in conducting media and the submarine-communication problem, is treated in detail in, for example, the intermediate-level text by Reitz, Milford, and Christy [RMC08].

The Schrödinger wave equation is treated in most introductory-level physics texts, all intermediate- and advanced-level texts on quantum mechanics, and all modern chemistry texts. The Klein-Gordon equation is featured in advanced-level texts on Einsteinian quantum mechanics, such as those of Bjorken and Drell [BD64, BD65] and that of Landau, Lifshitz, and Pitaevskii [LLP90].

Problems

1. A sufficiently intense beam of light can ionize the air. The energy of the beam is dissipated by the resulting currents, and the beam damps out.

   (a) Assuming the breakdown field of air is the same at optical frequencies as for static fields, viz., $E_{\max}^{(\text{air})} = 3.0 \times 10^6$ V/m, calculate the intensity of light required to do this, i.e., the maximum-intensity beam that can propagate freely in air.

   (b) What is the maximum magnetic-field amplitude for such a light wave?

   (c) Actually, the dielectric strength of air is likely to be somewhat lower at optical frequencies than for static fields. Why?

2. A plane-polarized plane electromagnetic wave impinges on a material. Treat a charge in the material as a damped oscillator driven by the electric field of the wave. Assume that this oscillator quickly attains steady-state oscillations.

   (a) Calculate the work done by the electric field on the charge in one cycle of the wave.

   (b) Calculate the impulse delivered to the charge by the magnetic field of the wave in the same time period. How is this related to result (a)?

3. Electromagnetic waves can be described via other forms than plane waves. The electric and magnetic fields of *spherical waves* from a "point" (i.e., small) source are described by an amplitude function $A(r)$ that varies slowly with radius times a function that oscillates with radius, time, and the angular coordinates.

   (a) How must the amplitude function $A(r)$ vary with radius?

   (b) How does this compare with the radial dependence of the electro*static* field of a point charge?

4. As described in Sec. B, if a third polarizing sheet is inserted between two polarizers oriented at right angles, the intermediate polarizer at 45° to the first, one-eighth of the unpolarized light intensity incident on the first polarizer will pass through the last, i.e., one-fourth of the polarized intensity passed through the first sheet.

(a) If two polarizers are inserted between the first two, each oriented at 30° to the one before it, what fraction of the intensity passed through the first sheet will emerge through the last?

(b) If $n - 1$ polarizers are inserted between the first two, each oriented at angle $\pi/(2n)$ radians to the one before it, what fraction of the intensity passed through the first sheet will emerge through the last?

(c) What is the limit $n \to \infty$ of result (b)? What does this imply?

5. The *photon*—a term introduced by Albert Einstein—is the quantum of excitation of the electromagnetic field, the "particle of light." With a proper skepticism for classical models of quantum objects, we can model the photon as a classical electromagnetic-wave packet of energy $E = \hbar\omega$, as suggested by the de Broglie relations.

(a) Photons are governed by Einsteinian, not Newtonian mechanics. Using this classical model, determine the *mass* of the photon.

(b) Photons are usually characterized by circular polarization. The *spin quantum number* of a particle is its angular momentum component along its spin axis divided by the reduced Planck constant $\hbar$. Use this classical model to calculate the spin quantum number of the photon.

6. As described in Ch. VII Sec. E, an electrostatic field can be written as the negative of the gradient of an electrostatic potential, i.e., as $\mathbf{E} = -\nabla V$. Using the appropriate Maxwell equation, construct a second-order partial differential equation for V. Known as the *Poisson*[23] *equation,* this result is the principal equation used in advanced treatments of electrostatics— and Newtonian gravitation, as a similar equation governs the Newtonian gravitational potential.

7. In addition to their dynamical content, the Maxwell equations are used, for example, to derive *boundary conditions* on electric and magnetic fields at interfaces between different media.

(a) Prove that in the absence of a surface charge layer, the normal (perpendicular) component of $\epsilon\mathbf{E} = \mathbf{D}$ is continuous at the boundary surface between two media. Similarly, prove that the normal component of $\mathbf{B}$ is always continuous at such a surface. *HINT*: Consider "pillbox integration," described in Ch. VII Sec. D.

(b) Prove that in the absence of a surface current layer, the tangential component of $\mathbf{B}/\mu = \mathbf{H}$ is continuous at the boundary between two media.

---

[23] Siméon Denis Poisson, 1781–1840

Similarly, prove that the tangential component of **E** is always continuous at such a surface. *HINT*: Use an infinitesimal rectangular loop of suitable shape.

8. Consider an electromagnetic wave with electric field of form (XI.23), propagating into a conducting medium with permittivity ϵ, permeability μ, and conductivity σ. The wave decays exponentially with depth, as its energy is dissipated by currents in the medium. Calculate the energy dissipation per unit volume in the medium, in terms of the relevant wave and medium properties, as a function of depth z.

9. Comparing the electromagnetic wave equation (XI.17) and the Klein-Gordon equation (XI.29b), determine the mass of the photon. Does this agree with the result of Problem 5(a)?

Chapter XII

A Pair of Binoculars

- *Reverse engineer an ordinary pair of binoculars: Analyze the performance of the prisms and objective and eyepiece lenses.*

- *Understand the function of the lens coatings.*

- *Examine the limitations on resolution imposed by diffraction.*

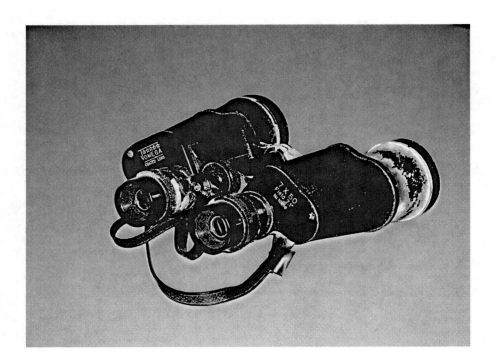

Figure XII.1: One of many devices with which we manipulate light to extend human vision. (Photograph by author.)

The science of *optics*—the study of light—is one of the oldest and most extensively developed branches of physics. Some of its principles were known to Euclid in the fourth century BCE, making them perhaps the oldest scientific laws still in use.

Optics is the basis of a diverse and sophisticated technology. "Quizzing glasses" and mirrors were known to the ancient Egyptians, Greeks, and Romans; spectacles appeared in medieval Europe. The introduction of the telescope in the seventeenth century completely transformed astronomy, while the microscope inaugurated a similar revolution in biology. Now our devices for manipulating light and other electromagnetic waves allow us to observe the universe from the most distant galaxies to the structure of living cells and to transmit images and messages all over the world. A comprehensive survey of optical technology is far beyond the scope of this text. Instead, we shall focus on a single device—an ordinary pair of binoculars, as pictured in Fig. XII.1—and explore the application of optical principles to as many aspects of its design as we can reach.

Like mechanics, optics is divided into three major branches: *Geometric* or *ray optics,* in which our principal concern is the direction in which light moves; *physical* or *wave optics,* in which the wave and electromagnetic properties of light come into play; and *quantum optics,* in which the behavior of light as excitations of the quantized electromagnetic field, interacting with other quantum systems, is central. The first two branches constitute *classical optics,* the subject of this chapter. Quantum optics must be deferred to another text.

Though it is the product of centuries of study, all of classical optics falls within the scope of the Maxwell equations. Maxwell's remarkable insight that light was a form of electromagnetic radiation incorporated the entire field within his theory. All of the principles and results we shall explore here can be derived by applying the Maxwell equations to suitable wave fields. Such breathtaking feats of analysis, however, must be left to more advanced treatments of electromagnetism.

A   Geometrical optics

It is usually said that geometrical optics applies when the scales on which we observe are much larger than the wavelengths of light or other electromagnetic waves—geometrical optics works perfectly well for radio, microwaves, or x-rays on suitable scales. But to be precise, some of the scales involved must be *smaller* than the relevant wavelength. For example, to function as a mirror, a metal surface must be smooth on wavelength scales. Its irregularities must be small compared to a wavelength. Hence, a layer of aluminum evaporated onto glass reflects images as a mirror, while a cast-iron frying pan does not. The frying pan, however, functions perfectly well as a mirror for microwaves with centimeter wavelengths.

Perhaps the longest-known principle in optics is the *Law of Rectilinear Propagation*: In the geometrical-optics limit, *light travels in straight lines* and casts sharp shadows.

Other important principles govern the behavior of light *reflected* and *refracted* at a smooth interface between different media, including vacuum. The standard depiction of this situation is shown in Fig. XII.2. The *incident, reflected,* and *transmitted* waves are identified by the corresponding rays; the directions of the rays are specified by angles θ_i, θ_r, and θ_t measured from the normal to the interface. (The term *transmitted* is used here in preference to *refracted* so that the subscripts are distinct.) The three rays lie in a plane—momentum conservation alone demands this—called the *plane of incidence.*

The three angles are related by two laws. One of these was also known to ancient Greek thinkers: The *Law of Equiangular Reflection* asserts the equality

$$\theta_r = \theta_i \ . \qquad\qquad (\text{XII.1})$$

The reflection from a smooth surface described by this law is termed *regular* reflection. The diffuse reflection from a rough surface, such as the aforementioned frying pan, is *specular* reflection.

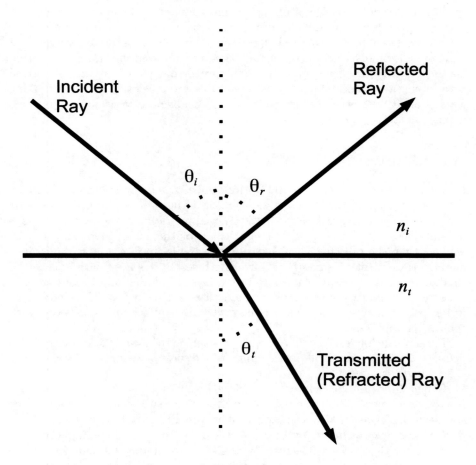

Figure XII.2: Incident, reflected, and transmitted (refracted) light rays at a smooth interface between two media.

The other law was not established for some twenty centuries. Willebrord Snell observed in 1621 that for given materials meeting at the interface, the angles θ_i and θ_t are not in fixed proportion but their sines are. He introduced *indices of refraction* for each material to specify the proportion—the *Snell Law of Refraction*—in the form

$$\frac{\sin \theta_i}{\sin \theta_t} = \frac{n_t}{n_i} \, , \tag{XII.2a}$$

or the alternative form

$$n_i \sin \theta_i = n_t \sin \theta_t \, , \tag{XII.2b}$$

where n_i is the index of the medium carrying the incident light and n_t that of the medium carrying the transmitted light. For example, vacuum has index $n_{\mathrm{vac}} = 1$, water $n_{\mathrm{H_2O}} = 1.333$, crown glass $n_{cg} = 1.520$, et cetera. Snell apparently envisioned that the index varied directly with the density of the medium, but this is not categorically true. As noted in Ch. XI Sec. B, it was some fifty years later that Huygens correctly connected the index of refraction to the speed of light in the medium, and it was more than two hundred years after Snell that Maxwell connected it with the medium's electric and magnetic properties.

Neither of these two laws specifies how much of the incoming radiation goes into each of the two outgoing beams. The Maxwell equations, applied to the wave fields, determine the relative intensities of the reflected and transmitted waves. These intensities depend on the angle of incidence θ_i, the indices of refraction n_i and n_t, and the polarization of the waves as well.

Most optical media are *dispersive*: They have an index of refraction $n(\omega)$ dependent on the frequency, i.e., color of the light. Hence, for a given angle of incidence, the angle of refraction differs for different frequencies. This is the basis of Isaac Newton's observation that a glass prism could separate white light into its component colors and the basis for the best-known example of dispersion in nature—the rainbow.

For light in a medium with a higher index of refraction incident on a medium with a lower index $(n_i > n_t)$, e.g., light traveling from glass into air or water into air, the Snell law can specify a value of $\sin \theta_t$ greater than unity. In that case there is no transmitted ray, and all the incident light is reflected back into the incident medium. This *Total Internal Reflection* occurs for incident angles θ_i greater than the critical angle

$$\theta_{\mathrm{crit}} = \mathrm{Sin}^{-1} \left(\frac{n_t}{n_i} \right) \, . \tag{XII.3}$$

This phenomenon is responsible for the sparkle of water droplets and diamonds, which have a very high n value.

Total internal reflection is not entirely total. It is not possible to have incident and reflected wave fields on one side of an interface and *nothing* on the other side—that would not be consistent with the Maxwell equations. When total internal reflection occurs, the fields on the other side of the interface constitute an *evanescent wave* that falls off exponentially with distance from the interface, rather than oscillating with distance as an ordinary wave. If the medium on the other side extends to infinity, the evanescent wave carries no energy away from the interface, and the intensity of the reflected beam equals that of the incident beam. But the evanescent wave is real. For example: It is occasionally necessary to excite fluorescence in molecules within the wall of a biological cell without exciting the molecules in the cytoplasm within, as that would overwhelm the desired signal. How can the illumination be focused so precisely? The sample is mounted near an interface, on the other side of which total internal reflection of the illuminating beam occurs. The evanescent wave will excite the molecules in the cell wall, but its exponential fall-off suppresses illumination of the interior. This very clever technique is known as *Total Internal Reflection Fluorescence.*

The evanescent wave can excite an ordinary wave if the "transmission" medium is not infinite, but a layer of finite thickness between two regions of the incident medium. For example, the two glass prisms with an air gap sketched in Fig. XII.3 form a beam splitter: Part of the incident beam is internally reflected at 90°, while part of the energy is "siphoned off" by the evanescent wave in the gap, appearing as a transmitted beam in the second prism. Because of the exponential fall-off of the evanescent wave, the distribution of intensity between the two beams can be adjusted via small changes in the width of the gap.[1] This phenomenon has the delightful name *Frustrated Total Internal Reflection.* It also occurs for the quantum waves that describe matter particles, which can pass through potential-energy barriers they lack the energy to surmount. In this context, the effect is known as *Schrödinger tunneling.* Although often presented as mysterious, quantum tunneling is simply an analogue of the classical optical effect.

These laws of propagation, reflection, and refraction can all be derived from a construction known as the *Huygens principle*: Each point of a wave front—a surface of constant phase—can be considered a point source of a new wave; the *envelope* of these waves constitutes the ongoing wave front. To reproduce the Snell law, Huygens needed to stipulate that the speed of the waves was *reduced* by the index of refraction of a medium,[2] i.e., given by $\tilde{c} = c/n$ as in Eq. (XI.4).

At an atomic level, both reflection and refraction are *scattering* processes. The reflected and refracted waves are, classically, reradiated by charges set oscillating by the incident wave fields. The directions these waves take are those in which the waves produced by individual atoms reinforce each other. This phenomenon, known as *constructive interference,* is described in Sec. D.

---

[1]This is not mere calculation. One of this author's graduate-school office mates built such a device. It worked.

[2]By contrast, Newton derived the Snell law by asserting that light particles *sped up* in media with larger n as particles entering a region of lower potential energy. The issue hung fire for the lifetimes of the two men, although Newton's prestige tipped opinion in his favor. The crucial experiment establishing the wave nature of light would not be performed until the turn of the nineteenth century, as described in Sec. D following. Explicit measurements of the speed to settle the matter (in Huygens' favor) were finally performed by Fizeau in the middle of the nineteenth century.

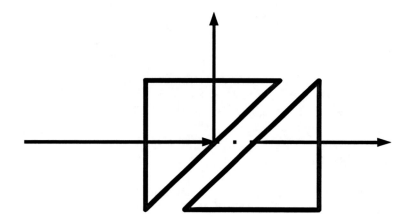

Figure XII.3: A beam splitter that works via Frustrated Total Internal Reflection.

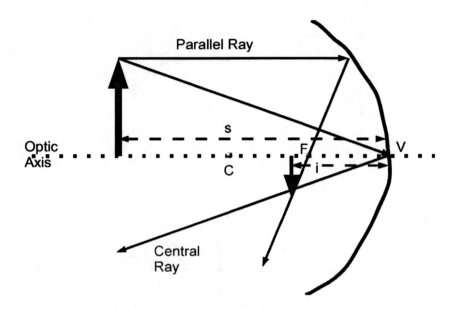

Figure XII.4: Reflection of light rays from a spherical mirror, shown in cross section.

B   Optical images

The most important optical technologies manipulate the direction of light to form images. Mirrors do this with reflection; refracting surfaces and lenses use refraction at interfaces.

Spherical and plane mirrors

A *spherical mirror* is a reflecting surface constituting a portion of a spherical surface. Light from a source is reflected from the mirror according to the Law of Equiangular Reflection—the only physics governing the device—to form an image, as illustrated in Fig. XII.4. In the *paraxial approximation*—in which it is assumed all rays are close to the *optic axis* and all relevant angles small enough that an angle (in radians), its sine, and its tangent can be treated as equal, its cosine treated as unity—the source distance s, image distance i, radius r, and focal length f can be shown to satisfy the relations

$$\frac{1}{s} + \frac{1}{i} = \frac{2}{r}$$

$$= \frac{1}{f} \, . \tag{XII.4}$$

All distances are measured from the vertex V at which the optic axis meets the mirror surface. The radius r refers not to the size of the mirror itself, but to the radius of the sphere of which the mirror surface is a portion—the distance from the vertex V to the center C. The focal length f is the distance at which rays parallel to the optic axis, i.e., rays from a very distant source, are brought together at the focal point or focus F. For a spherical mirror, in the paraxial approximation, this point is halfway between the vertex V and center C.

This single formula covers all cases of reflection at a spherical mirror once the algebraic signs of the distances s, i, r, and f are assigned and interpreted appropriately. The general rule is that *distances are positive where the light actually goes*. A standard *sign convention* for these quantities is given by

$$s = \begin{cases} + & \text{in front of mirror (real source);} \\ - & \text{behind mirror (virtual source);} \end{cases}$$

$$i = \begin{cases} + & \text{in front of mirror (real, inverted image);} \\ - & \text{behind mirror (virtual, erect image);} \end{cases} \tag{XII.5}$$

$$r, f = \begin{cases} + & \text{center in front of mirror (concave);} \\ - & \text{center behind mirror (convex).} \end{cases}$$

A "virtual source" (or "virtual object") is an image formed by another optical element, with rays converging on the mirror as if from a source behind the mirror.

The behavior of an optical system can be analyzed graphically by *ray tracing*, drawing the directions of individual rays to locate the image. Certain rays are particularly easy to identify. For a spherical mirror, these include:

- a *parallel ray*, parallel to the optic axis, reflected through the focus;

- a *focal ray*, passing through the focus, reflected parallel to the optic axis;

- a *radial ray*, passing through the center C, striking the mirror normally and reflected back on itself;

- a *central ray*, reflected from the vertex V at equal angle to the optic axis.

A ray that satisfies the Law of Equiangular Reflection but does not fit any of these special categories is called a *wild ray*.

Spherical mirrors can produce images smaller or larger than their sources, depending on the geometry. The *magnification* of the mirror is defined as the ratio of the (signed) height of the image to that of the source. This is given by

$$M = -\frac{i}{s} , \qquad (\text{XII.6})$$

as can be seen by noting that a central ray through source and image makes equal angles to each side of the optic axis at the vertex. The negative sign incorporates the fact that real images, with positive image distances, are inverted. As indicated by Eq. (XII.4), a concave ($f > 0$) mirror can produce enlarged or reduced real images or enlarged virtual images, depending on the position of the source. A convex ($f < 0$) mirror always produces reduced virtual images of real sources. All this can readily be demonstrated, e.g., with an ordinary spoon.

As indicated, these results are approximate. In particular, the set of points halfway from the center C of a sphere to its surface is not a single focal point, but a portion of a smaller sphere. Hence, for example, a bundle of parallel rays of finite cross section will be focused to a disk, not a point. This distortion is called *spherical aberration*.[3] For ordinary applications such as a makeup mirror or even a small telescope, this is unimportant. But for large astronomical instruments where high-precision imaging is required, spherical aberration is eliminated by using *parabolic* mirrors shaped as paraboloids of revolution.[4] Such surfaces do reflect all incoming parallel rays to a single point. The price paid for this advantage is that the parabolic figure is much harder to make than a spherical mirror.

The familiar plane mirror can be included in this analysis as the $r \to \infty$ limit of a spherical mirror. A plane mirror produces virtual images with $i = -s$ and magnification $M = +1$.

Spherical refracting surfaces

Refraction of light at a spherical boundary surface between media of different indices of refraction can likewise form images. This is illustrated in Fig. XII.5. Here, the indices of refraction of the media, as well as the geometry, are involved. The governing physics is the Snell Law of Refraction. Again, in the paraxial approximation, the source and image distances and the curvature radius of the surface are related by

$$\frac{n_1}{s} + \frac{n_2}{i} = \frac{n_2 - n_1}{r} . \qquad (\text{XII.7})$$

A sign convention appropriate to a spherical refracting surface is similar, but not

---

[3]The term *aberration* in optics refers to any aspect of an optical system that distorts the image.

[4]For example, the ten-meter mirrors of the Keck telescopes on Mauna Kea, Hawaii, each consist of thirty-six glass pillars, the top of each of which is a section of a paraboloid of revolution. The pillars have hexagonal cross sections and are fitted together like tiles, their heights precisely adjusted by computer to make the reflecting surface a paraboloid smooth to a fraction of the wavelength of light.

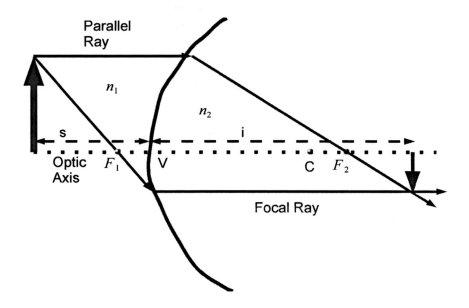

Figure XII.5: Refraction at a spherical interface between media with indices of refraction n_1 and n_2.

identical, to that for a spherical mirror. This relation covers all cases with signs assigned via

$$
s = \begin{cases} + & \text{incident side (real source);} \\ - & \text{transmission side (virtual source);} \end{cases}
$$

$$
i = \begin{cases} + & \text{transmission side (real, inverted image);} \\ - & \text{incident side (virtual, erect image);} \end{cases} \qquad \text{(XII.8)}
$$

$$
r = \begin{cases} + & \text{center on transmission side;} \\ - & \text{center on incident side.} \end{cases}
$$

A spherical refracting surface does not have a single focal length. For example, for $r > 0$ and $n_2 > n_1$, as sketched in the figure, focal rays passing through point F_1 at distance

$$
f_1 = \frac{n_1 r}{n_2 - n_1} \qquad \text{(XII.9a)}
$$

on the incident side are refracted parallel to the optic axis (i.e., to $i \to +\infty$). But parallel rays on the incident side (from $s \to +\infty$) are refracted through point F_2 at distance

$$
f_2 = \frac{n_2 r}{n_2 - n_1} \qquad \text{(XII.9b)}
$$

on the transmission side.

Ray tracing for spherical refracting surfaces is only slightly different from that for mirrors. Special rays include

- a *parallel ray,* incident parallel to the optic axis, refracted through point F_2 on the transmission side;

- a *focal ray,* incident through point F_1, refracted parallel to the optic axis;

- a *radial ray,* striking the surface normally, emerging undeflected;

- a *central ray,* striking the vertex, deflected as per the Snell law;

with *wild rays* not fitting any of these categories.

Because a central ray is deflected according to the Snell law, the magnification of a spherical refracting surface is a bit more complicated than that of a spherical mirror. In the paraxial (small-angle) approximation, it is given by

$$
M = -\frac{n_1 i}{n_2 s} \, , \qquad \text{(XII.10)}
$$

involving the indices of refraction as well as the source and image distances. As before, real images are inverted, with $M < 0$; virtual images are erect, with $M > 0$.

Refraction introduces another distortion into the imaging process: *chromatic aberration.* If either medium is dispersive, then light waves of different

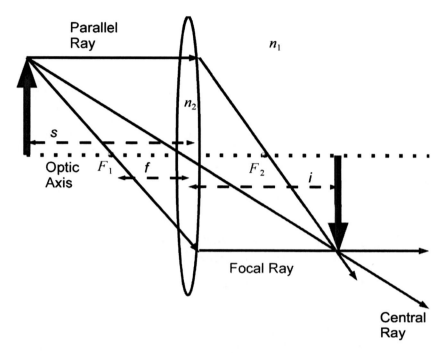

Figure XII.6: Focusing of light by a thin lens. A convergent lens, with focal length $f > 0$, is pictured.

wavelengths—colors—will be focused slightly differently. The image of a white-light source will be smeared out into its constituent colors. Again, for high-precision astronomical imaging, this is important. Isaac Newton introduced the *reflecting telescope,* in which the primary imaging element is a mirror[5] rather than a lens, in part to eliminate chromatic aberration. It is possible to eliminate this aberration in a *refracting telescope,* but a system of lenses is required.

Thin lenses

Perhaps the most widespread and important application of optics consists of two spherical refracting surfaces on the same optic axis that are close enough to-gether that the distance between them can be neglected—thus, the same source and image distances can be used for both. This is the *thin lens* sketched in Fig. XII.6. Applying Eq. (XII.7) to each of the surfaces in turn, we can re-late the source and image distances, the radii r_1 and r_2 of the two surfaces, and the indices of refraction n_2 of the lens and n_1 of the surrounding medium,

---

[5]Specifically, a *first-surface* mirror, in which the reflecting material is on the front side of the glass backing. An ordinary household mirror is a *second-surface* mirror, with the reflecting surface on the back of the glass for protection.

respectively, by the equations

$$\frac{1}{s} + \frac{1}{i} = \frac{1}{f}$$

$$= \left(\frac{n_2}{n_1} - 1\right)\left(\frac{1}{r_1} - \frac{1}{r_2}\right). \tag{XII.11}$$

These can be regarded as two equations for $1/f$. The first is the *Thin-Lens Equation,* describing the imaging properties of the lens, and the second the *Lens-Makers' Equation,* relating the focal length of the lens to its construction. These equations cover all cases, with sign convention

$$s = \begin{cases} + & \text{incident side (real source);} \\ - & \text{transmission side (virtual source);} \end{cases}$$

$$i = \begin{cases} + & \text{transmission side (real, inverted image);} \\ - & \text{incident side (virtual, erect image);} \end{cases} \tag{XII.12}$$

$$r_1, r_2 = \begin{cases} + & \text{center on transmission side of lens;} \\ - & \text{center on incident side of lens.} \end{cases}$$

The same sign convention applies to r_1 and r_2, so if both surfaces have the same figure, e.g., both convex, they will have radii of opposite signs. The negative sign in the Lens-Makers' Equation accounts for this.

A thin lens does have a single focal length f. Focal rays incident on the lens through the first focus F_1 emerge parallel to the optic axis (toward $i \to +\infty$). Parallel rays incident parallel to the axis (from $s \to +\infty$) are refracted through the second focus F_2, the same distance from the lens as the first.

Ray tracing for a thin lens is as simple as for a mirror. The special rays are:

- a *parallel ray,* parallel to the optic axis, refracted through the second focus;

- a *focal ray,* passing through the first focus, refracted parallel to the optic axis;

- a *central ray,* passing through the center of the lens undeflected;

all others are classed as wild rays. There is no radial ray, as the two surfaces of the lens have different centers.

The focal length of a lens depends on the index of refraction n_2 of the material of which it is made[6] and the index n_1 of the surrounding medium. Familiar lenses are constructed with index of refraction satisfying $n_2 > n_1$. But the Lens-Makers' Equation covers the opposite situation just as well: Air bubbles, or plastic bags filled with air, can serve as lenses under water. In that

---

[6]Modern *progressive* eyeglass lenses, replacing earlier bifocal and trifocal designs, have focal lengths made to vary continuously over the area of the lens by varying the index of refraction of the lens glass.

case, our usual intuition about the geometry of lenses, e.g., that convex lenses (with $r_1 > 0$ and $r_2 < 0$, say) are convergent (with $f > 0$), must be reversed.

Since central rays pass undeflected through a thin lens, the magnification of the lens is given by the simple relation

$$M = -\frac{i}{s} \,. \tag{XII.13}$$

As before, real images are inverted and virtual images are erect.

The thin lens is a convenient approximation, but in many modern applications, not all lenses can be considered thin. For example, the 75-205 mm zoom lens for this author's single-lens reflex (SLR) camera contains some fourteen different optical elements, some of them quite thick (compared to the dimensions and focal length of the assembly). A thick lens must be analyzed as two distinct refracting surfaces in two different positions, the image produced by one becoming the source for the other.

C Operation of binoculars

We can now confront the engineering problem of the design of the binoculars, such as those pictured in Fig. XII.1. Its features are chosen to meet performance requirements, subject to certain practical limitations.

Objective and eyepiece lenses

A pair of binoculars like that pictured is sketched in cross section in Fig. XII.7. The device is a pair of *simple telescopes,* one for each eye of the user. Each telescope contains a large *objective lens* and a smaller *eyepiece lens.* The objective lens forms a real image of a distant source, essentially at its focal point. The lenses are mounted so that this falls just inside the focal point of the eyepiece lens. (The eyepieces are mounted on screw columns so that the focus can be adjusted for objects at different distances. One eyepiece has a separate screw so that independent adjustments can be made for each eye.) In accord with the Thin-Lens Equation (XII.11), the eyepiece forms a large, distant, virtual image of the objective-lens image. It is this eyepiece image that the observer sees.

Magnification is one figure of merit for the binoculars. What matters here is not the linear size of the image, but the angle it occupies in the observer's field of view. That determines the size of the image formed on the observer's retina, i.e., the perceived size of the image. The relevant dimensions are sketched in Fig. XII.8. The height of the image is the angular size θ_s of the distant object times the focal length f_1 of the objective lens, i.e., it is given by

$$h = \theta_s f_1 \,, \tag{XII.14a}$$

in the paraxial approximation. The angular size θ_i of the eyepiece image is this height divided by the focal length f_2 of the eyepiece lens, viz.,

$$\theta_i = \frac{h}{f_2} \,. \tag{XII.14b}$$

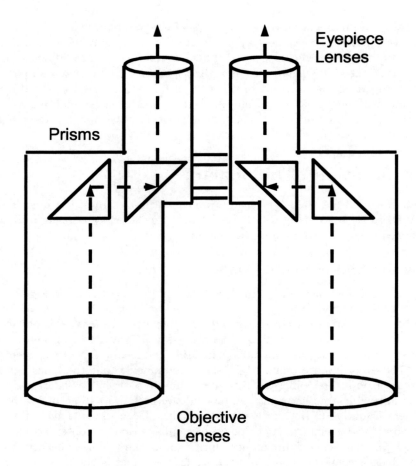

Figure XII.7: Simple schematic diagram of binoculars.

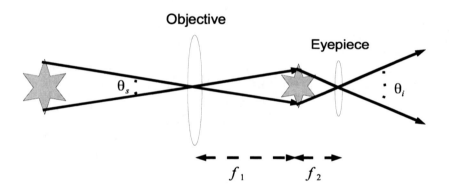

Figure XII.8: Dimensions and angles for a simple telescope or binoculars.

The *angular magnification* of the combination is the ratio

$$M = -\frac{\theta_i}{\theta_s}$$

$$= -\frac{f_1}{f_2} \; . \tag{XII.14c}$$

The negative sign indicates that the image the observer sees is upside-down. This is all right for an astronomical telescope, say, but for a spyglass or binoculars it is impractical. The sign is adjusted by an additional component to be described later.

The magnification of binoculars is constrained by practical considerations, as is typical of any engineering problem. The binoculars in Fig. XII.1 have magnification $M = 7.0$. Binoculars with $M = 20.0$ or even $M = 50.0$ are available, but if the magnification is too high, the task of holding the binoculars steady enough for viewing becomes prohibitive. High-magnification binoculars and telescopes are mounted on tripods (or for telescopes, more elaborate platforms) for this purpose. Also, higher magnification means a narrower field of view and a more stringent task of pointing the device. Furthermore, result (XII.14c) indicates that higher magnification calls for a longer objective focal length f_1. The *optical path* of the device is essentially the sum $f_1 + f_2$ of the two focal lengths. This, too, is constrained: A telescope can be tens of meters in length, but a spyglass or binoculars must be portable. The optical path of the binoculars in the photograph is approximately 30. cm. For magnification $M = 7.0$, this implies that the objective and eyepiece lenses have focal lengths

$$f_1 = \frac{7.0}{8.0} \, (30. \text{ cm})$$

$$= 26. \text{ cm} \tag{XII.15a}$$

and

$$f_2 = \frac{1.0}{8.0} \, (30. \text{ cm})$$

$$= 3.8 \text{ cm} \; , \tag{XII.15b}$$

respectively, to two significant figures.

A separate figure of merit for binoculars is *light-gathering power*. This determines the brightness of the image seen. Light-gathering power is proportional to the *area* of the objective lenses, i.e., to the square of their diameter. The size of the lenses is constrained by considerations of weight and cost. The binoculars pictured have objective lenses of diameter 50. mm. Binoculars with larger lenses are available, but a much greater size would be impractical. The binoculars here are labeled 7×50, indicating magnification 7 and objective lens diameter 50 mm. Most binoculars are labeled in this way.

Prisms

Small binoculars have eyepiece and objective lenses mounted in straight tubes, but larger versions are designed with the "double bend" shown in Figs. XII.1

and XII.7. Two carefully designed glass prisms are mounted in each bend. A total of four internal reflections (on each side) steer incoming light from the objective lenses around the bend and into the eyepiece lenses. This serves three purposes: The angles of the prism faces are chosen to invert the objective-lens image so that the image seen by the observer is erect. The double bend allows a longer optical path to be accommodated in a smaller device, enhancing both magnification and portability. And the bend allows wider objective lenses to be used for increased light-gathering power while keeping the eyepiece lenses at the same separation as the observer's eyes. This feature is implemented with prisms, rather than mirrors, because total internal reflection is not restricted by the reflectivity of mirrors, and the reflecting surfaces of the prisms are not subject to corrosion over time.

D   Physical optics: interference

To explore further the features of the binoculars, it is necessary to move from the geometrical-optics limit into the realm of physical or wave optics. The key phenomenon to be understood is *wave interference,* actually a "meta-phenomenon" or class of phenomena second in its sweep and significance, perhaps, only to resonance.

Superposition and coherence

A simple example of interference is that of two plane-polarized plane waves impinging on a common location. For simplicity, we can take the waves to have the same vector amplitude $\mathbf{E}_0$. As the Maxwell equations are linear, the electric field of the resulting wave is the vector sum of the individual contributions. It can be written

$$\mathbf{E} = \mathbf{E}_0 \left[\cos \phi + \cos(\phi + \delta) \right] , \qquad \text{(XII.16a)}$$

where ϕ is the phase of the first wave, and the second has phase differing from that by δ. These can be functions of position, time, or both. The squared magnitude of this field is given by

$$|\mathbf{E}|^2 = |\mathbf{E}_0|^2 \left[\cos^2 \phi + \cos^2(\phi + \delta) + \cos(2\phi + \delta) + \cos \delta \right] , \qquad \text{(XII.16b)}$$

the cross term transformed by suitable trigonometric identities. This is related to the intensity of the wave via Eq. (XI.5d). Averaged over many periods of the waves, the first two terms in the brackets contribute the same factor $\frac{1}{2}$ that they would in a calculation of the intensity of either wave individually. Hence, each of the first two terms contributes the intensity I_0 of the individual waves. The third term, not being squared, averages to zero, while the fourth contributes factor $\langle \cos \delta \rangle$ instead of $\frac{1}{2}$. The resulting intensity is

$$I = 2I_0 \left(1 + \langle \cos \delta \rangle \right) \qquad \text{(XII.16c)}$$

for the combined waves.

If the phase difference δ varies rapidly or randomly in time, the average of its cosine over many periods is $\langle \cos \delta \rangle = 0$. In this case, the two waves are said to be emphincoherent, as are, for example, the waves from different atoms in a hot light-bulb filament. The intensity of the combined sources is just $2I_0$, like streams of water from two garden hoses.

If the sources are linked in some way, so that the phase difference δ is constant in time, they are said to be *coherent*. In such cases, the intensity can be very different, taking the form

$$I = 2I_0 \left(1 + \cos \delta\right) . \tag{XII.17a}$$

The value of δ depends on how it arises. But in any circumstance in which it is an integer multiple of 2π, the intensity is enhanced, so:

$$\delta = 2\pi \, m \quad \Rightarrow \quad I = 4I_0 . \tag{XII.17b}$$

This phenomenon is *constructive interference*; the two original waves reinforce each other, crest to crest and trough to trough. If the phase difference is a half-integer multiple of 2π, the intensity is suppressed, so:

$$\delta = 2\pi \left(m + \tfrac{1}{2}\right) \quad \Rightarrow \quad I = 0 . \tag{XII.17c}$$

This is *destructive interference*; the two waves cancel each other,[7] crest to trough.

Streams of (classical) particles cannot do this—their fluxes can only add. Interference, wherever it appears, is the absolute signature of a wave phenomenon.

Two-slit interference

One way to obtain coherent sources of light is to use portions of a single wave front. Allowing such a wave front (obtained, for example, by allowing light to pass through a single narrow slit) to pass through two narrow, parallel slits and illuminate a distant screen creates *two-slit* interference. The geometry is illustrated in Fig. XII.9. The phase difference between the contributions of the two slits depends on the angle of propagation away from the central line and is constant in time. The path-length difference at angle θ for the two rays shown in the figure is $d \sin \theta$, where d is the slit spacing. The corresponding phase shift for waves of wavelength λ is

$$\delta = \frac{2\pi}{\lambda} d \sin \theta , \tag{XII.18a}$$

and the corresponding intensity distribution, from Eq. (XII.17a), is

$$I(\theta) = 4I_0 \cos^2 \left(\frac{\pi d \sin \theta}{\lambda} \right) . \tag{XII.18b}$$

This is plotted in Fig. XII.10. The pattern of light on the distant screen is not

---

[7] If the original waves are coherent but not of the same amplitude, constructive and destructive interference still occur, but the results are less extreme.

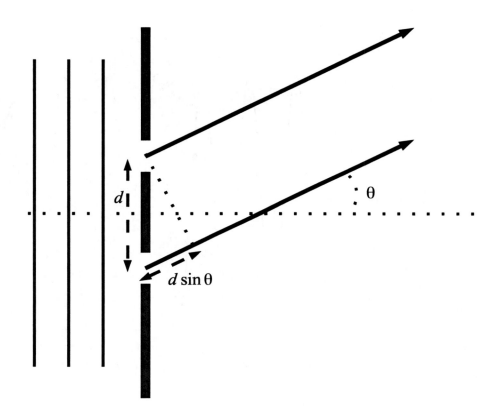

Figure XII.9: Geometry of two-slit interference.

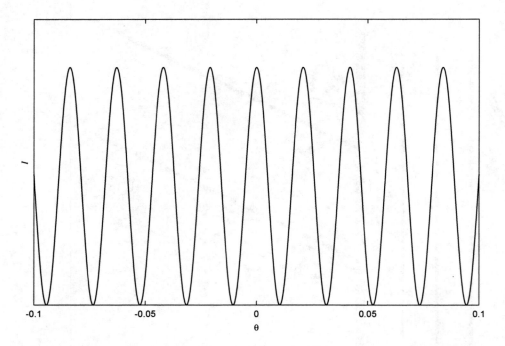

Figure XII.10: Two-slit interference intensity pattern. The intensity scale is arbitrary.

two spots or a single smudge, but a row of alternating bright and dark stripes extending far to both sides of the central line. The bright maxima of the pattern occur at angles $\theta_{\max}$ given by

$$\theta_{\max} = \mathrm{Sin}^{-1}\left(\frac{m\lambda}{d}\right) , \tag{XII.18c}$$

where the integer m is the *order* of the maximum. The minima—black spots, if the slits are equally illuminated—occur between, at angles

$$\theta_{\min} = \mathrm{Sin}^{-1}\left(\frac{(m + \frac{1}{2})\lambda}{d}\right) . \tag{XII.18d}$$

As with all interference phenomena, the scale of the pattern is set by the wavelength of the light. Two-slit interference can be used to measure the wavelength and can separate mixed illumination into its constituent wavelengths.

This experiment was first performed by Thomas Young[8] around 1801. It established incontrovertibly the wave nature of light.[9]

Similar interference patterns can be created with more than two slits. The resulting intensity peaks are higher and narrower, more cleanly separated. A modern *diffraction grating* consists of thousands of lines per centimeter engraved on a glass or plastic plate. The lines act as slits. The resulting interference pattern, with very sharply defined, bright peaks, is used for high-precision spectroscopic measurements. Certain types of *holograms* are made utilizing similar interference patterns. Laser light reflected from an object is allowed to interfere with a reference beam. The resulting pattern, recorded photographically, looks like an array of dark and clear stripes of varying spacings. When this is illuminated by another light source, the resulting "multiple slit" interference pattern reproduces the features of the original object in three dimensions.

Thin-film interference

A situation in which interference is more frequently encountered is that of *thin-film interference*. This is illustrated in Fig. XII.11. Waves reflected from interfaces on both sides of a thin layer combine and interfere. The beam reflected from the lower interface travels through the layer twice. Hence, for near-normal incidence, the phase difference between the two reflected waves is

$$\delta = \frac{4\pi t}{\lambda_2} + \delta_{23} - \delta_{12} , \tag{XII.19a}$$

with

$$\delta_{ij} = \begin{cases} \pi , & n_j > n_i ; \\ 0 , & n_j < n_i . \end{cases} \tag{XII.19b}$$

---

[8]He had to use lanterns and candles. Modern physics laboratory students enjoy the enormous advantage of being able to use lasers.

[9]The modern understanding that light is a *quantized* wave does not alter this.

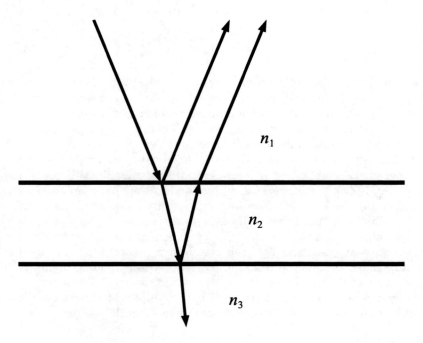

Figure XII.11: Thin-film interference with nearly normal incidence.

Here, t is the thickness of the layer and λ_2 is the wavelength of the light *in the layer*. The phase shifts δ_{ij} on reflection are determined by the boundary conditions imposed on the wave fields by the Maxwell equations.

Wavelengths for which this δ satisfies the constructive-interference condition (XII.17b) are seen strongly in reflection. The colors of oil layers on puddles of water are produced this way, as are the brilliant blue colors of the wings of *Morpho* butterflies. The reflection of waves for which δ satisfies the destructive-interference condition (XII.17c) is suppressed. All the energy of such waves is transmitted through the layer.

Newton observed thin-film interference caused by the layer of air between a convex lens and a flat glass blank, a pattern known as *Newton's Rings*. It would seem, however, that he did not fully realize the implications of this observation concerning the nature of light.

E   Lens coatings

Thin-film interference is put to work in high-quality lenses. For example, magnesium fluoride ($\mathrm{Mg\,F_2}$) has index of refraction $n_2 = 1.38$, intermediate between that of air and glass. A *quarter-wave* coating on a glass lens yields $\delta = \pi$, suppressing reflection and maximizing transmission at the corresponding wavelength.[10] For example, for minimum reflection and maximum transmission in the middle of the visible spectrum, wavelength $\lambda_{\mathrm{air}} = 550.$ nm, say, the required thickness of $\mathrm{Mg\,F_2}$ coating is

$$
\begin{aligned}
t_{1/4} &= \frac{\lambda_{\mathrm{air}}}{4n_2} \\
&= \frac{550.\ \mathrm{nm}}{4(1.38)} \\
&= 99.6\ \mathrm{nm}\ ,
\end{aligned}
\tag{XII.20}
$$

approximately one-tenth of a micron. The reduction of the wavelength from λ_{air} to $\lambda_{\mathrm{air}}/n_2$ in the coating is taken into account here.

Reflection of wavelengths at longer and shorter wavelengths creates the purplish color seen on coated lenses in binoculars and cameras. The lenses themselves, of course, are not tinted.

F   Physical optics: diffraction

We complete our exploration of the design and performance of the binoculars by examining another class of wave-interference phenomena, termed *diffraction*.

---

[10]Antireflective coatings on the glass of high-quality picture frames work the same way.

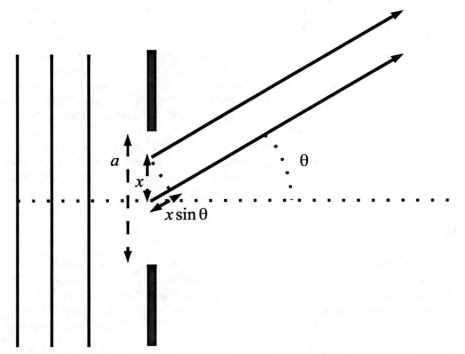

Figure XII.12: Geometry of single-slit Fraunhofer diffraction.

Diffraction

Diffraction is *deviation from rectilinear propagation due to interference from different portions of a wave front.* A diffracted wave is calculated by integrating the contributions of a continuous infinity of sources, in contrast to the discrete superpositions of the preceding section. If the incoming wave front is treated as planar, i.e., as from a distant source, the effects are termed *Fraunhofer*[11] *diffraction.* If the curvature of the incoming wave front is considered, as from a source at finite distance, the phenomena are labeled *Fresnel*[12] *diffraction.* The analysis of Fresnel diffraction is a challenge for a higher-level text. Here we can consider some simple but important examples of Fraunhofer diffraction.

Single-slit Fraunhofer diffraction

The geometry of Fraunhofer diffraction for a plane wave front impinging on a single rectangular slit of width a is illustrated in Fig. XII.12. The phase shift of wave contributions across the slit, relative to the central line, is

$$\delta(x) = kx \sin\theta , \qquad\qquad \text{(XII.21a)}$$

---

[11] Joseph von Fraunhofer, 1787–1826
[12] Augustin Jean Fresnel, 1788–1827

for $x \in [-a/2, +a/2]$, where $k = 2\pi/\lambda$ is the wave number (the magnitude of the wave vector) of the waves and θ the angle of the outgoing rays from the central line. It is simplest to consider the electric fields of the wave contributions, from "strips" of wave front across the slit, as the real parts of complex exponential functions, as was done in Ch. XI Sec. D. The combined wave field is given by

$$
\begin{aligned}
\mathbf{E} &= \frac{\mathbf{E}_0}{a} \, \Re \left(e^{i\phi} \int_{-a/2}^{+a/2} e^{ikx \, \sin\theta} \, dx \right) \\
&= \frac{\mathbf{E}_0}{a} \, \Re \left(e^{i\phi} \, \frac{e^{i(ka/2) \, \sin\theta} - e^{-i(ka/2) \, \sin\theta}}{ik \, \sin\theta} \right) \\
&= \mathbf{E}_0 \, \cos\phi \, \frac{\sin[(ka/2) \, \sin\theta]}{(ka/2) \, \sin\theta} \, ,
\end{aligned}
\tag{XII.21b}
$$

where ϕ is the phase of the wave contribution from the middle of the slit. Here the Euler relation is used to obtain the final expression from the result of the integral. This field corresponds [via Eq. (XI.5d)] to intensity

$$
I(\theta) = I_0 \, \frac{\sin^2\left(\dfrac{\pi a}{\lambda} \sin\theta\right)}{\left(\dfrac{\pi a}{\lambda} \sin\theta\right)^2} \, ,
\tag{XII.21c}
$$

with I_0 the intensity along the central line. This intensity distribution is plotted in Fig. XII.13. The pattern of light on a distant screen consists of a broad central maximum, flanked on both sides by alternating dark and bright bars, extending well beyond the width of the slit. The dark minima occur at angles

$$
\theta_{\min} = \mathrm{Sin}^{-1} \left(\frac{m\lambda}{a} \right)
\tag{XII.21d}
$$

for nonzero integers m, the bright maxima approximately halfway between.

For a real two-slit experiment such as that described in Sec. D preceding, this intensity distribution modulates the two-slit intensity pattern. An example of this is graphed in Fig. XII.14. The single-slit "envelope" is necessarily broader than the two-slit pattern, as the width a of the slits must be smaller than their separation d.

Circular-aperture Fraunhofer diffraction

Another important example of Fraunhofer diffraction is that of a plane wave front impinging on a barrier with a circular aperture of radius a. This is illustrated in Fig. XII.15. An exact calculation of the diffracted wave is quite complicated, sensitive to features such as the polarization of the incident wave. In the limits of normal incidence and a large aperture radius compared to the wavelength of the light ($a \gg \lambda$ or $ka \gg 1$), the intensity pattern is given by

$$
I(\theta) = 4I_0 \left(\frac{J_1(ka \, \sin\theta)}{(ka \, \sin\theta)} \right)^2 \, ,
\tag{XII.22a}
$$

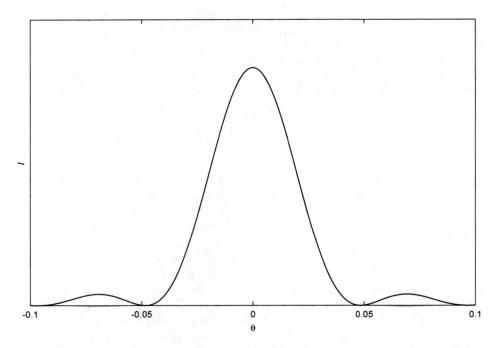

Figure XII.13: Single-slit Fraunhofer diffraction intensity pattern on an arbitrary intensity scale.

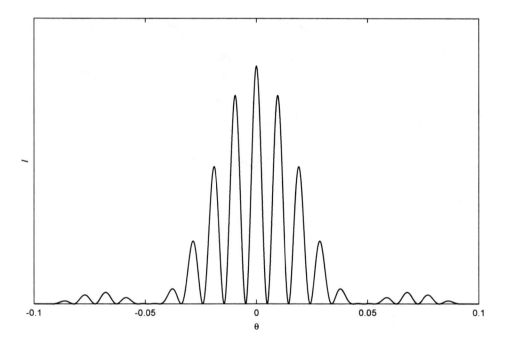

Figure XII.14: Two-slit interference pattern modulated by single-slit diffraction pattern. Again, the intensity scale is arbitrary. For this plot, the slit spacing d was taken to be 5.00 times the slit width a.

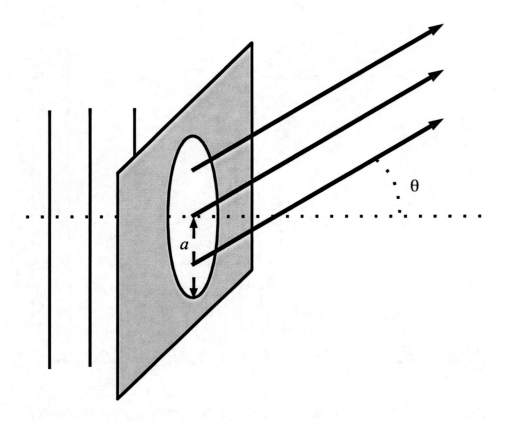

Figure XII.15: Geometry of Fraunhofer diffraction by a circular aperture.

where I_0 is the intensity along the central line (i.e., at $\theta = 0$) and J_1 is the *ordinary Bessel function of first order*. This is a special function similar, but not identical, to the sine function. The resulting pattern of light on a distant screen, say, is a bright central disk surrounded by alternating concentric dark and bright rings, the intensity of the bright rings decreasing with increasing ring radius. The first dark ring, i.e., the boundary of the central disk, is at angle to the central line

$$\theta_{\min} = \frac{j_{1,1}}{\pi} \frac{\lambda}{D}$$

$$\dot{=} 1.21967 \frac{\lambda}{D} \,, \tag{XII.22b}$$

where $j_{1,1} \dot{=} 3.83171$ is the first nonzero root of the function J_1 and $D = 2a$ the diameter of the aperture. This is a famous result.

G   Resolution of binoculars

These results allow us to examine another performance characteristic or figure of merit for the binoculars: *resolution,* the sharpness or amount of detail in the image. This characteristic is distinct from magnification. If the image were recorded photographically and enlarged, the magnification could be made as large as desired, but the detail discernible in the image would not be increased. The diffraction pattern (XII.22) limits the resolution available with objective lenses of a given size. One measure of this is provided by the *Rayleigh*[13] *Criterion*: Two point sources or features are distinguishable if the center of the diffraction pattern of one is separated from that of the other at least as far as the first diffraction minimum (dark ring). That is, the sources are distinguishable if their angular separation satisfies $\Delta\theta \geq \theta_{\min}$, with $\theta_{\min}$ given by Eq. (XII.22b). In fact, the human eye/brain combination can do a bit better than this (particularly if one knows in advance what one is looking at), but this is a useful guide.

For the binoculars of Fig. XII.1, with $D = 50.$ mm and visible wavelength $\lambda = 550$ nm, the Rayleigh Criterion requires

$$\Delta\theta \geq \frac{1.22\,\lambda}{D}$$

$$\geq \frac{1.22\,(5.5 \times 10^{-7}\text{ m})}{5.0 \times 10^{-2}\text{ m}}$$

$$\geq 1.3 \times 10^{-5}\text{ radian}$$

$$\geq 2''.8 \,. \tag{XII.23}$$

This is a small angle, representing high resolution. In actual use, the resolution of the binoculars is likely to be worse than this "diffraction-limited" value, determined by atmospheric seeing conditions, for example. Diffraction represents an irreducible lower limit to the resolution angle $\Delta\theta$.

---

[13]John William Strutt, Third Baron Rayleigh, 1842–1919

H   Matter waves and quantum mechanics

If a beam of electrons is made to pass through a thin metal foil—consisting of microcrystals with random orientations—and strike a fluorescent screen, the result is not a single spot or a smudge of electrons, but a central bright spot surrounded by concentric dark and bright rings. Similar experiments can be done with beams of atoms or neutrons. These are diffraction patterns, akin to the circular-aperture diffraction pattern described in Sec. F preceding. And like any diffraction or interference pattern, they are an absolute signature that *particulate matter has wave nature.* This is one of the foundational ideas of *quantum mechanics,* its complement being that wave fields have quantized, i.e, discrete, excitations—particle nature.

Quantum mechanics represents a far more radical transformation of physics than, say, Einstein's Special and General Theories of Relativity. Quantum mechanics is an *expansion* of physics; classical physics is a subset of it, a set of limiting cases. Quantum mechanics changes the conceptual framework within which questions are asked and answered. It has revolutionized physics and chemistry—not to mention philosophy—over the last century. The exploration of so vast an array of observations and ideas must await another text.

Classical mechanics is to quantum mechanics as geometrical optics is to physical optics. The parallel is very close, as both are consequences of constructive wave interference. But that is not for this text, either: It is the climactic conclusion of the first semester of an upper-division classical mechanics course.

The quest must go on.

References

This chapter treats several topics—geometrical optics, optical instruments, interference, and diffraction—which are covered by their own chapters in nearly all introductory physics texts. Optics, of course, is an extensively developed subject in its own right, with dedicated intermediate- and advanced-level texts. For example, Hecht [Hec01] is a detailed intermediate-level text.

The original classic optics text is Newton's *Opticks,* originally published (in English!) in 1704. A modern reprinting [New52] is available.

Problems

1. When light passes from one medium to another at an interface, its speed changes.

   (a) Does its frequency change, its wavelength, or both? Why?

   (b) Why, then, do the colors of objects not look radically different under water?

2. A light beam passes from point A in one medium, through a plane interface, to point B in another medium. The incident and refracted rays of

which its path is composed satisfy the Snell law of refraction. Prove that of all the two-segment paths—i.e., which bend somewhere on the interface— connecting A and B, this is the path along which the light takes *the least time*. This is known as the *Fermat*[14] *Principle* and is of great significance.

3. A 45°-45°-90° triangular block of crown glass, with index of refraction $n = 1.520$, can be used to reverse the direction of a light beam by means of two total internal reflections. Would this work under water, with index of refraction $n = 1.333$?

4. Consider the three classes of optical elements described in Sec. B.

   (a) For what range of source distances s will a concave spherical mirror produce 1) an enlarged real image, 2) a reduced real image, 3) an enlarged virtual image, and 4) a reduced virtual image?

   (b) Same question, for a convex mirror.

   (c) Same question, for a spherical refracting surface. Consider all possible cases.

   (d) Same question, for a converging ($f > 0$) thin lens.

   (e) Same question, for a diverging ($f < 0$) thin lens.

5. The classic question: The image of writing in a plane mirror is backwards; the image of a left-handed person is right-handed. Why does the mirror reverse left and right, not up and down? How does it know?

6. The *power* of a lens is its inverse focal length $1/f$, measured in *diopters* (inverse meters). If two thin lenses, of focal lengths f_1 and f_2, are placed together with negligible distance between them, what are the focal length and the power of the combination?

7. A nearsighted person uses contact lenses of power $P = -5.50$ diopters. If she wishes to replace these with spectacles mounted 1.00 cm from her eyes, what should be the prescription of the spectacles?

8. Calculate the intensity pattern produced by *three* equally spaced slits illuminated equally by a coherent wave front. Plot the intensity distribution. Compare these with the two-slit results of Sec. D of this chapter. Ignore single-slit diffraction effects.

9. Calculate and compare the diffraction-limited resolutions of the 10.0 meter Keck[15] telescope and the 2.40 meter Hubble[16] Space Telescope, which observe in visible light of wavelength 550. nm, and the 305 meter Arecibo radio telescope, which observes microwaves of wavelength 21.0 cm. Use the Rayleigh criterion.

---

[14] Pierre de Fermat, 1601–1665
[15] William Myron Keck, 1880–1964
[16] Edwin Powell Hubble, 1889–1953

10. Estimate the diffraction-limited resolution of a human eye using light of wavelength 550. nm. What is the greatest distance at which the headlights of a car, separated by 1.50 m, can be distinguished as two sources? How realistic is this estimate?

Bibliography

[ABS75]   Ronald Adler, Maurice Bazin, and Menahem Schiffer. *Introduction to General Relativity*. McGraw-Hill, New York, NY, 1975.

[And85]   John D. Anderson, Jr. *Introduction to Flight*. McGraw-Hill, New York, NY, second edition, 1985.

[BD64]   James D. Bjorken and Sidney D. Drell. *Relativistic Quantum Mechanics*. McGraw-Hill, New York, NY, 1964.

[BD65]   James D. Bjorken and Sidney D. Drell. *Relativistic Quantum Fields*. McGraw-Hill, New York, NY, 1965.

[BM70]   M. D. Burns and S. G. G. MacDonald. *Physics for Biology and Pre-Medical Students*, pages 419–422. Addison-Wesley, Reading, MA, 1970.

[BMSW13]   Roger R. Bate, Donald D. Mueller, William W. Saylor, and Jerry E. White. *Fundamentals of Astrodynamics*. Dover Publications, New York, NY, second edition, 2013.

[BO95]   Vernon Barger and Martin Olsson. *Classical Mechanics: A Modern Perspective*. McGraw-Hill, New York, NY, second edition, 1995.

[BW14]   Wolfgang Bauer and Gary W. Westfall. *University Physics with Modern Physics*. McGraw-Hill, New York, NY, second edition, 2014.

[FGT96]   Paul M. Fishbane, Stephen Gasiorowicz, and Stephen T. Thornton. *Physics for Scientists and Engineers*. Prentice Hall, Upper Saddle River, NJ, second edition, 1996.

[Gal54]   Galileo Galilei. *Dialogues Concerning Two New Sciences: Translated by Henry Crew and Alfonso de Salvio*. Dover Publications, New York, NY, 1954.

[Goo75]   David L. Goodstein. *States of Matter*. Prentice-Hall, Englewood Cliffs, NJ, 1975.

[GPS01]   Herbert Goldstein, Charles P. Poole, and John L. Safko. *Classical Mechanics*. Addison-Wesley, Cambridge, MA, third edition, 2001.

459

[Hal84]   Francis J. Hale. *Introduction to Aircraft Performance, Selection, and Design.* John Wiley & Sons, Inc., New York, NY, 1984.

[Har03]   James B. Hartle. *Gravity: An Introduction to Einstein's General Relativity.* Addison Wesley, San Francisco, CA, 2003.

[Hec01]   Eugene Hecht. *Optics.* Addison-Wesley, Reading, MA, fourth edition, 2001.

[Her71]   Samuel Herrick. *Astrodynamics: Orbit Determination, Space Navigation, Celestial Mechanics, Volume I.* Van Nostrand Reinhold, London, UK, 1971.

[Her72]   Samuel Herrick. *Astrodynamics: Orbit Correction, Perturbation Theory, Integration, Volume II.* Van Nostrand Reinhold, London, UK, 1972.

[HM94]   Mark A. Heald and Jerry B. Marion. *Classical Electromagnetic Radiation.* Cengage Learning, Stamford, CT, third edition, 1994.

[Hob73]   Russell K. Hobbie. "Nerve Conduction in the Pre-Medical Physics Course." *American Journal of Physics*, 41:1176–1183, 1973.

[HRK02]   David Halliday, Robert Resnick, and Kenneth S. Krane. *Physics, Volume 2.* John Wiley & Sons, New York, NY, fifth edition, 2002.

[Jac98]   John David Jackson. *Classical Electrodynamics.* John Wiley & Sons, New York, NY, third edition, 1998.

[Kit04]   Charles Kittel. *Introduction to Solid State Physics.* John Wiley & Sons, New York, NY, eighth edition, 2004.

[Koe90]   Arthur Koestler. *The Sleepwalkers: A History of Man's Changing Vision of the Universe.* Penguin Group (USA), New York, NY, 1990.

[LL76]   L. D. Landau and E. M. Lifshitz. *Mechanics.* Elsevier Science, Oxford, UK, third edition, 1976.

[LL80]   L. D. Landau and E. M. Lifshitz. *The Classical Theory of Fields: Volume 2.* Elsevier Science, Oxford, UK, fourth edition, 1980.

[LLP90]   L. D. Landau, E. M. Lifshitz, and L. P. Pitaevskii. *Quantum Electrodynamics.* Elsevier Science & Technology Books, Oxford, UK, second edition, 1990.

[LPL84]   L. D. Landau, L. P. Pitaevskii, and E. M. Lifshitz. *Electrodynamics of Continuous Media: Volume 8.* Elsevier Science, Oxford, UK, second edition, 1984.

[McK41]   Richard McKeon, editor. *The Basic Works of Aristotle*, pages 218–394. Random House, New York, NY, 1941.

[MEC74]   Howard V. Malmstadt, Christie G. Enke, and Stanley R. Crouch. *Electronic Measurements for Scientists*. W. A. Benjamin, Inc., Menlo Park, CA, 1974.

[Mer68]   N. David Mermin. *Space and Time in Special Relativity*. McGraw-Hill, Inc., New York, NY, 1968.

[MTW73]   Charles W. Misner, Kip S. Thorne, and John A. Wheeler. *Gravitation*. W. H. Freeman and Company, San Francisco, CA, 1973.

[New52]   Isaac Newton. *Opticks*. Dover Publications, Inc., New York, NY, 1952.

[New62]   Isaac Newton. *Philosophiae Naturalis Principia Mathematica: Andrew Motte's Translation, Revised by Florian Cajori*. University of California Press, Berkeley, CA, 1962.

[Res72]   Robert Resnick. *Basic Concepts in Relativity and Early Quantum Theory*. John Wiley & Sons, New York, NY, 1972.

[RHK02]   Robert Resnick, David Halliday, and Kenneth S. Krane. *Physics, Volume 1*. John Wiley & Sons, New York, NY, fifth edition, 2002.

[Rin06]   Wolfgang Rindler. *Relativity: Special, General, and Cosmological*. Oxford University Press, Oxford, UK, second edition, 2006.

[RMC08]   John R. Reitz, Frederick J. Milford, and Robert W. Christy. *Foundations of Electromagnetic Theory*. Addison-Wesley, Cambridge, MA, fourth edition, 2008.

[Rou12a] Edward John Routh. *The Advanced Part of a Treatise on the Dynamics of a System of Rigid Bodies: Being Part II of a Treatise on the Whole Subject*. Cambridge University Press, Cambridge, UK, fifth edition, 2012.

[Rou12b] Edward John Routh. *The Elementary Part of a Treatise on the Dynamics of a System of Rigid Bodies*. Cambridge University Press, Cambridge, UK, fifth edition, 2012.

[Sch58]   Julian Schwinger, editor. *Selected Papers on Quantum Electrodynamics*. Dover, New York, NY, 1958.

[Sch09]   Bernard Schutz. *A First Course in General Relativity*. Cambridge University Press, Cambridge, UK, 2009.

[Sta69]   Merriam-Webster Editorial Staff. *Webster's Biographical Dictionary*. G. & C. Merriam Company, Springfield, MA, first edition, 1969.

[Tay05]   John R. Taylor. *Classical Mechanics.* University Science Books, Sausalito, CA, 2005.

[Tip87]   Paul A. Tipler. *College Physics*, pages 496–500. Worth Publishers, New York, NY, 1987.

[TM04]   Stephen T. Thornton and Jerry B. Marion. *Classical Dynamics of Particles and Systems.* Thomson—Brooks/Cole, Belmont, CA, fifth edition, 2004.

[TW92]   Edwin F. Taylor and John Archibald Wheeler. *Spacetime Physics: Introduction to Special Relativity.* W. H. Freeman and Company, New York, NY, second edition, 1992.

[vM59]   Richard von Mises. *Theory of Flight.* Dover Publications, New York, NY, 1959.

[Wal84]   Robert M. Wald. *General Relativity.* University of Chicago Press, Chicago, IL, 1984.

[Wei72]   Steven Weinberg. *Gravitation and Cosmology: Principles and Applications of the General Theory of Relativity.* John Wiley & Sons, New York, NY, 1972.

[You71]   Louise B. Young, editor. *Exploring the Universe.* Oxford University Press, New York, NY, second edition, 1971.

Index

aberration
    chromatic, 436–437
    spherical, 434
absolute zero of temperature, 289
acceleration, 12–13, 16–20, 22, 26, 27, 32–34, 36, 39, 41, 49, 61, 63, 69, 89, 106, 112–113, 115, 118, 130, 131, 139, 146, 149, 156, 164, 167–169, 171, 180, 181, 203, 213, 215, 218, 221, 286, 291, 336, 338, 372, 403
    average, 13
    constant, 14, 62, 63, 96, 115, 124
    Einsteinian, 66
    instantaneous, 13
    piecewise-constant, 14
action at a distance, 156, 236, 274
action potential, 309–310, 323, 326
    depolarization, 310
    regeneration, 310
adiabatic (process), 221
aerobatics, 83n
Age of Exploration, 331n
air conditioner, 384
air hockey, 97, 111
aircraft carrier, 232
aircraft engine, 76, 81, 92, 383
    piston, 76, 78
    turbofan, 76, 77, 81
    turbojet, 76, 92
albatross, great, 88
Alpha Centauri (α Kentaurus), 141
α particle, 275
alphabet soup, 182, 316n
alternating current (AC), 290, 315, 326, 372, 383, 387n
    phase, 383n

alternator, 383
altitude, 65, 80
    airplane, 77, 78, 80–83, 87
    bullet, 65
    orbital, 173
    projectile with drag, 65
    rocket, 16, 18, 20, 24–26, 41–46
    tennis ball, 63, 65
aluminum, 311, 419, 427
amber, 228, 330
ammeter, 350
    analog, 341n
amount, 5, 277
Ampère Law, 347–356, 394
Ampère-Maxwell Law, 355–357, 377, 378, 399, 414, 416
Ampère, André Marie, 261n, 332, 333, 347
ampere, 5, 261, 282, 283, 286, 290n, 295, 311, 313, 340, 344, 345, 362, 364, 390
    definition, 350
Amperian loop, 350–353, 355, 356
amplifier, 316, 319, 320
    gain, 319
anaesthetic, 310
analog computation, 387
analytic geometry, 160
Anatolia, 330
AND gate, 322
angle of attack, 34, 36, 70
angle of incidence, 427, 429
angle of reflection, 427
angle of refraction (transmission), 427, 429
angular acceleration, 113, 118, 130, 131, 133, 136, 139

constant, 115
angular displacement, 131–132, 180
angular frequency, 184, 187, 190, 197,
        199, 200, 205, 208, 216, 225,
        335, 336, 373, 395, 401, 402,
        416, 417, 419–421, 423
angular impulse, 128, 130, 131, 142
angular magnification, 442
angular momentum, 134, 136, 138, 139,
        142, 163, 165, 166, 173, 186n,
        359, 360, 362, 363, 372, 373,
        411, 423
    conservation, 123, 136, 163, 165
    definition, 122
    dynamics, 122–123
angular speed, 113, 124, 132, 136, 138,
        139, 359, 373, 383
angular velocity, 112–115, 123, 128–134
anharmonicity, 184, 185
ant, 169
Antarctica, 331
antenna, 408, 419, 420
antiferromagnetism, 364, 371
    domain, 371
antimony, 311
antiparticle, 311n
antireflective coatings, 226
apastron, 165n
aphelion, 114, 165n
apocenter, 165n
apocynthion, 165n
apogee, 165n, 166, 173
Apollo, 146, 173, 323
archery, 63–64, 134n
    mounted, 63–64
arctangent, 40, 43
Arecibo radio telescope, 457
Aristarchus of Samos, 29
Aristotle, 2, 2n, 29, 31, 32n, 49, 62, 64,
        228, 230
Armstrong, Neil, 146
arsenic, 311
associativity, 133, 339
Aston, Francis William, 336
astrodynamics, 172
astronauts, 199

Astronomical Unit (AU), 114, 163
astronomy, 298, 426, 437
atom, 95, 101, 103, 123, 198, 216, 221,
        228, 232–235, 240, 250n, 256,
        257, 263, 269, 271, 283, 284,
        286, 288, 289, 311, 332, 340,
        341, 343, 360, 362, 364, 366,
        367, 369, 371–374, 403, 408,
        430, 444, 456
Atomic Age, 280
atomic nucleus, 95, 100–101, 103, 123,
        139, 193, 220, 232, 234, 263,
        272, 275, 334, 336, 361, 363,
        374, 403
Auckland, 68
aurora, 338
automobile, 376
    coil, 390
    distributor, 390
        points, 390
        rotor, 390
    electronic ignition, 390
    engine, 376, 383, 390
        cylinder, 390
        power stroke, 390
        spark plug, 390
    headlights, 458
    horn and headlights, 376
    starter, 376
Avogadro number, 233, 284
Avogadro, Amedeo, 233n
axon, see neuron

backspin, 128
ballistic theories, 400
barber pole, 348
Bardeen's Box, 316
Bardeen, John, 315, 316
barn
    structure, 103n
    unit, 103
Baroque physics, 55n
Barrow, Isaac, 31
basketball, 98, 140–141
Batavia, Illinois, 111
baton, 118

battery
  artillery, 293
  electric, 277, 291, 293–295, 332, 383, 390, 394, 404
Battery of Baghdad, 230n
beam splitter, 430
beats, 205, 224
  frequency, 207, 224
Bell Labs, 315
Bell, Alexander Graham, 222n
Bernoulli Principle, 70
Bernoulli, Daniel, 70n, 207, 216
Bessel functions, 217, 455
Bessel, Friedrich Wilhelm, 217
Bible, 261
  Genesis, 415n
bicycle, 139, 383
Big Bang, 315
big O, 238n
billiards, 94, 95, 111
binocular, 449
binoculars, 426, 439–443, 455
  eyepiece lens, 439–443
  field of view, 442
  light-gathering power, 442, 443
  magnification, 439–443
  objective lens, 439–443, 455
  optical path, 442, 443
  resolution, 455
    Rayleigh Criterion, 455
biological evolution, 280
biology, 298, 323, 426
Biot, Jean Baptiste, 344
Biot-Savart Law, 344–348, 353
birefringence, 407–408, 411
bismuth, 366
black hole, 87, 171
Block, Felix, 363
Bohr magneton, 362
Bohr radius, 257, 372
Bohr, Neils Henrik David, 235
Boltzmann constant, 241, 284, 313, 373
Boltzmann, Ludwig, 241, 399
Boole, George, 320
Boolean operation
  AND, 322, 323

NOT, 322
OR, 322
Boolean operaton
  XOR, 323
boron, 311
bottle brush, 348
Boulder, Colorado, 350
Boulliau (Bullialdus), Ismaël, 149, 153
Boulton, Matthew, 75, 75n
bound charge, *see* electric charge, bound
boundary condition, 417
boundary conditions, 215–217, 417, 423, 449
  Dirichlet, 215
  fixed end, 215–217, 220, 224, 226
  free end, 215–217
  Neumann, 215
boundary-value problem, 417
bowling, 97, 128n
  ball, hair on, 245n
Brahe, Tycho, 159, 161, 172
brain, 280, 323, 455
brass, 419
Brattain, Walter Houser, 315
Breguet range equation, 78, 79
Breguet, Louis-Charles, 78n
bremsstrahlung, 403
Brewster angle, 411
Brewster Law, 411
Brewster, David, 411
British units, 232
brushes (electric), 381, 384
bulk modulus, 221
bullet, 284
Bunsen burner, 298
Bunsen, Robert Wilhelm, 297, 298
Burroughs, Edgar Rice, 156
"butterfly" field pattern, 242, 340, 347, 403
butterfly, *Morpho*, 449

calcite, 407, 408
calcium carbonate, *see* calcite
Calculus, Fundamental Theorem of, 16, 88
California, 203

calorie, 74
    Large (Calorie), 74
Cambridge (UK), 146
camera, 439, 449
camera flash, 273
Canada, 331
cancer treatment, 336
candela, 5
capacitance, 263–269, 271–274, 276, 277, 305, 307, 308, 324, 326
    gravitational, 264
capacitor, 265, 271–273, 276, 277, 305, 306, 324, 386, 387, 389, 391
    charging, 306–307, 356–357, 389
    combination, 267, 293, 296–298
        parallel, 267, 269
        series, 267–269
    discharging, 305–306
    Earth, 264
    electrolytic, 267, 272
    energy, 273–274, 276, 277, 390, 391
    parallel-plate, 265–267, 269–271, 273, 276, 308, 353, 356
        fringing, 265, 276
    single sphere, 264–265
    ultracapacitor, 267, 272
Cape Canaveral, 146
Cape Kennedy, *see* Cape Canaveral
carbon, 295, 381
cardiac defibrillator, 273
cause and effect, 384, 416
Cavendish experiment, 150–153
Cavendish, Henry, 150, 153, 163, 230, 280
ceiling
    absolute, 83, 92
    cruise, 83
    service, 83
cell
    biological, 280, 426
        cell wall, 430
        cytoplasm, 280, 430
        nucleus, 280
        organelles, 280
    electrochemical, 230n, 291, 293–295

    in parallel, 293
    in series, 293
    photoelectric (photovoltaic), 293
Celsius, Anders, 241
center of gravity, 108, 140
center of mass, 116, 118, 121, 124, 128, 128n, 138, 140–142, 172, 264
    definition, 103–104, 109
    dynamics, 106
        kinetic energy, 106–107
        potential energy, 107–108
    kinematics
        acceleration, 106
        velocity, 104, 106, 107
    motion, 196
center-of-mass frame, *see* reference frame
center-of-momentum frame, *see* reference frame
Central Processing Unit (CPU), 322
centrifugal acceleration, 113n
centripetal acceleration, 113–115, 164, 335
centripetal force, 123, 366
chaos, 181, 182, 199
charge carriers, 272, 282, 283, 286, 288, 289, 311, 313, 316, 319, 341, 343–345, 348, 378, 380
chemical reaction, 233
chemistry, 49, 222, 233, 257, 269, 284, 291, 293, 295, 298, 310, 360, 384, 403, 422, 456
chlorine, 301
chocolate, 288
Cidenas or Kidinnu, 139
cinnamon bun, 386
circular motion, 112, 115, 122, 132
    kinematics, 112
        angular acceleration, 113, 118, 130, 131
        angular position, 112
        angular speed, 113, 124
        angular velocity, 112–115, 123, 128–131
        centripetal acceleration, 113–115, 335

linear acceleration, 112
linear velocity, 112
tangential acceleration, 113, 115
uniform, 113–114, 184, 204, 335
circular polarization, 411, 423
circulation, 347, 348, 353, 377, 378, 384, 413, 414
magnetic, 347
clarinet, 217
classical physics, 54–55, 55n, 198, 202
clocks, 56, 88, 184–185
coal
nn, 383
coefficient of restitution, 96–97, 100, 101, 109, 140–142
collider, 111
collision, 94–101, 108, 109, 126, 131, 140, 272, 334
elastic, 97–98, 101, 130, 141, 142
inelastic, 141
one-dimensional, 95–100, 129–131
Einsteinian, 98–100
Newtonian, 95–98, 100, 109, 140–142
perfectly inelastic, 97–100, 109, 111
two- or higher-dimensional, 100–103, 109
Einsteinian, 141
Newtonian, 100–101, 141
color
force, 234, 235
optical, 205, 212, 400, 403, 429, 437, 449, 456
comet, 95, 406
common-emitter circuit, 316–320
common-mode motion, 196
communicaton satellite, 309n
commutativity, 72, 132, 133
compressor motor, 384
computation
analog, 387
digital, 322–323, 376, 387
computer, 403
Computerized Axial Tomography, 220
conductance, 288n, 297n, 302, 326

conductivity, 285–286, 288, 288n, 310, 416, 419, 420, 424
conductor, 230, 248–253, 258–261, 263–265, 269, 271–276, 282–286, 288–291, 293, 297, 307, 310, 311, 313, 341, 343, 344, 350, 378, 380, 381, 416, 419, 424
cavity in, 250–251
cone, 104, 132, 138, 161
conic section, 161
apsides, 165, 173
eccentricity, 160–161, 164, 173
focus, 159–161
Conseil Européen pour la Recherche Nucléaire (CERN), 336n
conservation laws, 95, 298
Conservation of Angular Momentum, 123, 136
Conservation of Electric Charge, 285, 297
Conservation of Energy, 59, 60, 86–88, 123, 297, 305n, 378, 383, 384, 416
Conservation of Linear Momentum, 52, 59, 60, 98, 99, 123
Conservation of Mass, 60, 99, 285
conservative force, *see* force
contact lens, 457
continuity equation, 285, 297, 355
contour map, 257
Cooper, Leon N., 315
coordinates, 276, 360, 401, 414
curvilinear, 89, 413n
polar, 159
rectangular (Cartesian), 89, 261, 282n, 413n, 415n, 417
spherical, 422
Copernicus, *see* Koppernigk
copper, 283, 284, 288, 326, 331, 336, 343
corn syrup, 408
corona discharge, 258–261, 277, 291
cosine, 190, 205, 207, 213, 218, 238, 400, 432, 444
Cosmic Microwave Background Radiation, 315

cosmic string, *see* string, cosmic
cosmology, 171
coulomb, 232, 233, 236, 255, 264, 272, 282–284, 350, 405
Coulomb constant, 231–232, 234, 247, 263, 265
Coulomb Law, 230–236, 238, 245, 247, 274, 344, 345
couple, 123
coupled oscillators, *see* harmonic oscillator, coupled
Crookes radiometer, 406
Crookes, William, 406
cross product, *see* vector
cruise-climb flight, 77, 80, 81, 92
crystal, 363, 407, 408, 456
crystallography, 220
Curie Law, 364
Curie temperature, 369, 372
Curie, Marja Sklodowska (Marie), 364
Curie, Pierre, 364
curl, 413–415
current, 313, 316
current balance, 350
current density, 282, 283, 285, 286, 348, 350, 353, 355, 357, 414, 416
current divider, 324
curvature radius, 258
cyclotron, 336, 372, 390, 403
    dees, 336, 372
cyclotron motion, 335–338
    cyclotron frequency, 335, 336

d'Arsonval galvanometer, 341
d'Arsonval, Jacques Arsène, 341
da Vinci, Leonardo, 203
damped oscillator, *see* harmonic oscillator, damped
damping parameter, 186, 188, 200
de Broglie relations, 420–421, 423
de Broglie, Louis-Vector, 420
de Coulomb, Charles Augustin, 230, 232, 234, 235, 238, 242, 247, 255, 256, 280
de Fermat, Pierre, 457
de Laplace, Pierre-Simon, 166, 181

decibel, 222
declination, 372
degree Celsius, *see* kelvin
degree of freedom, 406
degrees of freedom, 84, 186, 198, 222
Dempster, Arthur Jeffrey, 336
dendrites, *see* neuron
density, 38, 64, 65, 70, 77, 79, 80, 80n, 81–83, 91, 103, 104, 116, 126, 128, 141, 142, 154, 156, 171, 213, 214, 221, 222, 224, 226, 234, 241, 242, 244, 247, 248, 250, 251, 253, 255, 258, 262, 263, 265, 270, 275–277, 282–284, 288n, 338, 343, 373, 404, 405, 413, 414, 429
    energy, 214
Denver, 80n
derivative, 12, 88, 89, 133, 176, 180, 181, 189, 215, 387
    partial, 89, 213, 309, 413, 414, 417
Descartes, René, 22, 160
desktop computer, 322
determinant, 133, 213, 380, 413
determinism, 181–182
deterministic dynamics, 181
dextrism, 132
Dialogues Concerning Two New Sciences, 62
diamagnetism, 364, 366–367
diamond, 429
dielectric, 269–272, 274, 274n, 276, 277, 355, 363, 366, 399, 407
    breakdown, 272, 277
        breakdown field, 272, 274, 276, 277, 391, 422
        of vacuum, 272
    constant, 270–272, 276, 277, 402
    strength, 272, 274, 276, 277, 391, 422
diesel engine, 384n
diesel-electric engine, 384n
differential, 91, 221, 240, 304, 309, 392, 393, 413
    first-order, 304
    second-order, 304, 309

diffraction, 449–458
  Fraunhofer, 450–455
    circular-aperture, 451–458
    single-slit, 450–451, 457
  Fresnel, 450
diffraction grating, 447
diffusion, 213, 309, 311, 313, 421
digital computation, 322–323, 376, 387
digital meter, 341n
diode, 313–315, 386
diopter, 457
dip angle, 372
dipole
  electric, 236–238, 240, 256, 269,
      275, 332, 340, 341, 403, 404,
      408
    dipole moment, 236, 240, 241,
        269, 270, 275, 371, 403, 404,
        408
    equipotentials, 258
    field lines, 242
  magnetic, 332, 340, 341, 347, 353,
      364, 366, 372, 394, 403, 404,
      408
    dipole moment, 332, 340, 341,
        345, 347, 357, 359, 361–364,
        366, 367, 369, 371–374, 383,
        393, 394, 403, 404, 408
    magnetic moment, 367
  mass, 238n
Dirac equation, 360, 362
Dirac, Paul Adrien Maurice, 49, 49n,
    87, 332, 360
direct current (DC), 290, 291, 315, 323,
    326
Dirichlet, Peter Gustav Lejeune, 207,
    215
disk, 104, 116
dispersion, 208–212, 214, 225, 402, 429,
    436
  dispersion relation, 208, 212, 214,
      225, 395, 401, 402, 416, 417
displacement, 5, 16, 18, 20, 22–27, 30,
    32, 55, 61, 63, 65, 72–74, 76,
    84–86, 115, 134, 136, 150, 176,
    178, 181, 186, 188, 191, 196,

197, 199, 200, 212, 214–217,
    220, 221, 224–226, 240, 255,
    339, 392
displacement (electric), 270, 355, 364,
    414n, 423
displacement current, 355–357, 377, 389n,
    404, 416
displacment current, 377
distributive law, 72, 133
divergence, 413, 415
Divergence Theorem, 414
dog, 315n
dogfight, 87
Doppler effect
  for light, 223–224
  for sound, 223
Doppler, Christian Johann, 223n
dot product, *see* vector
doughnut, 74, 250, 353
drag, 36, 38, 41, 43, 44, 46, 62, 83, 84,
    86, 91, 95, 97, 124, 130, 138,
    142, 186, 286
  induced, 70, 78
  laminar-flow, 38, 65, 286
  turbulent-flow, 38, 64, 65, 69, 70,
      72, 91
drag coefficient, 38, 38n, 64, 65, 70, 77,
    78, 91
  zero-lift, 78, 79
drag polar, 78, 82
drift velocity, 282–284, 338, 341, 343,
    348, 380
driven oscillator, *see* harmonic oscilla-
    tor, driven
dynamics, 2, 30–33, 56–61, 66, 69, 88,
    112, 131, 169, 171, 173, 245,
    276, 360, 400, 423
dyne, 232
Dyson sphere, 156
Dyson, Freeman, 156

efficiency, 76, 77, 92
Egyptians, 426
Einstein Field Equations, 171
Einstein mechanics, 422
Einstein's Elevator, 169

Einstein, Albert, 27, 27n, 30, 49, 54, 55, 55n, 58, 59, 59n, 61, 66, 98, 100, 168, 169, 235, 332, 400, 416, 423, 456
Einsteinian mechanics, 49, 59–61, 87, 98, 99, 109, 156, 224, 225, 263, 272, 311n, 360, 371, 400, 416, 421, 423
electret, 371
electric charge, 228, 230–236, 238, 240–242, 244, 245, 247, 248, 250, 251, 253, 255, 257, 258, 261–265, 267, 269–277, 280, 282–285, 290, 291, 293, 297, 305–308, 315, 324, 331–336, 338, 341, 343–345, 350, 355, 357, 359, 362, 371, 378, 380, 389, 390, 403, 405, 408, 413, 414, 422, 423, 430
  bound, 269–271
  conservation, 285
  cylinder, 248
  disk, 238, 357–359
  elementary, 233, 256, 257, 263, 313, 359
  free, 269–271
  line, 248, 338
  negative, 230–233, 238, 240–242, 253, 265, 269, 270, 283, 285, 311, 333, 341, 343, 378
  one-fluid model, 230
  plane sheet, 248, 253
  point, 234, 247, 256, 258, 263, 275, 276, 422
  positive, 230–232, 234, 238, 240–242, 253, 265, 269, 270, 275, 276, 283, 285, 306, 311, 333, 341, 343, 378
  ring, 238, 357–359
  sphere, 238, 245–247, 256, 277
    shell, 245–247, 250n, 256, 257
    solid, 247
    uniform density, 247, 256, 262, 275, 277, 330, 357–360, 373
  two-fluid model, 230
  units, 232–233, 264

electric circuit, 228, 267, 280, 282, 291–301, 305, 310, 313, 315, 316, 320, 322–324, 326, 327, 338, 345, 376, 380, 381, 383–387, 389, 390, 404
electric current, 5, 228, 261, 273, 280, 282–286, 288, 290, 291, 293, 295–298, 300–303, 305–309, 311, 313, 315, 316, 319, 320, 323, 324, 326, 332, 333, 338–341, 343–345, 348, 350, 353, 355–357, 372, 376, 377, 380, 381, 383–387, 389–391, 393, 394, 403, 404, 407, 408, 416, 419, 422–424
  long, straight wire
    magnetic field of, 345, 348–350
    magnetic force on, 350
  loop, 357, 372, 394
    magnetic field of, 345–347, 353, 372
    magnetic force on, 338–340, 347, 383
electric displacement, see displacement (electric)
electric field, see field, 411
electric force, 61, 84, 101, 228, 230–235, 248, 250, 253, 255, 261, 270, 275, 276, 291, 330, 331, 336, 340, 371, 380, 398, 405
electric generator, 378, 381, 383–384
  rotor coil, 383, 384
  stator coil, 383
electric motor, 335n, 338, 381, 383–384
  back EMF, 384
  burnout, 384
  commutator, 383
    split-ring, 383
  rotor coil, 383, 384
  series motor, 384
  shunt motor, 384
  stator coil, 384
electricity, 228, 230, 274, 280, 283, 289, 331, 343, 347, 362, 378, 394, 398, 399, 402, 422, 429
electrochemical cell, see cell

electrodynamics (Maxwellian), 398, 400

electrolyte, 310n

electromagnet, 322, 333, 336, 381, 383, 384, 390

electromagnetic forces, 61, 84, 154, 203

electromagnetic induction, 228, 377, 378, 380, 381, 383, 384, 386, 387, 389, 390, 393, 394, 404

electromagnetic radiation, *see* waves

electromagnetic spectrum, 403, 449

electromagnetic waves, *see* waves

electromagnetism, 228, 331, 371, 377, 394, 398, 427

electromechanical device, 376

electromotive force (EMF), 291–295, 377, 378, 380, 383–387, 389, 390, 393, 394

    ideal device or source, 293, 295, 298, 300, 324, 326, 394, 404

    photoelectric, 291

    real device or source, 293, 295

        internal resistance, 293, 295, 394

    thermoelectric, 291

electron, 198, 228, 232–234, 238, 247, 257, 258, 262, 272, 276, 282, 283, 283n, 284, 289, 289n, 291, 291n, 306, 311, 313, 316, 324, 330, 336, 338, 343, 357, 360–363, 366, 367, 371–374, 376, 403, 456

    classical radius, 262–263, 359, 373

    magnetic moment, 361–362, 364, 371–373

        classical, 357–360, 373

Electron Paramagnetic Resonance (EPR), 363, 373

electron volt (eV), 241, 255

electronics, *see* electric circuit

electrostatic stress, 253, 276, 277

electrostatic unit (esu), 232

electrostatics, 213, 228, 238, 245, 255, 262, 263, 274, 423

electrotonus case, 301, 304, 326

electroweak forces, 84, 235

elementary particle, 94, 95, 100, 101, 103, 111, 112n, 123, 193, 198, 220, 228, 233, 235, 263, 332, 360–362, 423

elephant, 63, 73–74

ellipse, 159–161, 163–167, 173

ellipsoid, 353, 355

elliptic functions, 185, 238, 345

elliptical polarization, 411n

e/m tube, 228, 338, 372

energy, 58–61, 74–77, 83, 86–89, 89n, 91, 92, 99–101, 109, 111, 116, 121, 122, 156, 158, 166, 168, 171, 173, 191, 200, 214, 215, 222, 255–257, 263, 272, 290, 291, 311, 313, 334, 336, 363, 373, 378, 380, 381, 383, 384, 392–395, 404, 406–408, 411, 417, 419, 421–424, 430, 449

    chemical, 76

    conservation, 88, 89, 91, 97, 100, 101, 123n, 124, 165, 185, 220, 240, 385, 406, 416

    heat, 74, 86, 91, 293, 369

    in capacitor, 273–274, 276, 277, 390, 391

    in electric field, 273–274, 277, 357, 390, 391, 395, 404, 405

    in inductor, 390–391, 394, 395

    in magnetic field, 390–393, 395, 404, 405

    kinetic, 59, 74, 75, 83, 84, 86, 87, 89, 91, 92, 97, 100, 101, 123–124, 141, 142, 172, 185, 200, 214, 222, 257, 275, 284, 335, 336, 372, 403

    of system, 106–107, 109, 121, 124, 142

    levels, 198, 216n

    potential, 83, 85–89, 124, 185, 199, 200, 214, 222, 240, 241, 341, 372, 394, 421, 430

        electric, 255, 257, 261, 262, 264

        gravitational, 157–158, 165, 172, 255, 264

        of system, 107–108

    self-energy

        electrostatic, 262, 263, 276, 277

gravitational, 276
  solar, 293
  total, 86–89, 165, 185, 186, 186n,
      189, 193, 199, 200, 257, 421
  of system, 109–111
energy height, 87
energy-momentum, *see* four-momentum
English, 128n
Enlightenment, 331
Eötvös, Baron Roland, 153
EötWash Experiment, 153
equation of motion, 38, 38n, 39, 42, 45,
      172, 178, 180–183, 185, 186,
      189, 191, 196, 199, 200, 215,
      225
equator, 372, 373
equilbrium point, 196
equilibrium, 69, 176, 178, 203, 213, 220–
      222, 226, 248, 251, 258, 269,
      271, 277, 286, 301, 311, 341,
      378, 384, 416
equilibrium point, 89, 176, 178, 188,
      189
  neutral, 89
  stable, 89, 176, 193, 198–200
  unstable, 89
Equinoxes, Precession of, *see* preces-
      sion
equipotential surface, 257–258, 269, 276,
      293
Erasmus of Formiae (St. Elmo), 258n
escape speed, 161, 165, 167
ether, *see* luminiferous ether
Euclid, 29, 426
Euclidean inner product, 73n
Euler relation, 182–183, 190, 451
Euler, Leonhard, 182, 207, 216
evanescent wave, 429–430
exchange interaction, 367, 369, 371
existence theorem, 180, 182
exoplanet, 141
exponential function, 178n, 182, 188,
      190, 250n, 304, 306, 307, 310,
      313, 373, 389, 390, 417, 419,
      424, 429, 430, 451
eye, 310, 403, 439, 443, 455, 457

pixels, 310
  resolution, 458
  retina, 403, 439

F-104 *Starfighter*, 70n
farad, 264, 265, 267, 272, 306
faraday, 233, 264
Faraday field tensor, 415n
Faraday induction, *see* electromagnetic
      induction
Faraday Law of Induction, 356, 377–
      381, 383, 384, 386, 394, 399,
      414, 416
Faraday's Ice Pail, 250
Faraday, Michael, 233n, 250, 263, 264,
      377
Faustus, 399
Fermat Principle, 457
Fermilab, 111, 336n
ferroelectricity, 371
  domain, 371
  electret, 371
  hysteresis, 371
  memory, 371
  permanent polarization, 371
ferromagnetism, 367–372, 376, 392
  domain, 367, 369
  domain wall, 367
  hard, 369
  hysteresis, 369
  memory, 369
  permanent magnetization, 369, 372
  remnant magnetization, 369
  soft, 369
  superparamagnets, 369
Feynman, Richard P., 361, 362n
fiber optics, 309n
fictitious force, *see* force, inertial
field, 156–157, 198, 203, 241, 255, 270,
      274, 357, 360, 364, 389, 456
  electric, 236, 238, 240–242, 244,
      245, 247, 248, 250, 251, 253,
      255–258, 261, 262, 265, 269–
      272, 274–276, 284–286, 290,
      334, 336, 340, 341, 344, 347,
      353, 357, 360, 363, 377, 378,

380, 381, 391, 395, 399–401, 403–405, 407, 408, 411, 413–417, 419, 420, 422–424, 427, 429, 430, 443, 449, 451

energy density, 273–274, 277, 357, 390, 391, 395, 404, 405

electromagnetic, 423, 427

gravitational, 156–158, 164, 171, 203, 236, 247, 274

magnetic, 331–336, 338–341, 343–345, 347, 348, 350, 353, 357, 360, 363, 364, 366, 367, 369, 371–374, 376–378, 380, 381, 383, 384, 386, 391–395, 399–401, 403–405, 407, 414–417, 419, 422, 423, 427, 429, 430, 449

Earth's, 372, 395

energy density, 390–391, 393, 395, 404, 405

quantum wave, 198, 343

scalar, 156

vector, 156, 241, 242, 244, 258, 270, 283, 347, 377, 378, 413, 415

field lines, 241–242, 244, 257, 258, 331, 347, 357, 363, 378, 403

figure skating, 123

fine-structure constant, 361

first integral, 186, 189, 199

First Law of Thermodynamics, 87

First Rocket Equation
Einsteinian, 60
Newtonian, 52–54, 91

fish, 286

FitzGerald, George Francis, 56

Fizeau, Armand Hippolyte Louis, 402, 430n

fluid, 38, 70, 112, 115, 156, 233, 241, 242, 244, 280, 282, 285, 343, 344, 347

fluorescence, 430

flux, 242, 283, 285, 348, 353, 355, 384, 444

electric, 242–245, 247, 248, 251, 411, 413

energy, 404, 405

magnetic, 344, 347, 348, 377, 378, 381, 383, 385, 386, 392, 393, 414

flux density, 378

quantization, 344

units, 378

mass, 244, 282

Poynting, 404

fly, 310

flyswatter, 310n

focal length, 439

spherical mirror, 433

spherical refracting surface, 436

thin lens, 438–442, 457

force, 31–34, 36–38, 38n, 39, 41, 43, 44, 52, 61, 64, 65, 69, 70, 72–75, 83, 85–89, 95, 106, 107, 115, 116, 122, 123, 126, 128–131, 136, 138, 139, 141, 142, 150, 153–154, 156, 158, 169, 171, 172, 176, 184, 186, 189, 190, 193, 200, 203, 213–216, 218, 228, 232–234, 236, 240, 275, 286, 291, 334, 335, 338, 376, 406

action/reaction pairs, 33, 49, 52, 95, 106, 123

central, 84, 154, 157, 161, 163–165, 230, 231

centripetal, 123, 366

conservative, 83–86, 89, 157, 185, 186, 255, 378

constant, 84, 85

elastic, 95, 96, 126, 129, 130, 270

external, 52, 95, 106–108, 121, 381

force law, 33, 34, 37, 38, 43, 49, 50, 61, 70, 84, 91, 150, 153, 154, 156, 161, 176

fundamental, 84, 157, 234, 235

impulsive, 126

inertial, 50, 168–169

internal, 52, 95, 106, 107, 121–123, 131

inverse-cube, 149n, 238

inverse-square, 149–150, 153–154,

156, 161, 164, 165, 231, 234, 242, 345
nonconservative, 389
nonconservative (dissipative), 84–86, 186, 191, 200, 286, 288, 290, 378
force field, 156, 198, 203
force law, 230, 231, 235, 236, 240, 248, 331–334
Fosbury flop, 108
Fosbury, Richard Douglas, 108
four-force, 61
four-momentum, 58–61, 99, 421
four-scalar, 57–58
four-vector, 57–58, 61
four-velocity, 58, 61
Fourier analysis, 207–208, 216, 217
Fourier, Jean Baptiste Joseph, 190, 207, 216
frame of reference, *see* reference frame
Frankenstein, 280
Franklin, Benjamin, 230, 232, 261, 280, 283, 289, 343
    kite experiment, 261n
Fraunhofer diffraction, 450–455
    circular-aperture, 451–458
    single-slit, 450–451, 457
free will, 181, 182
frequency, 113, 184, 187, 191, 193, 198, 199, 204, 205, 207, 208, 216, 217, 220, 221, 223, 224, 297, 335, 336, 363, 372, 373, 400, 402–403, 419, 420, 422, 429, 456
Fresnel diffraction, 450
Fresnel, Augustin Jean, 450
friction, 31, 32n, 37, 62, 84, 86, 95, 128, 128n, 129–131, 138, 142, 196, 199, 215, 240, 383
    dry-surface, 37
    fluid, 38
    kinetic, 37, 131, 142
        coefficient of, 37, 131, 142
    static, 37, 124
        coefficient of, 37, 141
    tractive or rolling, 37, 91, 124, 130

coefficient of, 37, 91
Frustrated Total Internal Reflection, 430
frying pan, 427
fugacity, *see* velocity parameter

galaxy, 87, 95, 115, 168, 171, 223n, 333, 426
Galilean kinematics, 11, 55, 57, 400
Galilean transformation, 29, 30, 32, 49, 55, 56, 98, 107, 168
Galilei, Galileo, 2, 13, 20, 23, 27, 27n, 29–32, 32n, 49, 55, 62–64, 149, 184, 331, 401
Galvani, Luigi (Aloisio), 341n
game of catch, 29
gas, 38, 203, 220–222, 226, 313, 336, 338, 373, 374
    ideal, 226, 284
    real, 226
gasoline, 91
gauss, 333
Gauss Law, 244–253, 262, 274, 285, 347, 348, 355, 399, 413, 415, 416
    gravitational, 247–248
    with dielectric, 271–272
Gauss Law for Magnetism, 347, 348, 399, 413, 416
Gauss, Karl Friedrich, 4, 333
Gaussian probability distribution (bell curve), 4
Gaussian surface, 245–248, 250, 251, 272, 347, 348, 411
Gaussian units, 232, 333, 378
Geneva, 111, 336n
Genghis Khan (Temujin), 63
geocentric model, 29
geodesic deviation, 169
germanium, 288, 311
Gibbs overshoot, 208
Gibbs, Josiah Willard, 208
Gilbert, William, 331
Glashow, Sheldon, 235
glass, 427, 429, 430, 443, 447, 449, 457
glider, 91
Global Positioning System (GPS), 171

gluon, 362
Gödel Incompleteness Theorems, 323
Gödel, Kurt Friedrich, 323
Gordian knot, 168
Gordius, 168n
Gordon, Walter, 225n
Gothic cathedral, 2
gradient, 89, 158, 261, 423
Grand Unified Theories (GUTs), 235, 332
gravitation, 33, 61, 84, 95, 101, 107, 141, 143, 150, 153–154, 156, 158, 163, 164, 166, 167, 228, 233–235, 264, 275
 artificial, 50n
 Earth's, 17, 18, 20, 22, 30, 33, 34, 43, 44, 61, 63–65, 70, 85, 87, 107, 114, 124, 140, 149, 156, 157, 178, 186, 191, 200, 203, 232
 Einsteinian, 154, 168–171
 Law of Universal Gravitation, 31, 85, 146, 156, 157, 159, 161, 164, 168, 171, 230, 231, 233, 248
 Newtonian, 84, 85, 154, 158, 168, 171–172, 213, 234, 247, 255, 423
 of spherical source, 155–156, 165, 247–248
gravitational field, see field
gravitational potential, see potential
gravitational waves, see waves
graviton, 228n
Greek tragedy, 400
Greeks, 426, 427
ground (electrical), 298, 390
ground state, 257
group velocity, see waves, kinematics
gun, 293
gymnasium, 332
gyromagnetic ratio, 359, 361, 363, 371
 reduced (g-factor), 359, 362
gyromagnetic ration
 reduced (g-factor), 361
gyroscope, 136–140, 363

Hall effect, 341–344
 fractional quantized, 344
 Hall field, 341
 Hall potential, 341, 343
 Hall resistance, 343–344
 quantized, 343–344
Hall, Edwin Herbert, 341, 343
Halley, Edmund, 31, 164
harmonic analysis, see Fourier analysis
harmonic oscillator, 176, 180, 198, 225
 coupled, 193–198, 225
 damped, 186–189, 198, 200, 422
  critically damped, 188–189
  overdamped, 188–189
  underdamped, 187–190, 193, 200
 driven, 189–193, 198, 200, 422
  steady-state response, 190, 191, 193, 200, 422
  transient response, 190, 191, 200
 quantum, 198
 simple, 178, 180, 183–186, 191, 193, 196, 198–200, 204, 216
harmonics, 208, 216–217, 326
 fundamental, 216, 224
hartree, 257
Hartree, Douglas Rayner, 257n
Hawking, Stephen, 31
heart, 273, 290n, 293, 309, 320
heat, 84, 91, 213, 221, 240, 277, 288–290, 293, 309, 364, 369, 383n
Heisenberg Uncertainty Relations, 181
Heisenberg, Werner, 87
helical motion, 338
heliocentric model, 29
helium, 374
 atom, 374
Helmholtz coils, 372–373
hematite, 331
henry, 385, 386
Henry, Joseph, 377, 385
hertz, 403
hertz (Hz), 113, 204, 205, 224
Hertz, Heinrich Rudolph, 113, 357n, 399
hexadecapole, 238
Hipparchus (Hipparchos), 139

Hohmann Transfer, 167, 172–174
Hohmann, Walter, 167, 168
hole, 311, 316
Hollywood, 172
hologram, 447
Homo sapiens sapiens, 280
Hong Kong, 68
Hooke Law, 85, 176, 178, 184, 186, 189, 222
Hooke, Robert, 85n
horse, 75n
horsepower, 75
horsepower-specific fuel consumption (HSFC), 76
Hubble Space Telescope, 457
Hubble, Edwin Powell, 457
Huntingdon Library (San Marino, CA), 62
Huygens principle, 430
Huygens, Christian, 184, 402, 429, 430
hydrogen, 240, 295n
    atom, 176, 233, 234, 257, 372–373
    bonds, 240
    molecule, 373
hyperbola, 161, 165
hyperbolic functions, 40–41, 41n, 57
hyperfine splitting, 372
hysteresis, 369, 371

ideal gas, *see* gas, ideal
Ideal Gas Law, 226, 277
image, 427, 432–443
    enlarged, 457
    erect, 433, 436, 438, 439, 443
    inverted, 433, 434, 436, 438, 439, 442
    mirror, 275, 276
    real, 433, 434, 436, 438, 439, 457
    real source (object), 433, 434, 436, 438
    reduced, 457
    virtual, 433, 434, 436, 438, 439, 457
    virtual source (object), 433, 436, 438
imaginary unit, 182, 421

impedance matching, 295
impulse, 126–131, 422
    angular, 128, 130, 131, 142
incandescent light bulb, 290–291, 293, 444
incident ray, 408, 427, 429, 430, 456
inclined plane, 124
index of refraction, 402, 407, 408, 429, 430, 434–439, 449, 457
indium, 311
inductance, 385–391, 393, 394
    definition, 385
    mutual, 385, 386
    self, 385, 387
    solenoid, 386
    units, 385
inductor, 386–391
    energy, 390–391, 394, 395
inertia, 29, 31, 32
    magnetic, 385
inertia tensor, 134
inertial force, *see* force
inflation (cosmic), 332n
initial conditions, 180–184, 187–189, 191, 197, 199, 215, 225
initial-value problem, 180–184, 186, 188, 189, 225
insulator, 269, 271, 272, 276, 280, 286, 288–289, 291, 295, 310, 313, 326, 381
integrable system, 186
integrated-circuit (IC) chip, 295, 313, 316, 322, 386, 390
    Very Large Scale Integration (VLSI) chip, 322
integration, 39–40, 74, 88, 126, 245, 387
    constants of, 16
    definite, 16, 200, 277
    dummy variable, 16
    line integral, 73, 345
    Riemann integral, 73, 84
    surface integral, 242, 244, 377
intensity, 103, 222, 405–407, 420, 422, 423, 429, 430, 443–444, 451, 455, 457

interference, 193n, 205n, 443–450, 456
    constructive, 430, 444, 449, 456
    destructive, 444, 449
    multiple-slit, 447, 457
    thin-film, 447–449
    two-slit, 444–447, 451, 457
intergalactic space, 333
internal resistance, 293, 295, 394
interstellar space, 333
inverse-square force, *see* force
inverse-square law, 406
invertebrate, 310
inverter, 322
ion, 199, 228, 271, 282, 301, 302, 310,
    336, 372
ionization, 258, 261, 272, 324, 336, 338,
    422
    energy, 257
    potential, 293
I^2R loss, 290, 391, 404
iron, 331, 367, 369, 386, 427
isotropy
    conductor, 285
    dielectric, 270, 271, 285, 407
    radiation, 406

Jacobi, Karl Gustav Jakob, 185
Japan, 203, 261
jerk, 13, 63
joule, 73, 74, 116, 255, 362, 364
Joule heating, 290, 391, 404
Joule, James Prescott, 73, 74, 86
Jupiter, 141, 160, 402

kaon (K meson), 407n
Keck telescope, 434n, 457
Keck, William Myron, 457
kelvin, 5, 241, 289, 289n, 291
Kelvin, Baron, *see* Thomson, William,
    1st Baron Kelvin
Kepler's Laws of Planetary Motion, 159–
    164, 172
    First Law, 159–161
    Second Law, 161–163
    Third Law, 163–164, 167

Kepler, Johannes, 159–163, 166, 172–
    173, 186n
Keplerian orbit, 235
kettledrums, *see* tympani
kilogram, 5
kinematics, 2–13, 27, 29, 30, 55–57, 61,
    62, 66, 96, 109, 114, 131, 169,
    360, 400
kinetic energy, *see* energy
kinetic friction, *see* friction
kinetic theory, 284
Kirchhoff rules, 297–302, 308
    junction rule, 5, 297–298, 300, 303,
    305n, 308, 316, 389
    loop rule, 297, 298, 300, 305, 306,
    309, 387, 389
Kirchhoff, Gustav Robert, 297, 298
kissin' cousins, 41
Klein, Oskar, 225n
Klein-Gordon equation, 225, 421, 424
Klein-Gordon equations, 422
knot, 202, 217, 218, 220, 226
Koppernigk (Copernicus), Niklas, 29

Labrador Retriever, 280
Laplace-Runge-Lenz (LRL) vector, 166,
    173, 186n
Laplacian operator, 415
laptop computer, 322
Large Hadron Collider (LHC), 111, 336n
laser, 447, 447n
Laughlin, Robert B., 344
Law of Equiangular Reflection, 408, 427,
    432, 433
Law of Rectilinear Propagation, 427,
    450
Lawrence, Ernest Orlando, 336
LC circuit, 387n
Le Châtelier's Principle of Spite, 269,
    384
Le Châtelier, Henry Louis, 269
lead, 293
lead sulfate, 293
Leaning Tower of Pisa, 64
left-hand rule, 283n
left-handed coordinates, 133n

Leibniz rule, 73, 133
length, 5
length contraction, 56
lens, 432, 437, 449
    coated, 449
    convex, 449
    eyepiece, 439–443
    objective, 439–443, 455
    progressive, 438n
    thick, 439
    thin, 437–439, 457
       center, 438
       convergent, 439, 457
       convex, 439
       divergent, 457
       focal length, 438–442, 457
       focus (focal point), 438–439
       image, 437–443
       magnification, 439, 457
       optic axis, 438
       power, 457
       ray tracing, 438
       sign convention, 438
Lens-Makers' Equation, 438
Lenz Law, 384–385, 387, 393, 416
Lenz, William, 166
lever arm, 115, 122, 123, 128, 130, 134,
    136
lift, 69–70, 72, 77–80, 80n, 82, 87, 91
    Bernoulli, 70
    impact, 70
lift coefficient, 70, 77, 78, 80, 91
light, 86, 101, 156, 169, 171, 212, 222,
    258, 273, 297, 326, 403, 404,
    407, 408, 411, 422, 423, 426,
    427, 429, 432–434, 436, 437,
    442–444, 447, 449, 451, 455–
    458
    particles, 430n
    speed of, 55, 57, 59, 60, 87, 98,
       109, 171, 223–225, 232, 263,
       284, 361, 371, 399, 401–402,
       429, 430, 456
light ray, 402
light waves, see waves
lightning, 233, 251n, 272, 280, 390

lightning rod, 261
limit, 11–13, 16, 55, 59, 88, 103, 157,
    158, 236
linear momentum, 50–52, 59–61, 74,
    91, 95, 97–100, 104, 108, 109,
    111, 122, 126, 128, 130, 131,
    134, 163, 171, 225, 405–406,
    411, 420, 421
    conservation, 52, 91, 95, 98–101,
       106, 108, 111, 406, 427
    definition, 50
    dynamics, 52
    of system, 107n
linearity
    conductor, 285
    dielectric, 270, 271, 285
    force, 154–155, 234, 245
    waves, 204, 213–214, 216, 217, 224,
       443
lines of force, see field lines
lipid, 280, 301
liquid, 38, 203, 220, 240, 264, 271, 408
liquid crystal, 408
lithium, 374
    atom, 373
little o, 238n
locomotive, 384
lodestone, 331
logarithm, 40, 44, 222, 233
logic gate, 323, 387
London, 68
longitudinal waves, see waves
Lorentz force law, 334, 399
Lorentz transformation, 55–58, 61, 99,
    109, 111, 168, 191, 223, 332
Lorentz, Hendrik Antoon, 55, 60, 191,
    332
Lorentz-FitzGerald contraction, see
    length contraction
Lorentzian kinematics, 55, 61, 62, 66,
    360, 400
Lorentzian peak, 191
Los Angeles, 68
LR circuit, 387–390, 394–395
    bleeder resistor, 390
    LR time constant, 389, 390, 395

LRC circuit, 387n
Lucasian Professorship of Mathematics, 31
Lucent Technologies, 315
luminiferous ether, 400
luminous intensity, 5
lumped components, 308, 387
Luna spacecraft, 168
Lunar Orbit Insertion (LOI), 146, 164–168, 173

Maclaurin series, 176, 180, 183
Maclaurin, Colin, 176
MacPherson strut, 189
MacPherson, Earle Steele, 189
Magnesia, 330
magnesium fluoride, 449
magnet, 330–332, 334, 340, 364, 371, 381
    permanent, 340, 369, 381, 383, 384
magnetar, 334
magnetic charge, 331, 332, 340, 347
magnetic compass, 331, 332, 341
magnetic field, *see* field
magnetic force, 61, 235, 270, 330–335, 338–341, 350, 363, 366, 367, 371, 372, 376, 378, 380, 384, 398, 405
magnetic inertia, 385
magnetic intensity, 364, 369, 414n, 423
magnetic materials, 330, 331, 341, 363, 364, 371, 373, 386, 391, 392, 399
    antiferromagnetic, 364, 371
        domain, 371
    diamagnetic, 364, 366–367
    ferromagnetic, 364, 367–372, 376, 392
        domain, 367, 369
        domain wall, 367
        hard, 369
        hysteresis, 369
        memory, 369
        permanent magnetization, 369, 372
        remnant magnetization, 369
    soft, 369
    superparamagnets, 369
    paramagnetic, 364–367, 369
magnetic memory, 369
magnetic monopole, 332, 347
magnetic poles, 331–332, 340, 364, 366, 381
    North-seeking (N) pole, 331–332
    South-seeking (S) pole, 331–332
Magnetic Resonance Imaging (MRI), 363
magnetism, 228, 274, 282, 330–333, 340, 341, 343, 344, 347, 357, 363, 367, 369, 371, 385, 394, 398, 399, 402, 422, 429
magnetization, 331, 341, 364, 366, 367, 369
    curve, 364, 366, 369
    permanent, 369, 372
    remnant, 369
    saturation, 366, 369
    vector, 364, 366, 369, 373
magneto, 383
magnetostatics, 228n
magnetostriction, 350
magnification, 457
    angular, 442
    binoculars, 439–443
    plane mirror, 434
    spherical mirror, 434, 436
    spherical refracting surface, 436
    thin lens, 439
magnitude (astronomical), 222
Malibu^{TM} lamps, 326–327
Malibu, California, 326
Malus Law, 407
Malus, Étienne Louis, 407
mantissa, 4
Mars, 159, 173–174
mass, 5, 32–34, 59, 60, 65, 66, 76–81, 88, 91, 98, 103–104, 106–108, 124, 126, 136, 138, 141, 142, 150, 153, 154, 158, 165, 172, 244, 282
    active gravitational, 33, 150, 153, 155–157, 163, 164, 169, 171,

172, 203, 234, 248, 276
chemical, 33, 153, 226, 284, 343
conservation, 98–100, 285
inertial, 33, 50, 52, 54, 58–61, 65,
    66, 74, 81, 91, 92, 95, 97–101,
    111, 116, 118, 124, 128, 130,
    134, 140–142, 153, 154, 163,
    165, 168–169, 173, 178, 193,
    196, 199, 200, 213–215, 217,
    218, 221, 224–226, 228, 262,
    284, 335, 336, 338, 359, 362,
    371, 421, 423, 424
negative, 154
passive gravitational, 33, 34, 64,
    70, 73, 140, 150, 153, 155, 156,
    158, 168–169, 173, 178, 200,
    214, 232, 234, 276
relativistic, 59
rest, 59
mass spectrometer, 336–338
massé shot, 126n
matrix, 133, 134, 270
    diagonalization, 134n, 198n
matter waves, see waves, quantum
Mauna Kea, Hawaii, 434n
maxwell, 378
Maxwell Equations, 449
Maxwell equations, 245, 347, 399–402,
    411, 416, 417, 420, 422, 427,
    429, 443
    differential form, 411–415, 423
    integral form, 399, 414, 423
    Lorentzian tensor form, 415n
Maxwell's Rainbow, see electromagnetic
    spectrum
Maxwell, James Clerk, 54, 55n, 228,
    270, 355–357, 377, 398–400,
    415, 427, 429
mechanical waves, see waves
mechanics, 2, 228, 427
Meissner(-Ochsenfeld) effect, 367
Meissner, Fritz Walther, 367
Meissner, Fritz Walther Meissner, 367
memory, 280
Mercury, 161
mercury, 338

meridian of longitude, 169
meter, 5, 280, 282, 284, 288, 291, 296,
    304, 313, 316, 322, 340, 350,
    362, 364, 378, 402, 403, 427,
    442, 447, 457
meter loading, 324
Method of Images, 275–276
metric system, 333
Mexico City, 108
mho, 286
Michell, John, 171
Michigan State University, 336, 390
microcomputer, 390
micron, 280, 316, 403, 419, 449
micropulsations, 405
microscope, 301, 367, 426
microwave oven, 240–241
mileage, 77, 91
Millikan, Robert Andrews, 232
Ming dynasty, 331n
Minkowski, Hermann, 58, 169
mirror, 402, 426, 427, 432, 437, 443
    first-surface, 437n
    parabolic, 434
    plane, 275, 434, 457
        image, 434, 457
        magnification, 434
    second-surface, 437n
    spherical, 432–434
        center, 433–434
        concave, 433, 434, 457
        convex, 276, 433, 434, 457
        focal length, 433
        focus (focal point), 433–434
        image, 432–434
        magnification, 434, 436, 457
        optic axis, 432–434
        radius, 432–434
        ray tracing, 433, 436, 438
        sign convention, 433, 436
        spherical aberration, 434
        vertex, 433–434
mirrors
    reflectivity, 443
Möbius, August Ferdinand, 244
Möbius strip, 244

modern physics, 54, 55n

molasses, 284

mole, 5, 226, 233, 284, 343

molecular bonds, 84, 240

molecule, 95, 101, 123, 198, 222, 258, 269, 270, 272, 280, 291, 301, 313, 363, 364, 366, 373, 403, 407, 430

    diatomic, 222, 226

    monatomic, 222

    triatomic, 222

moment, 116

    first, 104

    second, 104, 116, 134

    zeroth, 104

moment of force, *see* torque

moment of inertia, 116–118, 122–124, 128, 134, 136, 138, 141–143

momentum, *see* linear momentum

Mongols, 63–64

Monkey and Hunter, 30

monopole

    electric, 238

morality, 416

motional EMF, 378–381

motorcycle, 139

myelin, *see* neuron

Mylar, 271

nabla (∇), 89

NASA (National Aeronautics and Space Administration), 50, 70n, 146n

National Institute of Standards and Technology (NIST), 350

National Superconducting Cyclotron Laboratory, 336

natural gas, 383n

nawabari, 63

nebbish, 172n

neoclassical physics, 55n

neon

    gas, 324

    lamp, 324

nerve, 280, 304, 309, 310, 326

nerve cell, *see* neuron

nervous system, 280, 282, 310

Neumann, Carl Gottfried, 215

neuron, 280, 282, 301, 304, 309, 323

    axon, 280, 301–305, 308–310, 323–326

        axoplasm, 280, 301, 302, 310

        interstitial fluid, 280, 301, 302, 310

        membrane, 302–305, 308, 310, 326

        myelin, 280, 282, 301, 310

        nodes of Ranvier, 282, 310

        Schwann cells, 282

    cell body, 280, 304

    dendrites, 280

    synapse, 280, 323

    terminal fibers, 280

neurophysiology, 280, 301

neurotransmitter, 72n, 280

neutron, 234, 374, 456

    magnetic moment, 362–363

neutron star, 142, 334

New Delhi, 68

New Horizons, 146n

New York, 68

newton, 32–34, 116, 236, 272, 291, 350, 405

Newton's Constant, 150–153, 163, 172, 234

Newton's Cradle, 97

Newton's law of resistance, 38, 62

Newton's Laws of Motion, 30, 31, 33, 34, 49, 52, 87, 106, 123, 164, 171, 181

    First Law, 31–32, 49

    Galilean covariance, 49

    Second Law, 32–34, 38, 49, 50, 52, 69, 75, 87, 95, 101, 106, 122, 126, 136, 139, 150, 153, 161, 168, 178, 185, 186, 189, 196, 212, 213, 218, 221, 230, 286

        Einsteinian version, 61, 334

        rotational version, 118, 139, 180

    Third Law, 33, 49, 52, 70, 106, 123, 153

        strong form, 123, 136, 142, 154

        weak form, 123

Newton's Rings, 449

Newton, Isaac, 12, 13, 29–33, 38, 49,
        52, 54–56, 59, 62, 74, 146–
        150, 153, 154, 156, 159, 164,
        171, 230, 429, 430n, 437, 449,
        456
    and the apple, 146–150

Newtonian mechanics, 49, 50, 55n, 59–
        61, 86, 87, 98, 99, 103, 104,
        109, 136, 139, 141, 153, 169,
        224, 225, 285, 360, 372, 400,
        421, 423

nitrogen
    liquid, 289

Nobel Prize, 315, 344, 361, 363

Nobel, Alfred Bernhard, 315

nodes of Ranvier, see neuron

nonconservative force, see force

nonorientable surface, 244

nonrelativistic mechanics, 49

normal force, 37

normal mode, 198, 216–217, 224, 225,
        360

North America, 301

North Magnetic Pole, 331

North Pole, 169, 372

Nuclear Magnetic Resonance (NMR),
        363, 371, 373
    spectrometer, 373

nuclear magneton, 362

nuclear power, 383n

nuclear weapons, 267n

nucleon, 362, 363

nucleus
    cell, 280
    of atom, see atomic nucleus

nutation, 139

oblateness, 143, 157

observer, see reference frame

Ochsenfeld, Robert, 367

Ockham (Occam), William of, 323

Ockham's Electric Razor, 323

Ockham's Razor, 323

octave, 403

octupole, 238

ohm, 286, 288, 295, 306, 344

Ohm Law, 285–289

Ohm law, 344

Ohm, Georg Simon, 285

Ohmic heating, 290, 391, 404

ohmic materials, 285–291, 416, 419

ohmmeter, 291

oil, 383n, 449

oil-drop experiment, 232

Olympic Games, 108

onion, 247

Onnes, Heike Kamerlingh, 289

optic
    geometric (ray)
        image, 455

optic axis, 432–434, 436–438

optical activity, 408

optical lever, 150

optics, 205, 220, 226, 228, 398, 401,
        422, 426–457
    geometric (ray), 427–443, 456
        aberration, 434, 436–437
        image, 427, 432–443, 455
        magnification, 434, 436, 439–443,
            455, 457
        paraxial approximation, 432–434,
            436, 439
        reflection, 427, 429, 430, 432–
            434, 437, 443, 447–449
        refraction, 427, 429, 430, 432,
            434–439
    physical (wave), 427, 443–456
        diffraction, 449–458
        interference, 430, 443–450, 456
        resolution, 455, 457–458
    quantum, 427

OR gate, 322

orbital speed, 164

order of magnitude, 4–5, 37, 79, 220,
        221, 234, 263, 272, 286, 288,
        295, 313, 343, 362, 391, 403,
        420

ordinary differential equation (ODE),
        180, 186, 301, 304–307, 389
    coupled, 196
    homogeneous, 189, 306

inhomogeneous, 189, 306
  homogeneous solution, 189, 191, 200
  particular solution, 189, 191, 200
  linear, 180, 182, 189
  second-order, 180, 189, 196
orientable surface, 244, 245, 347
Örsted, Hans Christian, 332, 333
Örsted Medal, 332
orthogonal (rotation) matrices, 112n, 132
oscillation, 89, 139, 184, 198–200, 203, 216, 220, 221, 224, 240, 315, 336, 400, 401, 403–405, 408, 416, 422, 429, 430
  amplitude, 184–186, 188, 191, 200, 216, 225
  damped, 187–188
  phase shift, 184, 193, 225
oscilloscope, 336
Oswald span efficiency, 78, 79
Oswald, W. Bailey, 78n
oxygen, 240, 364

pacemaker, 293
Pacific Ocean, 203
pagoda, 261
parabola, 22, 24, 29, 106, 161, 165, 167
paraboloid of revolution, 434
Parallel-Axis Theorem, 118
paramagnetism, 364–367, 369
Parthians, 64
partial differential equation (PDE), 215, 225, 411
  coupled, 414
  first-order, 309, 414
  linear, 213
  second-order, 213, 309, 414, 423
    elliptic, 213, 416
    hyperbolic, 213, 415, 417
    parabolic, 213, 309, 421
particle accelerator, 111, 324, 403
pascal, 221
Pascal, Blaise, 221n
past and future, 384
pearl, 247, 262

Pellucidar, 156
pendulum, 153, 184–185
  simple, 178–180, 184, 185
Penzias, Arno Allan, 315
periastron, 165n
pericenter, 165n, 167
pericynthion, 165n
perigee, 165n, 166, 173
perihelion, 114, 159, 165n
period, 113, 138, 139, 141–143, 149, 163, 167, 173, 184, 185, 187, 199, 200, 215, 223, 405, 443, 444
  spatial, 204
  temporal, 204
periodic motion, 113, 184, 186, 188, 197, 200, 204
permeability, 364, 386, 391, 393, 401, 419, 420, 424
  vacuum, 344, 345, 348, 350, 364, 393, 395, 402, 420
permittivity, 401, 419, 420, 424
  dielectric, 271
  of free space (vacuum), 232, 245, 267, 357, 395, 402
perpetual-motion machines, 87
perturbations, 361
phase transition, 289, 369
phase velocity, *see* waves, kinematics
Philosophiae Naturalis Principia Mathematica, 31, 62, 146
philosophy, 456
phosphorus, 311
photon, 101, 228n, 423–424
Physical Science Study Committee (PSSC), 224
physics, defined, 2
piano, 217
piano tuner, 207
pillbox integration, 251–253, 423
pistachio, 288
Planaria, 280
Planck's (reduced) constant, 198, 225, 361, 362, 372, 373, 423

Planck, Max Karl Ernst Ludwig, 198, 228
plane of incidence, 408, 427
planet, 95, 159–161, 163, 168, 193, 220, 233, 234, 265, 338
Plexiglas, 276
plumbing, 282
Pluto, 146n
plutonium, 293
pn junction, 311–313, 315, 316, 319
    depletion layer, 311
    diffusion current, 311, 313
    forward biased, 311–313, 316, 319
    intrinsic current, 311, 313
    reverse biased, 313, 316
    unbiased, 311
point particle, 157, 202
Poisson equation, 423
Poisson, Siméon Denis, 423
polarization
    dielectric, 269–270, 363, 371
        permanent, 371
        vector, 270, 364
    wave, 204, 269, 406–411, 429, 451
        circular, 411, 423
        elliptical, 411n
        plane, 407, 423
        regeneration, 407
polarizing material, 407–408, 411, 423
pool, 94, 97, 111, 126, 128n, 142
position, *see* displacement
positron, 272
postmodern physics, 55n
potassium, 301, 310
potential, 264, 421
    electric, 213, 255–262, 264, 265, 267, 271–277, 286, 288, 290, 291, 293, 295–298, 300, 301, 303, 304, 306, 307, 309–311, 313, 316, 319, 322, 324, 326, 327, 336, 338, 341, 343, 372, 378, 380, 383, 387, 389, 390, 394, 408, 423
        units, 255–256
    gravitational, 158, 165, 213, 257, 423

potential energy, *see* energy
power, 75–76, 81–82, 92, 121, 215, 219, 290, 295, 298, 313, 315, 320, 323, 324, 326, 327, 380, 381, 383, 390, 391, 393, 394
    thin lens, 457
power line, 290
Poynting vector, 404
Poynting, John Henry, 404
Prague, 172
Prandtl, Ludwig, 38, 70
precession, 136–139, 363, 373
    of the Equinoxes, 139, 143
pressure, 37, 70, 86, 221, 222, 226, 253, 277, 374, 406
    atmospheric, 221
    radiation, *see* radiation pressure
principal axes, 134, 142
Principia Mathematica, *see Philosophiae Naturalis Principia Mathematica*
Principle of Equivalence of Gravitation and Inertia, 168
Principle of Superposition, *see* linearity
prism, 429, 430, 443
probability, 103, 103n
probability amplitude, 203
Project Seafarer, 420n
projectile, 56, 83, 142
projectile motion, 23, 46, 106
    ground-to-ground, 26–27
    seen from projectile, 30, 169
    with drag, 65
proper time, 57, 61, 66
proton, 99–100, 111, 233, 234, 257, 324, 335, 336, 363, 372–374
    magnetic moment, 362–363, 372
pseudoforce, *see* force, inertial
Ptolemaeus (Ptolemy), Claudius, 29
pulsar, 334
Purcell, Edward Mills, 363
pyramid, 2

quadrupole, 238
quality factor (*Q* factor), 191, 193

quantities, dissected, 2–5
quantum chromodynamics (QCD), 84, 234, 235, 362n
quantum electrodynamics (QED), 331, 360, 361, 427
quantum field theory, 87, 198, 272, 331, 343, 360, 371, 427, 456
quantum mechanics, 49, 54, 55, 55n, 87, 89n, 123, 181–182, 198, 202–204, 213n, 216n, 228, 235n, 257, 289, 311n, 331, 343, 344, 359, 360, 362, 363, 367, 371, 373, 420–422, 427, 430, 456
quantum number, 423
quantum waves, *see* waves
quark, 233n, 234, 362
quarter-wave coating, 449
quarter-wave plate, 411
quartz, 288
quizzing glass, 426

racquetball, 140–141
radar, 220, 223, 398
radiation pressure, 406
    absorption, 406
    reflection, 406
radio, 336, 357n, 398, 403, 408, 416, 420, 427
    frequency, 335, 336, 363, 373
        AM band, 335, 420
        FM band, 335
    telescope, 457
    tuner, 265n
radioactivity, 234, 336
radioisotope, 293
Radioisotope Thermal Generator (RTG), 293
rainbow, 212, 403, 429
range
    airplane, 68, 76–80
    arrow, 63, 64
    projectile with drag, 65
    rocket, 20, 24–27
Ranvier, Louis Antoine, 282
rapidity, *see* velocity parameter
ray tracing

spherical mirror, 433, 436, 438
    central ray, 433, 434
    focal ray, 433
    parallel ray, 433, 434
    radial ray, 433
    wild ray, 433
spherical refracting surface, 436
    central ray, 436
    focal ray, 436
    parallel ray, 436
    radial ray, 436
    wild ray, 436
thin lens, 438
    central ray, 439
    focal ray, 438
    parallel ray, 438
    radial ray, 438
    wild ray, 438
Rayleigh Criterion, 455, 457
RC circuit, 305–307, 387, 389
    RC time constant, 306, 307, 324
real gas, *see* gas, real
rectifier, 315
recursion, 302–303
reduction to quadrature, 40, 43, 45
reference frame, 27, 29, 30, 32, 49, 50, 55–60, 64, 66, 87, 98, 109, 168–169
    center-of-mass (CM) frame, 108–109, 141
    center-of-momentum (CM) frame, 109–111
    inertial, 32, 49, 55–57, 59, 60, 66, 98, 99, 108, 109, 128n, 168–169
    LAB frame, 97, 109, 111, 141
reflected ray, 408, 427, 429, 430
reflection, 216, 217, 219, 220, 223, 226, 297, 406, 408, 411, 417, 427, 429, 430, 432–434, 437, 443, 447–449
    angle of, 427
    reflected ray, 427, 429, 430
    regular, 427
    specular, 427

refracted (transmitted) ray, 427, 429, 430

refracted ray, 456

refracting surface, 432
  spherical, 434–439, 457
    center, 436, 438
    chromatic aberration, 436–437
    convex, 438
    focal length, 436
    focus (focal point), 436
    image, 434–437, 439
    magnification, 436, 457
    optic axis, 436–437
    radius, 434–439
    ray tracing, 436
    sign convention, 434–436
    vertex, 436

refraction, 226, 427, 429, 430, 432, 434–439
  angle of refraction (tramsmission), 427
  angle of refraction (transmission), 429
  index, 402, 407, 408, 429, 430, 434–439, 449, 457
  refracted (transmitted) ray, 427, 429, 430
  refracted ray, 456

refrigerator, 384

relative permeability, 364, 402

relativistic mass, *see* mass

relativistic mechanics, 49

relativity, 55
  Einsteinian, 54, 62, 156, 331, 416
    general theory, 30, 61, 62, 87, 91, 123, 153, 168–172, 203, 456
    special theory, 49, 54–62, 87, 98, 99, 109, 123, 168, 169, 224, 225, 263, 272, 311n, 360, 371, 400, 421–423, 456
  Galilean, 27–30, 32, 49, 62, 99, 168, 400

relativity of simultaneity, 56

relaxation oscillator, 324

renormalization, 361

resistance, 286, 288, 290, 291, 293, 295–297, 300–303, 305, 315, 320, 324, 326, 327, 344, 384, 387, 389–391, 394

resistance ladder, 302–304, 326–327

resistive/capacitive ladder, 308–309

resistivity, 288
  temperature coefficient of, 289n

resistor, 286, 290, 291, 295–298, 300, 301, 305, 306, 313, 315, 319, 322, 386, 387, 390, 394, 395, 404
  color code, 295–296
  combination, 298
    parallel, 296–298, 303, 324
    series, 296–298, 303, 324

resonance, 191–193, 198, 200, 224, 336, 363, 373, 403, 443
  bandwidth, 403
  resonance peak, 191, 193, 200, 403
  resonant amplification, 191

resonator, 419

rest energy, 59

rest mass, *see* mass

restitution, coefficient of, *see* coefficient of restitution

Riemann, Georg Friedrich Bernhard, 73n

Riemann, Georg Friedrich Bernhard, 235

right-hand rule, 132, 133, 283n, 333, 340, 348, 353, 381, 411

right-handed coordinates, 133n

rigid body, 115, 118, 123, 139

ring, 116

Rio de Janiero, 68

Rockefeller Library (Cambridge, UK), 62

rocket engine, 14, 34, 50, 66, 81, 91–92, 166, 173
  Einsteinian, 59–62, 66
  Newtonian, 52–54, 61, 62

rocket flight, 34
  angled, 46
    drag-free, 20–27
    free-fall, 20, 22, 199
    shape of trajectory, 20, 22–24

under power, 20–22
with drag, 38–39
drag-free, 106
vertical
  coasting, 42–46
  drag-free, 14–20, 46
  free-fall, 14, 17–19, 23
  under power, 14–17, 39–42
  with drag, 39–46, 186
rod, 118
roller coaster, 87
roller-ball mouse, 311n
rolling, 37, 94, 124, 128
  without slipping, 124, 126, 128–131, 141, 142
rolling friction, *see* friction
Rolling Race, 124–126
  cheating, 126n
Romans, 426
Romantic physics, 55n
Römer, Ole (Olaus), 402
root-mean-square (rms) value, 284, 290
rotation, 111–112, 128, 128n, 130–132, 140, 142, 150, 157, 168, 222, 240, 330, 347n, 357, 360, 383, 384, 408, 411
  axis, 112, 115, 116, 118, 121–124, 128, 131, 132, 138, 143, 357, 362, 372, 423
  defined, 112
  differential, 115
  dynamics
    Einsteinian, 139
    fixed axis direction, 115–123, 134
    in three dimensions, 134–139
  kinematics
    fixed axis direction, 114–115
  rigid, 115, 121, 122, 124, 141, 359
rotation axis, 359
Runge, Carl David Tolmé, 166
Rutherford, Ernest, 235
rydberg, 257
Rydberg, Johannes Robert, 257n

Salam, Mohammad Abdus, 235
satellite, 160, 161, 163–165, 171–173

Saturn V rocket, 146, 173
Savart, Félix, 344
scalar, 11–12, 57, 72, 74, 87, 98, 101, 131, 134, 158, 270, 283, 285, 360, 401
  Lorentz, *see* four-scalar
scattering, 94, 101–103, 108, 109, 193, 220, 263, 288, 289, 408, 430
  angle, 109
  cross section, 103
    differential, 103
    total, 103
  Rutherford, 235
  wave, 202, 217–220, 224
    phase shift, 217, 219, 220, 226
    reflection coefficient, 219, 220, 226
    transmission coefficient, 219, 220, 226
Schieffer, John Robert, 315
Schrödinger equation, 421, 422
Schrödinger tunneling, 430
Schrödinger equation, 49, 213n, 225, 360
Schrödinger tunneling, 89n
Schrödinger, Erwin, 49n, 87
Schwann cells, *see* neuron
Schwann, Theodor, 282
Schwinger, Julian, 361
sea urchin, 242
seagull, 87, 91
seawater, 280, 301, 416, 420
second, 5, 282–284, 306, 324, 405
Second Rocket Equation
  Einsteinian, 60, 61
  Newtonian, 54, 60, 91
semiconductor, 288–289, 310, 313, 316, 343
  doped (extrinsic), 311
    n-type, 311, 313, 316
    p-type, 311, 313, 316
  pure (intrinsic), 310, 311, 313
shed (unit), 103
Shelley, Mary Wollstonecraft, 280
shock wave, *see* waves
Shockley equation, 313

Shockley, William, 315
Shockley, William Bradford Jr., 313
short (electrical), 295, 322
siemens, 286, 288n
significant figures, 4, 9, 18, 19, 36, 74,
        76, 114, 149, 150, 164–167,
        188, 253, 256, 263, 295, 442
silicon, 288, 311
silver, 286, 288, 419
sine, 180, 190n, 205, 213, 216, 218, 238,
        400, 429, 432, 455
sinusoidal function, 189, 190, 200, 205,
        207, 214–217, 222, 224–226,
        290, 400, 401, 416
skin depth, 419–420
smartphone, 323
Snell Law of Refraction, 408, 429, 430,
        434, 436, 457
Snell van Royen (Snellius), Willbrord,
        402
Snell van Royen (Snellius), Willebrord,
        429
snooker, 94, 111
soap bubble, 277
sodium, 301, 310
Solar Constant, 405
solar day, 113, 167
solar prominences, 338
solenoid, 376, 387, 391
        end effects, 350, 353, 386, 392, 393
        force, 376, 392–395
        magnetic field, 350–353, 376, 393
        mutual inductance, 386
        self inductance, 386
        toroidal, 353
solenoid relay, 322, 376
solid, 203, 220, 310, 369, 371
        band structure, 289
        mechanics, 399n
solid angle, 103
solid-state circuitry, 316, 323
Sommerfeld, Arnold Johannes Wilhelm,
        235
sonar, 220
sonic boom, 223
sound, 86, 224

loudness, 222
pitch, 205
speed of, 38, 54, 221–222, 226, 400
sound waves, *see* waves
South Magnetic Pole, 331
South Pole, 372
Space Chase, 172–173
spacetime, 58, 203, 235, 384, 416
        curved, 169–171
        flat, 169
specific heat, 222
specific impulse, 66
spectacles, 426, 457
spectroscopy, 297–298, 447
speed, 11–12, 19, 25–26, 29, 31, 32, 38,
        42, 43, 46, 54–56, 58–61, 63–
        66, 69, 70, 74, 76–79, 81, 82,
        87, 91, 92, 96–100, 109, 111,
        113, 124, 129, 142, 156, 161,
        164–167, 169, 171, 173, 186,
        207, 223, 224, 283, 284, 291,
        333, 335, 336, 371, 373, 384,
        400
        average, 12, 124
        instantaneous, 12
        supersonic, 38
        transsonic, 38
speedometer, 11–12
sperm whale, 310
sphere, 141–143
spin, 94, 112n, 128n, 130, 131, 136,
        138, 139, 359, 360, 362–364,
        366, 367, 372–374, 421, 423
spinal cord, 280
spinor, 360
spool, 118–121
spoon, 434
spring, 85, 86, 176, 178, 186, 188, 196,
        199, 200, 203, 214, 341
        spring constant, 86, 178, 196, 199,
        200
spyglass, 442
square wave, 208, 217
squid, 301, 304, 310
St. Elmo's Fire, 258
St. Louis, Missouri, 326, 372

standard atmosphere, 80
Standard Model (of Particle Physics), 235
standing wave, 216–217, 224, 225
    antinode, 216
    node, 216, 217
star, 298, 334, 406
static friction, *see* friction
statics, 2, 139, 140
    static equilibrium, 139–140, 248, 250, 258, 285
statistics, 4n, 103
steam turbine, 383
steradian, 103
stimulus-response relation, 270, 285
Stokes Theorem, 414
Stokes' law of resistance, 38, 62, 65, 286
Stokes, George Gabriel, 38
Stormer, Horst L., 344
strawberry, 288
streamlines, 241
string, 202–204, 207, 212–218, 220, 224–226
    cosmic, 224
string theory, 235
strong nuclear force, 84, 154, 234, 235
Strutt, John William, 3rd Baron Rayleigh, 455
submarine, 398, 416, 417, 420, 422
sulfur, 286
sulfuric acid, 295n
Super Unified Theory, 235
supercomputer, 322
superconductor, 289, 367, 391, 394
    BCS model, 315
    critical temperature, 289, 367
    high-T_c, 289
surface tension, 277
susceptibility, 364, 366, 367, 373, 374, 392, 393
swan, carnivorous, 63
switch, 284, 293, 295, 298, 305–307, 315, 320, 322, 323, 376, 378, 380, 387, 389, 390, 395
    combination

    parallel, 320, 322
    series, 320–322
    ideal, 315
Sydney, 68
synapse, *see* neuron
synchrocyclotron, 336n
synchrotron, 336n, 403
synchrotron radiaton, 403
Système International des Poides et Mesures (SI), 5, 32, 33n, 73–75, 116, 232, 236, 264, 282, 286, 333, 344, 364, 378, 385

table tennis, *see* ping pong
tangent, 57, 213, 432
Taylor series, 176, 238, 372
Taylor, Brook, 176
teeth, 280
Teflon, 272
telegraph, 309, 326
Telegraphers' Equation, 309
telephone, 309
telescope, 426, 434, 442
    radio, 457
    reflecting, 437
    refracting, 437, 439
    resolution, 457
television, 267, 315, 398, 403
    picture tube, 336
temperature, 5, 222, 226, 241, 277, 284–286, 288, 289, 291, 293, 298, 313, 364, 366, 367, 369, 372, 373
    room, 226, 284, 291, 366, 374
temperature coefficient of resistivity, 289n
tennis, 63, 65, 98, 128n
tension, 118, 213, 214, 224, 226
tensor, 134n, 171, 270
    contravariant, 57n
    rank, 57n
terminal fibers, *see* neuron
terminal speed, 40–43, 45, 46, 62, 64, 65, 286
terminal velocity, *see* terminal speed
tern, common, 88, 88n

tesla, 334, 343, 362, 364, 405

Tesla, Nikola, 333

Tevatron, 111, 336n

Thatcher, Margaret, 31

Theory of Everything, 235

thermocouple, 293

thermodynamics, 222, 373, 399n

thin lens, *see* lens

Thin Lens Equation, 439

Thin-Lens Equation, 438

Thomson, Joseph John, 228, 338

Thomson, William, 1st Baron Kelvin, 241

thorium, 275

thought, 280

thrust
    airplane, 69, 70, 72, 76, 81–83, 91, 92
    rocket, 34, 36, 39, 41, 42, 50, 52, 54, 66, 166, 173

thrust-specific fuel consumption (TSFC), 76, 79

thundercloud, 261

tidal stabilization, 140n

tides, 169

time, 5, 27, 32, 55, 58, 61, 65, 75, 180, 457

time dilation, 56, 58

time of flight, 18–20, 24–26, 43–46, 65

time-translation invariance, 88

timescales, 95

toaster, 290

Tokyo, 68

Tomonaga Shin-ichiro, 361

topspin, 128

torque, 115, 116, 118, 121–123, 128, 128n, 136, 138–140, 142, 143, 163, 166, 178, 240, 269, 332, 339–341, 363, 383, 384
    external, 123, 136
    internal, 122, 123

torsion balance, 150, 230

torus, 353

total energy, *see* energy

Total Internal Reflection, 429–430, 443, 457

evanescent wave, 429–430

Frustrated Total Internal Reflection, 430

Total Internal Reflection Fluorescence, 430

tractive friction, *see* friction

transformer, 369

transistor, 310n, 315–320, 322, 323, 327, 386
    active region, 319
    base, 316, 319, 320
    Bipolar Junction Transistor (BJT), 316
    collector, 316, 319, 320
    dc current gain (β), 319
    emitter, 316, 319, 320
    Field Effect Transistor (FET), 316n
    Metal Oxide Semiconductor Field Effect Transistor (MOSFET), 316n
    npn transistor, 316
    pnp transistor, 316
    saturation, 319, 320, 327

Transistor-Transistor Logic (TTL), 322

translation, 115, 124, 142

transmitted ray, 408

transoceanic cable, 309
    repeater, 309

transverse waves, *see* waves

triode, 316

tripod, 442

Tsui, Daniel C., 344

tuning fork, 207

Turkey, 330

turning point, 89

two-body problem, 172

tympani, 217

U.S. Air Force, 50, 68

U.S. Navy, 420

uncertainty, 4, 295, 296

Unified Field Theory, 235

uniqueness theorem, 180, 182

units, 4, 32, 34, 36, 38, 55, 73, 75, 79, 95, 103, 113, 116, 122, 158,

186, 232, 236, 255–257, 263–265, 267, 272, 282, 286, 333, 334, 340, 344, 362, 364, 378, 385, 391, 405
    basic, 5, 282
Universal Gas Constant, 226
universal quantum of action, *see* Planck's (reduced) constant
universe, 41n, 88, 91, 156, 157, 168, 171, 181, 223n, 233, 280, 304, 332n, 426
Upper Peninsula (Michigan), 420
uranium, 275

vacuum, 20, 37n, 46, 50, 55, 203, 244, 263, 265, 270–274, 284, 350, 361, 386, 391, 395, 399, 402, 403, 405, 414, 419, 427, 429
vacuum tube, 310n, 313, 315, 316, 322
valve, 319
vampire tap, 326
Van de Graaff generator, 291
Van de Graaff, Robert Jemison, 291
vanilla, 288
vector, 5–12, 32, 41, 50, 57, 57n, 61, 72, 73, 87, 89, 100, 101, 122, 126, 131–133, 136, 138, 142, 150, 153, 154, 158, 161, 163, 165, 228, 231, 238, 242, 282, 331, 335, 338, 339, 345, 359, 360, 362, 377, 380, 401, 404, 413, 443
    addition, 5, 9, 32, 34–36, 50, 52, 106, 131–134, 154, 283, 340, 443
    analysis, 414
    antisymmetrized tensor product, 133n
    area, 244, 283, 340, 348
    aree, 413
    box product, 380, 381
    components, 36, 39, 56, 57, 57n, 58–61, 70, 72, 73, 88, 89, 95, 99, 100, 111, 113, 115, 116, 122, 133, 136, 141, 157, 214, 235, 238, 241, 242, 244, 245,

248, 251, 257, 258, 262, 333, 335, 338, 343, 345, 347, 348, 359, 361, 362, 372, 373, 380, 407, 411, 413, 415, 423, 424
    integration, 16
    resolution into, 9, 13
    scalar, 8–9, 11
    vector, 8–9
    direction, 5, 9, 11, 13, 72, 73, 89, 101, 109, 123, 131–133, 136, 142, 153, 154, 163, 231, 241, 258, 261, 275, 282, 283, 333, 339, 340, 348, 353, 359, 362, 371–374, 394, 395, 401, 403, 404, 406, 407, 411
    Lorentz, *see* four-vector
    magnitude, 5, 9, 11, 13, 34, 39, 72, 73, 76, 89, 109, 113, 115, 123, 131, 132, 136, 143, 153, 163, 230–233, 240–242, 257, 261, 262, 272, 274, 275, 286, 333–335, 343, 348, 371–373, 383, 391, 393, 395, 404–407, 411, 420, 422, 443, 451
    multiplication, 11, 134
    scalar (dot or inner) product, 72–73, 84, 133, 242, 413
    spacetime, *see* four-vector
    tensor (outer) product, 72
    unit vectors, 8, 13, 38, 39, 70, 73, 112, 113, 133, 138, 157, 236, 242, 244, 247, 248, 251, 253, 265, 282n, 348, 350
    vector (cross) product, 133–134, 136, 163, 240, 333, 335, 338, 339
        BAC-CAB identity, 415
velocity, 11–12, 16, 18–20, 22, 23, 25–27, 29–33, 38, 39, 41–43, 45, 46, 49, 50, 52, 54–61, 63, 65, 66, 69, 70, 74, 75, 91, 95, 97, 98, 100, 101, 106, 108, 109, 111, 113, 124, 128–132, 134, 141, 142, 156, 158, 166, 181, 186, 199, 214, 223, 225, 226, 244, 282, 284, 286, 298, 333–335, 338, 344, 345, 378, 380,

405
   average, 11–12
   instantaneous, 11–12
velocity parameter, 57, 59, 60
velocity selector, 334, 336, 341
ventricular fibrillation, 273
Venus, 161
vertebrate, 280, 301, 310
Vikings, 408
vis viva, 74
volt, 255–257, 261, 264, 272, 286, 291,
   295, 324, 343, 344
Volta, Alessandro, 255
voltage, *see* potential, electric
voltage divider, 324
voltaic pile, 293
voltmeter, 324
   analog, 341n
von Fraunhofer, Joseph, 450
von Goethe, Johann Wolfgang, 399
von Helmholtz, Hermann Ludwig, 86
von Klitzig constant, 344
von Klitzig, Klaus, 344
von Leibniz, Gottfried Wilhelm, 12
von Siemens, Werner, 286n
Voyager, 68

water molecule, 240, 241, 269, 271, 275
waterfall, 383
watt, 75, 290
Watt, James, 75, 75n
wave, 450
wave equation, 204, 207, 212–213, 224,
   225, 309, 360, 416, 420
   electromagnetic, 420
     in nonconducting medium, 415–
     417, 424
     in ohmic medium, 417
   for sound, 221
   for stretched string, 204, 212–214,
     225
   linear, 204
wave equations, 421
wave front, 451
wave function, 181, 204–205, 215–217,
   221, 222, 224, 225, 401, 421

waveguide, 419
waves, 202, 213, 214, 224–226, 427, 429,
   430, 444–451, 456
   antinode, 221
   coherent, 444, 457
   crest, 401, 444
   defined, 203
   diffraction, 449–458
     Fraunhofer, 450–455
     Fresnel, 450
   dispersion, 208–212, 214, 225, 402,
     429, 436
     dispersion relation, 208, 212, 214,
     225, 395, 401, 402, 416, 417
   dynamics, 204, 207, 212–217, 220,
     400, 411
   electromagnetic, 86, 203, 398–411,
     415, 416, 422–424, 426, 427,
     429, 430, 450–451
     circularly polarized, 411, 423
     elliptically polarized, 411n
     gamma radiation, 403
     in ohmic medium, 416–420, 422,
     424
     infrared, 403, 404
     micropulsations, 405
     microwaves, 240, 315, 402, 403,
     419, 427, 457
     plane wave, 401, 405, 422, 443–
     444, 450, 451
     plane-polarized, 400, 401, 405,
     407, 417, 422–423, 443–444
     radio waves, 403, 420, 427
     ultraviolet, 403
     visible light, 403, 404, 407, 408,
     411, 422, 423, 426, 427, 429,
     432–434, 436, 437, 442–444,
     447, 449, 451, 455–458
     x-rays, 403, 427
   gravitational, 171, 203
   gravity, 203, 203n
   incoherent, 444
   interference, 443–450, 456
     constructive, 430, 444, 449, 456
     destructive, 444, 449
     multiple-slit, 447, 457

thin-film, 447–449

two-slit, 444–447, 451, 457

kinematics, 204–212, 220

  amplitude, 204, 205, 215, 216, 219–221, 224, 401, 405, 406, 417, 419, 422, 443

  angular frequency, 205, 208, 214, 216, 395, 401, 402, 416, 417, 419–421, 423

  angular spatial frequency, 205n

  frequency, 204, 205, 207, 208, 216, 217, 220, 221, 223, 224, 400–403, 419, 420, 422, 429, 456

  group velocity, 207, 208, 212, 214, 225

  period, 204, 405, 443, 444

  phase, 204, 207, 208, 217, 219, 220, 224, 401, 416, 430, 443, 451

  phase shift, 205, 411, 416, 443, 444, 447, 449, 450

  phase velocity, 204, 207, 208, 212, 214, 225, 401

  polarization, 204, 269, 406–411, 423, 429, 451

  shape, 204, 208, 212, 214, 216, 326, 401, 416

  speed, *see* waves, kinematics, velocity

  velocity, 204, 207, 214, 215, 221, 224, 399–402, 407, 430, 456

  wave number, 205, 208, 214, 216, 326, 417, 451

  wave vector, 395, 401, 404, 416, 420, 451

  wavelength, 204, 205, 213, 216, 217, 223, 400, 402–403, 408, 427, 436, 444, 447, 449, 451, 455–458

light, 212, 223, 291, 399, 400, 403, 404, 407, 408, 411, 422, 423, 426, 427, 429, 432–434, 436, 437, 442–444, 447, 449, 451, 455–458

longitudinal, 203, 220, 225

mechanical, 203, 224, 400

monochromatic, 205, 207, 208, 214–216, 222, 224–226, 400, 401

node, 221

periodic, 204, 205, 207, 215

quantum, 203, 220, 225, 360, 420, 421, 430, 456

shock, 223

sound, 86, 203, 205, 207, 220–223, 400

  threshold of hearing, 220–222, 406

  threshold of pain, 221

spherical, 422

transverse, 203–204, 212, 213, 224, 401, 406, 416

trough, 401, 444

wave front, 430, 444, 450, 451, 457

wave packet, 205, 207, 406, 411, 423

wave train, 203

weak nuclear force, 84, 154, 234, 235, 361

weber, 378

Weber, Wilhelm Eduard, 378

weight, 33, 34, 41, 66, 69, 72, 73, 87, 118, 131, 136, 138, 140, 150, 153, 169, 178, 200, 232, 264, 275, 394, 442

weightlessness, 199

Weinberg, Steven, 235

Wilson, Robert Woodrow, 315

wind tunnel, 70

windmill, 383

wine, 182, 316n

wing, 70, 70n, 77–79

  aspect ratio, 78

  wingspan, 78, 79

wire gauge, 283n, 288, 326

work, 72–77, 83–86, 91, 107, 116, 121–124, 166, 189, 200, 240, 255, 257, 262, 264, 273, 277, 290, 335, 341, 369, 376, 380, 381, 383, 384, 390–394, 405, 422

Work-Energy Theorem, 74–75, 86, 87, 107, 121, 128, 335

workstation, 301, 320
world scalar, *see* four-scalar
world vector, *see* four-vector
Worshipful Guild of Scientific Instru-
          ment Makers, 153n

x-ray source, 171
x-rays, 363, 403, 427

year, 114
Young, Thomas, 74, 447

ziggurat, 2
zoom lens, 439
zooming, 83, 87

CPSIA information can be obtained at www.ICGtesting.com
Printed in the USA
LVOW09s0223210516

489291LV00001BA/1/P